Second Supplements to the 2nd Edition of

RODD'S CHEMISTRY OF CARBON COMPOUNDS

ELSEVIER SCIENCE B.V.
Sara Burgerhartstraat 25
P.O. Box 211, 1000 AE Amsterdam, The Netherlands

ISBN: 0-444-82229-1

Second Supplements to the 2nd Edition of

RODD'S CHEMISTRY OF CARBON COMPOUNDS

VOLUME I

ALIPHATIC COMPOUNDS
★

VOLUME II

ALICYCLIC COMPOUNDS
★

VOLUME III

AROMATIC COMPOUNDS
★

VOLUME IV

HETEROCYCLIC COMPOUNDS
★

VOLUME V

MISCELLANEOUS

GENERAL INDEX
★

Second Supplements to the 2nd Edition of

RODD'S CHEMISTRY OF CARBON COMPOUNDS

A modern comprehensive treatise

Edited by
MALCOLM SAINSBURY
School of Chemistry, The University of Bath,
Claverton Down, Bath BA2 7AY, England

Second Supplement to

VOLUME III AROMATIC COMPOUNDS

Part F: Polybenzenoid Hydrocarbons and their Derivatives:
Hydrocarbon Ring Assemblies, Polyphenyl-substituted
Aliphatic Hydrocarbons and their Derivatives
(Partial: Chapter 24 in this Volume)
Part G: Monocarboxylic Acids of the Benzene Series: C_7-C_{13}-
carbocyclic Compounds with Fused-ring Systems
and their Derivatives
Part H: Polycarbocyclic Compounds with more than Thirteen
Atoms in the Fused-ring System

1995
ELSEVIER
Amsterdam – Lausanne – New York – Oxford – Shannon – Tokyo

Contributors to this Volume

J.M. MELLOR

Department of Chemistry, University of Southampton,
Southampton SO9 5NH, U.K.

M. SAINSBURY

School of Chemistry, The University, Claverton Down, Bath BA2 7AY, U.K.

M.M. COOMBS

Department of Chemistry, University of Surrey, Guildford,
Surrey GU2 5XH, U.K.

N.H. WILSON

Department of Pharmacology, University of Edinburgh, 1 George Square,
Edinburgh EH8 9JZ, Scotland, U.K.

R. BOLTON

Department of Chemistry, University of Surrey, Guildford,
Surrey GU2 5XH, U.K.

Preface to Volume III F (partial), G and H

The most important feature of the supplements to the second edition of *Rodd's Chemistry of the Carbon Compounds* is the way in which these volumes provide a carefully crafted survey of the *whole* of organic chemistry. Nowhere else is this done in such depth and yet in such a readable form.

This sub-volume is a typical illustration. It surveys a very large subject area of aromatic chemistry: phenylalkylbenzenes, benzoic acids and related compounds, polycyclic aromatic hydrocarbons and their derivatives, together with a range of other systems including benzocyclopropenes and benzocyclobutenes. All these subjects are dealt with thoughtfully. Thus, although topics of current interest, such as new synthetic methodology and control of stereochemistry, are highlighted, less familiar areas are certainly not neglected and older ones are brought up to date.

The authors have used their considerable collective skills and industry to take the output of factual literature surveys, to analyse them and to bring these subjects to life. As the editor of the work I am much in their debt and I hope the reader will find the contents as fascinating and as useful as I have.

Malcolm Sainsbury

Bath
March 1995

Contents
Volume III F(partial), G and H

Aromatic Compounds: Polybenzenoid Hydrocarbons and their Derivatives; Hydrocarbon Ring Assemblies, Polyphenyl-substituted Aliphatic Hydrocarbons and their Derivatives (partial)

Monocarboxylic Acids of the Benzene Series: C_7–C_{13}-carbocyclic Compounds with Fused-ring Systems and their Derivatives

Polycarbocyclic Compounds with more than Thirteen Atoms in the Fused-ring System

Chapter 24. Bis- and Tris-(phenylalkyl)benzenes and Linear Oligo(phenylenealkyl)s, their Derivatives and Oxidation Products
by J.M. MELLOR

Chapter 25. Monocarboxylic Acids of the Benzene Series
by M. SAINSBURY

Chapter 26. *Benzocyclopropene, Benzocyclobutene and Indene, and their Derivatives*
by M.M. COOMBS

Chapter 27. Aromatic Compounds with Condensed Nuclei: Naphthalene and Related Compounds
by N.H. WILSON

Chapter 28. Anthracene, Phenanthrene and Derivatives
by R. BOLTON

Chapter 29. Polyannular aromatic compounds containing one or more
five-membered rings
by R. BOLTON

Chapter 30. Polyannular aromatic compounds containing four or more
six-membered rings
by R. BOLTON

xiv

List of Common Abbreviations and Symbols Used

A	acid
Å	Ångström units
Ac	acetyl
a	axial
as, asymm.	asymmetrical
at.	atmosphere
B	base
Bu	butyl
b.p.	boiling point
c, C	concentration
CD	circular dichroism
conc.	concentrated
D	Debye unit, 1×10^{-18} e.s.u.
D	dissociation energy
D	dextro-rotatory; dextro configuration
d	density
dec., decomp	with decomposition
deriv.	derivative
E	energy; extinction; electromeric effect
*E*1, *E*2	uni- and bi-molecular elimination mechanisms
E1cB	unimolecular elimination in conjugate base
ESR	electron spin resonance
Et	ethyl
e	nuclear charge; equatorial
f.p.	freezing point
G	free energy
GLC	gas liquid chromatography
g	spectroscopic splitting factor, 2.0023
H	applied magnetic field; heat content
h	Planck's constant
Hz	hertz
I	spin quantum number; intensity; inductive effect
IR	infrared
J	coupling constant in NMR spectra
J	Joule
K	dissociation constant
k	Boltzmann constant; velocity constant
kcal	kilocalories
M	molecular weight; molar; mesomeric effect
Me	methyl
m	mass; mole; molecule; *meta-*
m.p.	melting point
Ms	mesyl (methanesulphonyl)

[M]	molecular rotation
N	Avogadro number; normal
NMR	nuclear magnetic resonance
NOE	Nuclear Overhauser Effect
n	normal; refractive index; principal quantum number
o	*ortho-*
ORD	optical rotatory dispersion
P	polarisation; probability; orbital state
Pr	propyl
Ph	phenyl
p	*para-*; orbital
PMR	proton magnetic resonance
R	clockwise configuration
S	counterclockwise configuration; entropy; net spin of incompleted electronic shells; orbital state
S_N1, S_N2	uni- and bi-molecular nucleophilic substitution mechanism
S_Ni	internal nucleophilic substitution mechanism
s	symmetrical; orbital
sec	secondary
soln.	solution
symm.	symmetrical
T	absolute temperature
Tosyl	*p*-toluenesulphonyl
Trityl	triphenylmethyl
t	time
temp.	temperature (in degrees centrigrade)
tert	tertiary
UV	ultraviolet
α	optical rotation (in water unless otherwise stated)
$[\alpha]$	specific optical rotation
ϵ	dielectric constant; extinction coefficient
μ	dipole moment; magnetic moment
μ_B	Bohr magneton
μg	microgram
μm	micrometer
λ	wavelength
ν	frequency; wave number
χ, χ_d, χ_μ	magnetic; diamagnetic and paramagnetic susceptibilities
(+)	dextrorotatory
(−)	laevorotatory
−	negative charge
+	positive charge

Second Supplements to the 2nd Edition of Rodd's Chemistry of Carbon Compounds, Vol. III F(Partial), G and H, by M. Sainsbury

1

Chapter 24

BIS- AND TRIS-(PHENYLALKYL)BENZENES AND LINEAR
OLIGO (PHENYLENEALKYL)S, THEIR DERIVATIVES AND
OXIDATION PRODUCTS

J. M. MELLOR

1. Introduction

Interest in this area is dominated by the chemistry
of cyclophanes (cyclic phenylenealkyls) (F. Vogtle,
"Cyclophane Chemistry",J. Wiley, Chichester 1993;
and F. Diederich, "Cyclophanes",Royal Society of
Chemistry, 1991) and the chemistry of calixarenes
(cyclic phenol-formaldehyde oligomers) (C. D.
Gutsche, "Calixarenes", Royal Society of Chemistry,
Cambridge,1989). The establishment of simple
synthetic routes to both series of compounds has
permitted the recent thorough exploration of many
facets of their chemistry. The linear poly-
(phenylenealkyl)s continue to have important coating
applications, and their isocyanates have widespread
adhesive and fabric applications following their
polymerisation with alcohols. However the synthesis
and applications of such poorly defined polymers is
not a subject to be discussed here. Following a
review of the important cyclophane and calixarene
areas, a brief update is given of the rather limited
types of natural products having structures relevant
to this chapter.

2. Cyclophanes

The positions of bridges in cyclophanes are
indicated by the prefix ortho, meta or para.
Alternatively a numerical indication of bridging can
be appropriate, e.g. as in paracyclophane or
(1,4)cyclophane.In a multibridged cyclophane the

bridging is easily defined if the bridges are alike,
e.g. as in superphane, [2,2,2,2,2,2]cyclophane.
Beyond the simple single bridged [n]cyclophanes and
the double bridged [m,n]- and multibridged-
cyclophanes, e.g. [2,2]metacyclophane interest now
extends to more complex architectures. A sub-group
of cyclophanes are the [1n]metacyclophanes,known as
calixarenes, which command much interest.
Multilayered cyclophanes, e.g [2,2][2,2]para-
cyclophane, are a further significant subgroup. The
chemistry of the cyclophanes is divided into
sections concerned with [n]cyclophanes,[m,n]cyclo-
phanes, multilayered cyclophanes and other multibridged cyclophanes. Calixerenes are discussed
separately.

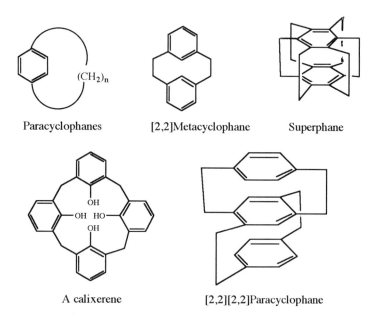

Paracyclophanes   [2,2]Metacyclophane   Superphane

A calixerene   [2,2][2,2]Paracyclophane

(a) [n]Cyclophanes

Although the synthesis of [n]cyclophanes (n>6) is
well established (see F. Vogtle),there has been a
recent interest in the synthesis of smaller cyclophanes. Thermolysis or photolysis of Dewar benzenes
permits the synthesis of [n]paracyclophanes

(F. Bickelhaupt, Pure Appl.Chem.,1990, _62_, 373).
When n=6 thermolysis at 60^{0}C affords [6]paracyclo-
phane. When n=5 similar thermolysis does not allow
isolation of the unstable [5]paracyclophane. However
photolysis at -20^{0}C leads to a photostationary
equilibrium between the Dewar benzene and the
cyclophane. Evidence for generation of
[4]paracyclophane by photolysis at -20^{0}C has been
given (F. Bickelhaupt et al.,J. Am. Chem. Soc.,
1987, _109_, 2471) and at 77K [4]paracyclophane is
sufficiently stable to permit the uv-spectrum
(λ340nm.) to be observed (T. Tsuji and S. Nishida,
J. Am .Chem. Soc.,1988,_110_,2157).

The instability of the strained boat shaped benzene
rings of the smaller cyclophanes is reflected in
their high reactivity with acids and alcohols. This
high reactivity is observed in both the meta- and
para- series. The metacyclophanes are somewhat less
strained than the paracyclophanes and this is
reflected in the ability to promote rearrangement of
the latter with acids to give the former.
Thus [6]paracyclophane gives [6]metacyclophane (Y.
Tobe et al., Tetrahedron, 1986, _42_, 1851).

The strain in [6]paracyclophane explains the unusual
addition of alkyllithiums (Y. Tobe et al., J. Org.
Chem., 1993, _58_, 5883). The strain in [4]meta-
cyclophane is so great that a remarkable

dimerisation by a Diels Alder reaction gives both [4,4]metacyclophane and [4,4]paracyclophane.(F. Bickelhaupt and W. H. de Wolf, Recl. Trav. Chim. Pays-Bas, 1988, 107, 459). [6]Paracyclophane is the smallest stable member of the [n]paracyclophane series. It is sufficiently strained that reactions with electrophiles are anomalous. Although p-diethylbenzene with strong bases reacts at

a benzylic centre, [6]paracyclophane after metallation by a superbase, and reaction with electrophiles, affords both mono- and di-substituted products, thus effecting overall an aromatic electrophilic substitution. The lack of reactivity at a benzylic site in [6]paracyclophane is attributed to strain preventing stabilisation of the benzylic anion in the bent aromatic system (Y. Tobe et al., Tetrahedron Lett., 1993, 34, 4969).

Chiral derivatives of simple cyclophanes have been used to achieve enantioseparations. An enantiomer of [10]paracyclophane-13-carboxylic acid has been resolved and used, following linkage to a silanized silica gel via an amide linkage, as a chiral stationary phase. Moderate hplc separations of enantiomeric pairs were achieved (S. Oi and S. Miyano, Chem. Lett. 1992, 987).

(b) [m,n]Paracyclophanes

Routes to [2,2]paracyclophanes based on dimerisation
of polyenes, generated by elimination reactions, are
well established (see F.Vogtle). The [2,2]paracyclo-
phanes have been well known for their potential as
precursors of thin film polymers known as parylenes
(W. F. Gorham, J. Polym. Sci., Part A, 1966, 4,
3027). Octafluoro[2,2]paracyclophane, might be
expected to give particularly stable coatings. An
interesting reductive elimination using $TiCl_4$ and
$LiAlH_4$ provides an effective route to this
cyclophane (W. R. Dolbier, et al., J. Org. Chem.,
1993, 58, 1827).

Cheletropic eliminations have always played a part
in the synthesis of cyclophanes. A new route is
based on pyrolytic decomposition (at 625^0C) of
ketones (F. Vogtle et al., Chem. Ber., 1993, 126,
97).

Functionalisation of [2,2]paracyclophane at both the
bridge and the aromatic sites is possible.
Sulfonation gives mainly the 4,15-disulfonic acid,
but milder conditions permit monofunctionalisation
in high yield (H. Cerfontain et al., Recl. Trav.

Chim. Pays-Bas, 1992, 111, 379). Photobromination
affords bridge substituted bromides. By subsequent
treatment with potassium tert-butoxide, and
reduction of the resultant vinyl bromides with
LiAlH$_4$, the paracyclophane-1,9-diene is obtained in
25% overall yield from [2,2]paracyclophane. (A. de
Meijere et al., Chem Ber, 1987, 120, 1667). The
diene, acting as a dienophile undergoes Diels Alder
additions characterised by inverse electron demand
(A. de Meijere and B. Konig, Helv. Chim. Acta, 1992,
75, 901). Derivatives of [2,2]paracyclophane have
been prepared and used as chiral auxiliaries.
Thus 2-formyl-3-hydroxy[2,2]paracyclophane is a
chiral equivalent of salicylaldehyde (Y. Belokon et
al., Angew. Chem. Int. Ed. Engl., 1994, 33, 91).

[2,1]Paracyclophanes are unknown. Surprisingly
however [1,1]paracyclophane has been prepared and
has a moderate stability at -20^{0}C (T. Tsuji et al.,
J. Am. Chem. Soc., 1993, 115, 5284).

(c) [m,n]Metacyclophanes

The chemistry and conformational analysis of
[2,2]metacyclophanes have been particularly well
studied. Recently two new routes have been developed
to [2,2]metacyclophane. Elimination of N$_2$O from
nitrosamines is effective (H. Takemura, et al.,
Tetrahedron Lett., 1988, 29, 1031). Extrusion of
selenium is also efficient (H. Higuchi et al., Bull.
Chem. Soc. Jpn., 1987, 60, 4027).

The chemistry of other [m,n]metacyclophanes is less studied because the necessary synthetic entries have not been developed. However the synthesis and conformational analysis have been reported for [3,2], [3,3], [4,2], and [4,3]metacyclophanes (D. Krois and H. Lehner, Tetrahedron, 1982, 38, 3319). Ring inversion characteristics are also reported for higher members of the series [5,2], [6.2]. and [8,2] metacyclophanes (T. Yamato et al., J. Org. Chem., 1992, 57, 5243).A decreasing length of the bridge increases the barrier to ring inversion relating anti- and syn-conformers. In the case of [2,2]meta-cyclophane the anti and syn conformers readily interconvert above 0^0C. However syn[2,2]meta-cyclophane was first synthesized in 1985. The final step requires decomplexation of the syn complex at -45^0C (R. H. Mitchell et al., J. Am. Chem. Soc., 1985, 107, 3340).

The presence of large substituents increases the energy barrier to flipping. Hence in a series of [n,2]metacyclophanes carrying tert-butyl substituents, both anti- and syn-isomers can be individually isolated (T. Yamata et al., Chem. Ber. 1992, 125, 2443). The ethers can be cleaved with $AlCl_3$-CH_3NO_2 to afford phenols without loss of stereochemical integrity. Hence anti- and syn-[3,2]metacyclophanequinones have been prepared. The attempt to effect a similar oxidation in the [4,2] series failed to give the syn-quinone. The

R = t-Bu

n=2: 85%	0%
n=3 59%	19%
n=4 51%	19%
n=5 41%	39%
n=6 41%	42%

larger chain length facilitated ring flipping. Hence both anti- and syn-phenols afforded the anti-quinone (T. Yamato et al., J. Org. Chem., 1992, 57, 5154).

An interesting route to a [7,7]metacyclophane is based on generation of metallophanes and their subsequent decomposition in moderate yield (E. Lindner et al., Angew. Chem. Int. Ed. Engl., 1994, 33, 321).

The electrophilic functionalisation of
[n,2]metacyclophanes is complicated by the isolation
of substituted tetrahydropyrenes, formed by an
addition elimination mechanism. Formylations using
Cl_2CHOMe catalysed by $TiCl_4$ permit high yields of
mono- or di- formyl[n,2]metacyclophanes to be
isolated. The order of reactivity is [2,2]> [3,2] >
[4,2] (T. Yamato et al., J. Chem. Res., 1993, 44).
An alternative strategy for functionalisation of
metacyclophanes is via prior formation of
bis(tricarbonylchromium) complexes. Thus a one pot
route to a bisethoxycarbonyl[2,2]metacyclophane is
based on this approach (F. Vogtle et al., J. Chem.
Soc. Perkin Trans 2, 1992, 2095).

(d) Other [m,n]cyclophanes

A variety of mixed cyclophanes e.g. metapara-,
orthometa- and orthopara- are known. Some of the
more important examples are shown in TABLE 1.
The original preparation of [2,2]metaparacyclophane
was made by acid catalysed rearrangement of
[2,2]paracyclophane. Another convenient route to
mixed cyclophanes is based on cheletropic
elimination of sulfur or sulfur dioxide from

dithiacyclophanes. This method has been used in the preparation of tert-butyl substituted [2,2]metapara-cyclophanes (T. Yamato et al., J. Chem. Soc. Perkin Trans. 1, 1992, 2675 (see over).

TABLE 1

OTHER [m,n]CYCLOPHANES

	mp^{0}C	Reference
[2,2]Metaparacyclophane	79-80	1
[3,3]Metaparacyclophane	90-91	2
[2,2]Orthometacyclophane	oil	3
[2,2]Orthoparacyclophane	168-170	4
[3,3]Orthoparacyclophane	oil	5
[3,3]Orthometacyclophane	oil	5
[4,4]Orthoparacyclophane	oil	5

1. See Supplement to 2nd edition, vol. 3F.
2. T. Otsubo et al., Bull. Chem. Soc. Jpn., 1979, 52, 1515.
3. H. Hopf et al., Angew. Chem. Int. Ed. Engl., 1989, 28, 455.
4. Y. Tobe et al., J. Am. Chem. Soc., 1993, 115, 1173.
5. H. A. Staab et al., Tetrahedron Lett., 1991, 32, 2117.

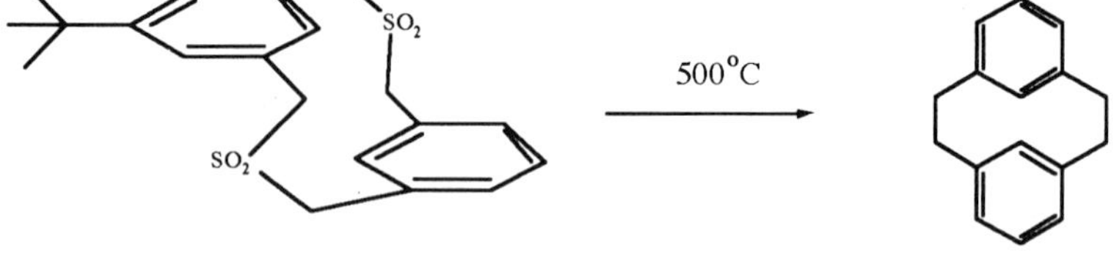

(e) Multilayered cyclophanes

Although some multilayered compounds are known
having features of mixed meta- and para-cyclophanes,
the vast majority of studies have been concerned
with either multilayered metacyclophanes or
multilayered paracyclophanes. Triple layered and
quadrupole layered paracyclophanes are readily
accessible. Similar procedures have been used to
give access to quintuple- and sextuple-layered
paracyclophanes (see F. Vogtle, "Cyclophane
Chemistry", J. Wiley, Chichester 1993).

A different approach has been used in the synthesis
of metacyclophanes. Photoextrusions of sulfur are
effective and have been used in the synthesis of
both triple and quadrupole layered metacyclophanes
(see F. Vogtle p.295).

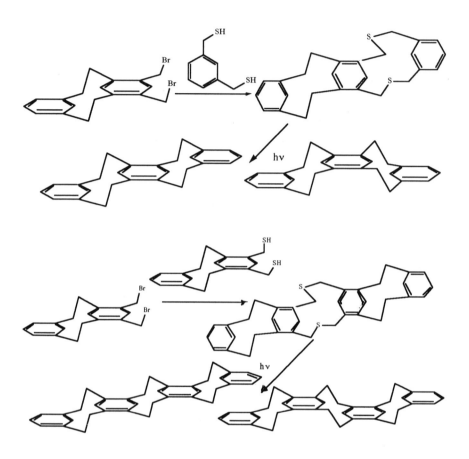

The conformational isomerism has been studied for both the above multilayered metacyclophanes and the mixed systems shown here and over the page (see F. Vogtle p. 305).

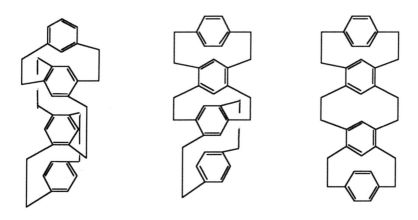

A feature of multilayered cyclophanes is the transmission of electronic effects from one ring to another. Addition of successive layers facilitates electron donation. Hence in complexes with TCNE the wavelength maxima are shifted to longer wavelength on increasing the number of layers. Internal charge transfer is observed within donor acceptor phanes.

14

The increased electron availability resulting from addition of extra layers leads to faster bromination of multilayered cyclophanes. Triple layered paracyclophanes are prone to acid catalysed rearrangement to afford metacyclophanes.

An unusual synthetic route to triple layered cyclophanes is based on dimerisation and rearrangement (H. Hopf et al., Angew. Chem. Int. Ed. Engl., 1992, 31, 1073).

(f) Multibridged cyclophanes

The early history of these cyclophanes concerned the [2n]phanes and was crowned by the synthesis of superphane by Boekelheide in 1979 (see R. Gleiter and D. Kratz, Acc. Chem. Res.,1993, <u>26</u>, 311).

The 8 possible 'symmetrical' [2_n]cyclophanes are known but only 2 'unsymmetrical' [2_n]cyclophanes are known. Emphasis has recently moved to a study of the [3_n]cyclophanes. Data for these cyclophanes are collected in TABLE 2.

TABLE 2

[3_n]CYCLOPHANES

	mp^{0}C	Reference
[3_2] (1,3)Cyclophane	79–80	1
[3_3] (1,3,5)Cyclophane	140	2
[3_4] (1,2,3,5)Cyclophane	254.5–255	3
[3_4] (1,2,4,5)Cyclophane	>300	3
[3_5] (1,2,3,4,5)Cyclophane	265–266.5	4

1. T. Shinmyozu et al., Chem. Lett. , 1976, 1405; M. F. Semmelhack et al., J. Am. Chem. Soc., 1985, <u>107</u>, 7508.
2. A. J. Hubert, J. Chem. Soc C., 1967, 6; T. Otsubo et al., Bull. Chem. Soc. Jpn., 1979, <u>52</u>, 1515; T. Meno et al., Canad. J. Chem., 1990, <u>68</u>, 440;
3. T. Shinmyozu et al., Chem. Ber., 1993, <u>126</u>, 1815.
4. T. Shinmyozu et al., Chem. Lett., 1994, 669.

The ultimate target in this series [3_6](1,2,3,4,5,6)cyclophane has yet to be synthesized. The satisfactory procedure for introduction of the 3-carbon bridges is illustrated on the next page.

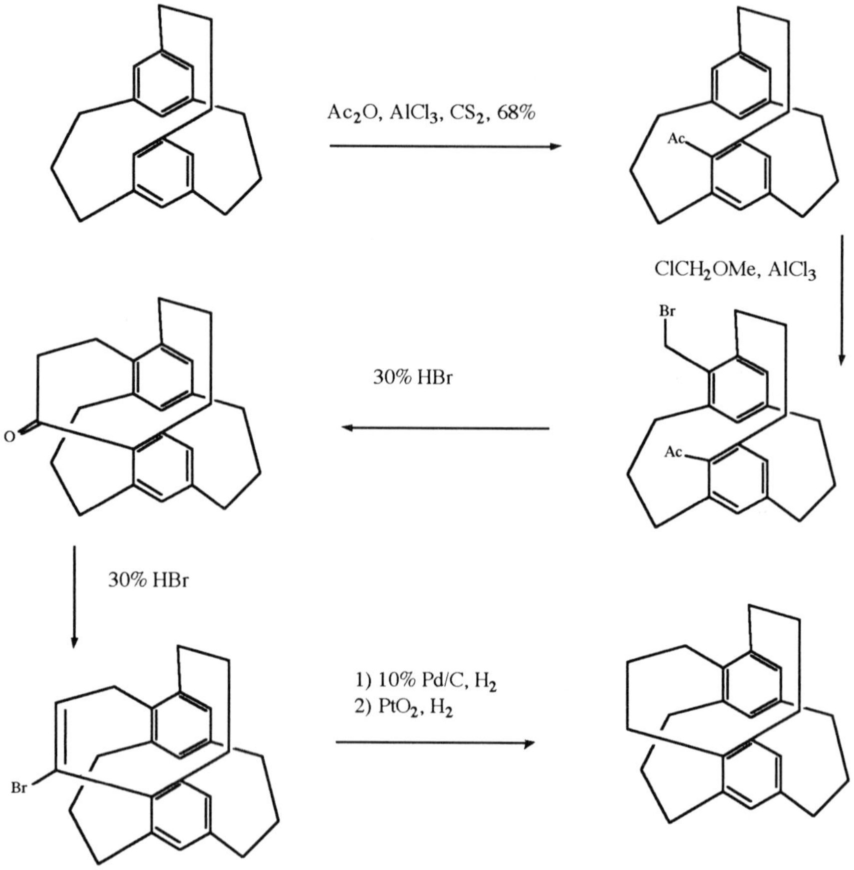
Ac2O, AlCl3, CS2, 68%
Ac
ClCH2OMe, AlCl3
Br
Ac
30% HBr
O
30% HBr
Br
1) 10% Pd/C, H2
2) PtO2, H2

(g) $[m_n](m,n)$Cyclophanes

The series of $[0_n]$cyclophanes, e.g. $[0_6]$orthocyclophane and $[0_5]$metacyclophane are outside the scope of this chapter. Of the remaining skeletal types it is particularly the calix[n]arenes, a series of $[1_n]$metacyclophanes,, e.g. calix[4]arene, or $[1_4]$metacyclophane which dominate recent studies. They are discussed in section 3.

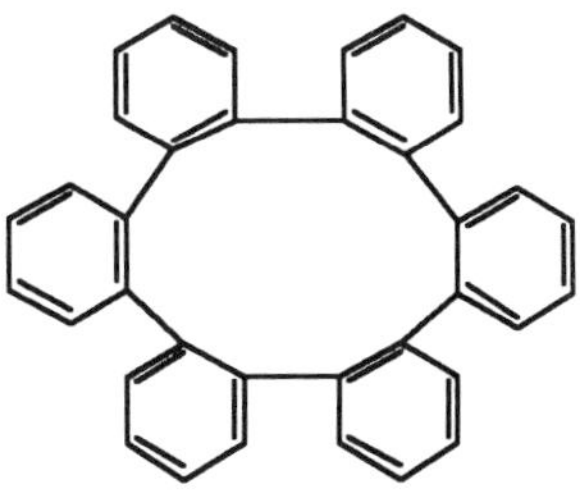

[0$_6$]Orthocyclophane

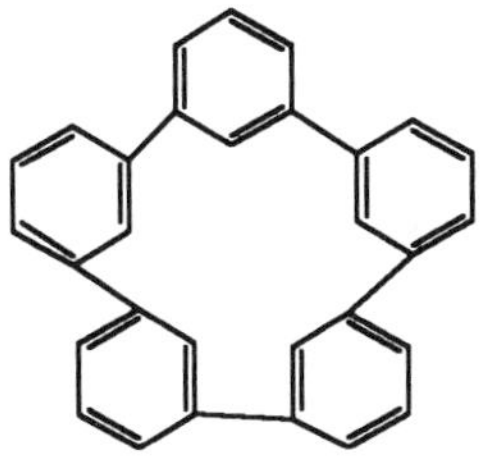

[0$_5$]Metacyclophane

Similarly some aspects of the chemistry of [1$_3$]-orthocylophanes, the cyclotricatechylenes and the cyclotri- and tetra-veratrylenes are mentioned alongside the calixarenes.

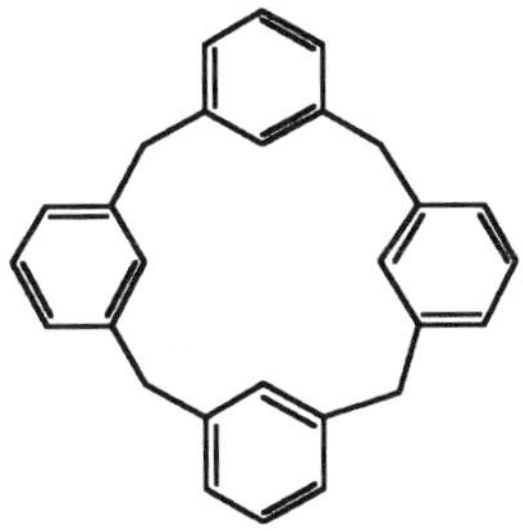

[1$_4$]Metacyclophane

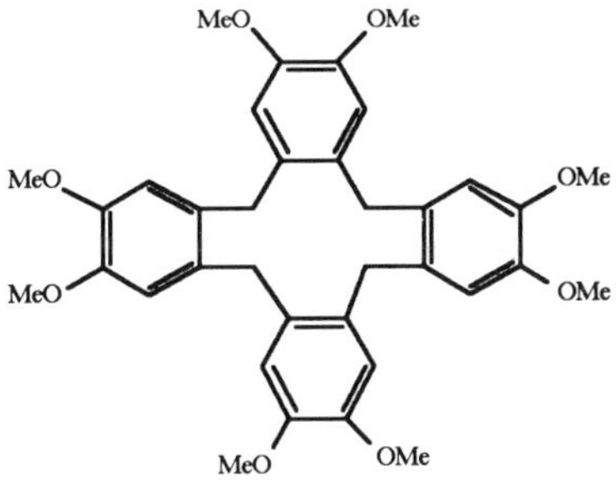

Cyclotetraveratrylene

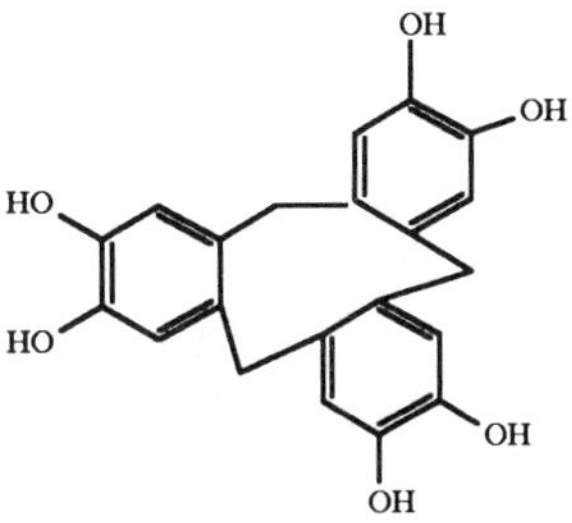

A [1$_3$]Orthocyclophane cyclotricatechylene

The synthesis of the less studied ortho- and para [1$_n$], and [2$_n$]cyclophanes was commenced some time ago, but as shown in TABLES 3 and 4 many examples have only recently been reported.

18

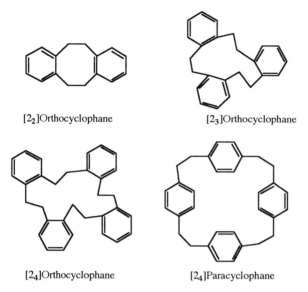

[2₂]Orthocyclophane   [2₃]Orthocyclophane

[2₄]Orthocyclophane   [2₄]Paracyclophane

TABLE 3

ORTHOCYCLOPHANES

	mp^{0}C	Reference
[1$_3$] Orthocyclophane	277–278	1
[1$_4$] Orthocyclophane	345–346	2
[1$_5$] Orthocyclophane	257–258	2
[1$_6$] Orthocyclophane-1,3-dione1	288–289	3
[1$_6$]Orthocyclophane-1,4-dione **2**	>290	3
[2$_3$] Orthocyclophane	184.5	4
[2$_4$] Orthocyclophane	205	5
[2,1,2,1] Orthocyclophane	195	5
[1$_4$](1,2)(1,2)(1,2)(1,3)Cyclophane	160	6

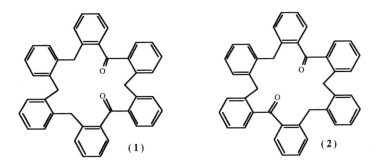

(1)   (2)

1. W. Y. Lee et al., J. Chem. Soc. Perkin. Trans. 1,
   1992, 881.
2. W. Y. Lee et al., J. Org. Chem., 1992, 57, 4074.
3. W. Y. Lee and C. H. Park, J. Org. Chem., 1993,
   58, 7149.
4. W. Baker et al., J. Chem. Soc., 1945, 27.
5. E. D. Bergmann et al., J. Am. Chem. Soc., 1953,
   75, 4281.
6. W. Y. Lee et al., J. Chem. Soc. Perkin Trans. 1,
   1993, 719.

TABLE 4

PARACYCLOPHANES

	mp^{0}C	Reference
[2_3] Paracyclophane	162–163	1
[2_4] Paracyclophane	179–181	1
[2_5] Paracyclophane	170–172	2
[2_6] Paracyclophane	200–202	2

1. H,G. Wey et al., Chem. Ber., 1990, 123, 93
   L. A. Errede et al., J. Am. Chem. Soc., 1960, 82,
   5218
   W. Baker, J. Chem. Soc., 1951, 1114.
2. F. Vogtle and W. Kissener, Chem. Ber., 1984, 117,
   2538.

3 Calixarenes

The calixarenes are so named because of the similarity of their molecular shape to that of a Greek chalice. The size of the internal cavity is determined by the number of arene units and the preferred conformation. The calixarenes act as excellent hosts for guest molecules and this area of molecular recognition has been much studied recently (S. Shinkai, Tetrahedron, 1993, 49, 8933; C. D. Gutsche, 'Calixarenes', Royal Society of Chemistry, 1989 Cambridge; J. Vicens and V. Bohmer (eds) 'Calixarenes: A Versatile Class of Macrocyclic Compounds', Kluwer, Dordrecht, 1991). The calixarenes are a family of compounds related to the phenol formaldehyde resins originally investigated by von Baeyer in the last century. As $[1_n]$meta-cyclophanes the more important members of the series are the calix[4]arenes and the calix[6]arenes. However all members of the series n=4-14 are now known.

A calix[4]arene

A calix[5]arene

The isolation of calixarenes by direct reaction of phenol with formaldehyde is unsatisfactory, due to excessive formation of polymers. However by blocking the para position with alkyl or aryl substituents, isolation of calixarenes is greatly facilitated. The most popular entry has been by use of p-tertbutyl-phenol. Procedures are well described for the direct transformation of p-tertbutylphenol to calix-[4]arene, calix[6]arene and calix[8]arene. Thus initial heating of p-tertbutylphenol with 37%

formaldehyde and 0.045 equivalents of sodium hydroxide at 110-120°C for 2h, followed by heating the reaction mixture in refluxing diphenyl ether for 2h on cooling affords p-tertbutylcalix[4]arene in 50% yield.

p-tertButylcalix[6]arene is better obtained (yield 80%) by a similar procedure using potassium hydroxide. Removal of the tertbutyl groups with $AlCl_3$ permits the parent calixarenes to be isolated. However in order to prepare diversely substituted calixarenes a stepwise synthetic approach is to be preferred. Typically condensation of phenolic fragments with benzylic halides provides a satisfactory solution.

The calixarenes are mobile systems and therefore are characterised by an interesting conformational isomerism. Cornforth originally proposed that calix[4]arenes might be characterised by four discrete conformations, the cone, the partial cone and 1,2- and 1,3-alternating conformations. With the parent calix[4]arene having four hydroxyl groups the cone conformer is preferred at room temperature, but on heating to 45^{0}C interconversion between the conformers occurs. Substitution of a hydroxyl group by a more bulky alkyloxy or acyloxy substituent greatly increases the barriers to conformational isomerism. Thus tetrabenzylcalix[4]arene is locked in the cone conformation. The tetraacetate exists as the 1,3-alternating conformation. An illustration of the conformational complexity of calix[4]arenes is provided by an example in which three conformational isomers can be independently isolated. Extraction studies establish that the cone has a strong metal ion affinity. By comparison the partial cone has a poor metal ion affinity. The noticeable metal ion affinity of the 1,3-alternate conformer is attributed to complexation at the ester sites (S. Shinkai et al., J. Org. Chem., 1992, 57, 1516).

The selective functionalisation of the calixarenes is the key to permitting their chemistry to be fully explored. Monoalkylated calix[4]arenes are obtained by alkylation with potassium carbonate in acetonitrile or caesium fluoride in dimethyl-formamide. These conditions minimize overalkylation (D. N. Reinhoudt et al., Tetrahedron, 1991, 47, 8379). Surprisingly tetraalkylation under more vigorous conditions affords 1,3-alternate conformers in good yield (D. N. Reinhoudt et al., J. Org. Chem., 1992, 57, 5394). Conditions for dialkylation lead to 1,3-products having cone conformations (D. N. Reinhoudt, Pure Appl. Chem., 1993, 65, 387) and by use of NaH in DMF, 1,2-products, again having cone conformations, are obtained. By use of potassium tertbutoxide in tetrahydrofuran the tetraethoxy p-tertbutylcalix[4]arene has been obtained as the 1,2-alternate conformer.

In a cone conformation the region of the hydroxyl groups is called the lower rim. Extensive structural modifications about the lower rim have been achieved, e,g, the alkylation described above. However the upper rim is also susceptible to considerable structural variation. This may be achieved by stepwise synthesis. Alternatively removal of the tertbutyl groups using aluminium chloride, can be followed by a new functionalisation at these para positions. At the top rim both amino groups (C. D. Gutsche and K. Y. Nam, J. Am. Chem. Soc., 1988, <u>110</u>, 6153) and very bulky groups (S. K. Sharma and C. D. Gutsche, Tetrahedron Lett., 1993, <u>34</u>, 5389) have been introduced to calix[4]arenes.

The interest in calixarenes as molecular hosts has developed later by comparison with crown ethers and cyclodextrins. However it is now clear that their supramolecular chemistry offers many possibilities. They have the capacity to bind both neutral and ionic guests. The cation binding ability of calixarenes has been applied to the recovery of caesium from aqueous solutions of nuclear waste, and of uranium from sea water. PVC membranes incorporating calix[4]arenes have been used in the development of ion-selective electrodes useful in determination of sodium in human blood plasma (R. Forster et al., Anal. Chem., 1991, <u>63</u>, 876). By judicious incorporation of suitable ligating functional groups the molecular architecture of calixarenes can be modified to permit not only selective binding of specific metal ions, but also to permit the development of switching devices. Incorporation of 1,3-bispyridyl- or 1,3-bipyridyl-units permits the formation of copper complexes (P. D. Beer et al., Tetrahedron, 1993, <u>48</u>, 9917).

$R_1 = CH_2COOEt$

$R_2 =$

A calix[4]arene, efficient in capture of sodium ions is capped by two anthracene units. On irradiation a photochemical addition between the two anthracene units results in a closed cavity. The sodium ion is squeezed out, but thermal reversion can regenerate the original calixarene, which has a high affinity for sodium ions. Hence the ionophoric properties can be photochemically controlled with implications for ion extraction and ion sensing. (S. Shinkai et al., Tetrahedron Lett., 1992, 33, 2163). Suitable disposition of a fluorophore (a pyrene residue) and a quencher (a nitrobenzene residue) permits ion binding to be monitored by observation of fluorescence intensity (S. Shinkai et al., J. Chem. Soc. Chem. Commun., 1992, 730)

The useful host characteristics of calixarenes are not limited to capture of metal ions. Selective uptake of neutral molecules has been developed. Suitably substituted calixarenes can separate the isomeric xylenes (R. Perrin et al., Pure Appl. Chem., 1993, 65, 1549).

An extensive patent literature is developing concerning the applications of calixarenes. Although developments are limited by the ease of synthesis, their potential in electronic devices, in coatings and in catalysis, is recognised. Certain calixarenes act as efficient phase transfer catalysts (comparable to crown ethers) (S. Shinkai et al., Tetrahedron, 1993, 49, 6763). Certain steroids are known to act as gelators of organic fluids. This property is also associated with calixarenes having long chain para substituents (S. Shinkai et al., J. Chem. Soc. Perkin Trans. 2, 1993, 347). A consequence of the conformational rigidity of substituted calixarenes is their chirality. Potentially the applications of calixarenes as chiral hosts are considerable. The progress in making chiral calixarenes is still relatively modest. However examples are well known of complete resolution, by HPLC on chiral columns, of enantiomeric calix[4]arenes. With the larger calixarenes greater conformational mobility is observed, and conformational studies on calix[6]arenes and calix[8]arenes are less advanced.

Alkylation of calixarenes with αω-dihalides or ditosylates offers the possibility of synthesis of double calixarenes, characterised by a very large central cavity. Although it is known that intramolecular 1,3-capping is possible, e.g. in a calix[5]arene (V. Bohmer et al., Tetrahedron, 1993, 49, 6019), successful conditions have been established for linkage of two separate calix[4]arenes by two, or three bridges (J. Vicens et al., Pure Appl. Chem., 1993, 65, 585; and Synlett, 1993, 719.) Further elaboration of calixarenes permits the creation of cavitands and carcerands, having large central cavities. Similarly cyclotriveratrylenes have been extended to give cryptophanes. Much of this interesting chemistry is outside the scope of this chapter.

The study of metacyclophanes has been dominated by those regular structures, generated by direct synthesis, having one carbon bridges. Stepwise synthesis is now making available a series of irregular structures. Thus incorporation of biphenyl units affords a series of [1,0,1,0,1,0]metacyclo-phanes. (T. Yamato et al., Chem. Ber., 1993, 126, 1435). Similarly units, having alternative hydroxyl group locations, have been constructed (V. Bohmer et al., Tetrahedron Lett., 1992, 33, 769).

A [1,0,1,0,1,0]Metacyclophane

A [1$_4$]Metacyclophane

Other regular metacyclophanes than calixarenes have been synthesized, e.g. members of the [2,2,2]- and and [2,2,2,2]-series (M. Tashiro et al., J. Org. Chem., 1989, 54, 2632) and members of the [3,2,3,2]- and [3,3,3,3]-series (D. H. Burns et al., J. Org. Chem., 1993, 58, 6526).

A [2,2,2,2]Metacyclophane   A [2,2,2]Metacyclophane   A [3,3,3,3]Metacyclophane

4 Naturally occurring products

The occurrence of phloroglucinol derivatives as fern constituents has been known for some time (see next page). Systematic investigation has established their occurrence as dimers-hexamers (T. Murakami and N. Tanaka, Fortschritte der Chemie organischer Naturstoffe, Springer Verlag, 1988, 54, 1).

Although diarylmethanes are frequently found in ferns, diarylethanes are also known. Notholaenic acid is found in fern frond exudates, but the related lunularic acid is found both in liverworts and algae.

Notholaenic acid   Lunularic acid

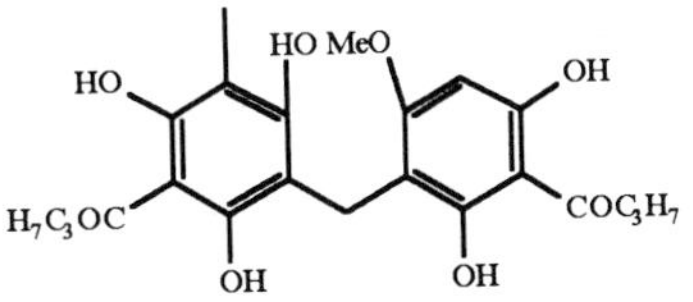

Phloraspin

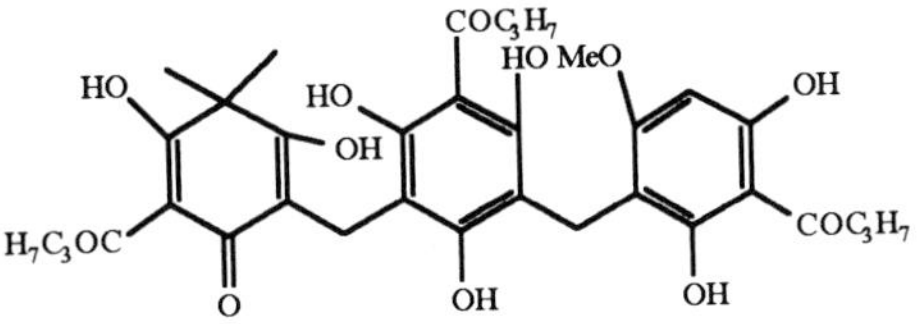

Trisdesaspidin

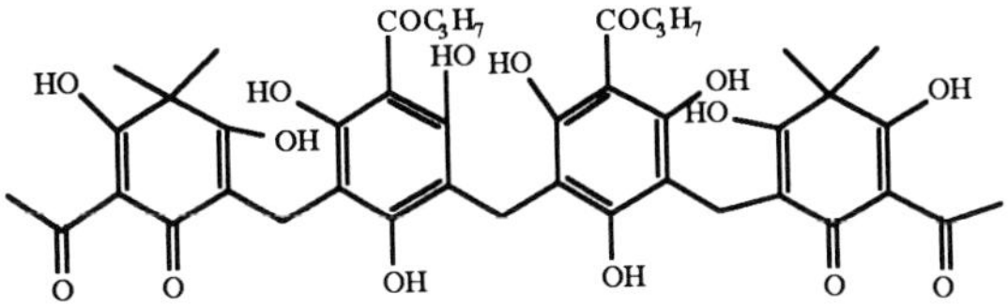

Dryocrassin

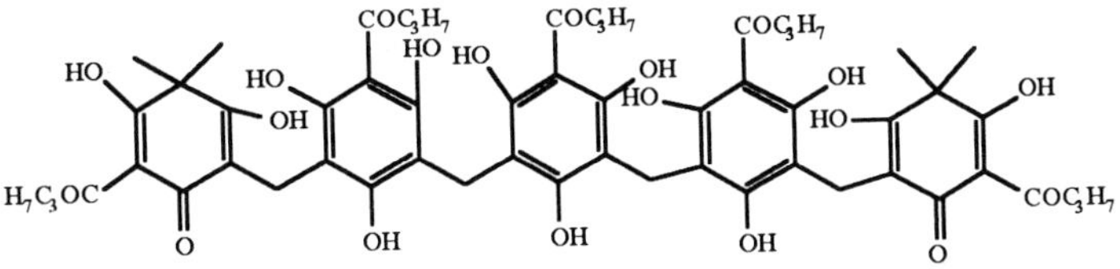

Pentaalbaspidin

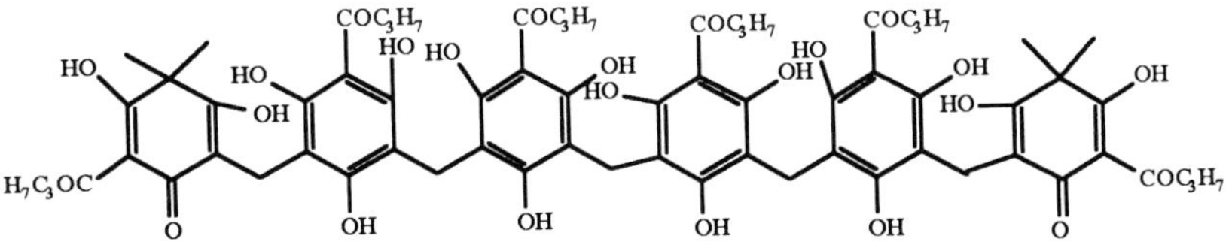

Hexaalbaspidin

Stilbenes, chalcones and some 1,4-diarylbutanes
occur via a shikimate biosynthesis pathway (P. M.
Dewick, Nat. Prod. Rep. 1994, 11, 173) Some
dihydrochalcones display antimicrobial activity, and
also have potential as sweetening agents, and
uvaretin has cytotoxic activity. In particular
liverworts are rich in bibenzyls (J. Gorham, Proc.
Phytochem. Soc. Eur., 1990, 29, 171).

Uvaretin

The lignans afford a rich variety of structural type
(R. S. Ward, Nat. Prod. Rep., 1993, 10, 1). 1,4-
Diarylbutanes are prominent but the
dibenzocyclooctadiene derivatives represent an
important subgroup. Gomisin A is found in Korean red
ginseng and the cytotoxic properties of a range of
steganacin derivatives have been reported (K.
Tomioka et al., J. Med. Chem., 1991, 34, 54). The
biological properties of these compounds have
stimulated extensive synthetic work. An important
route to both steganes and isosteganes is by non-
phenolic oxidative coupling of diarylbutanes.

Gomisin A   Steganacin

Ruthenium salt oxidations are particularly efficient
(Y. Landais et al., Tetrahedron Lett., 1986, 27,

5377). An alternative oxazoline mediated coupling has been used in the synthesis of (−) schizandrin (A. M. Warshawsky and A. I. Meyers, J. Am. Chem. Soc., 1990, 112, 8090).

A number of naturally occurring metacyclophanes having a 7-membered alkyl chain are known, e.g. porson.

Porson

A well documented class of antibiotics are the ensamycins, characterised by an alkyl chain with meta bridging of a benzenoid ring. Many including

the antileukemic maytansine have an amide
funtionality in the chain. Now a further series of
antibiotics, typified by garuganin-2, having meta-
bridging through an all carbon 7-membered chain,
have been discovered (S. Krishnaswamy et al, Acta.
Cryst., 1987, 43C, 527). Myricanol, which shows
insecticidal activity has been synthesized (D. A.
Whiting and A. F. Wood, J. Chem. Soc. Perkin Trans.
1, 1980, 623).

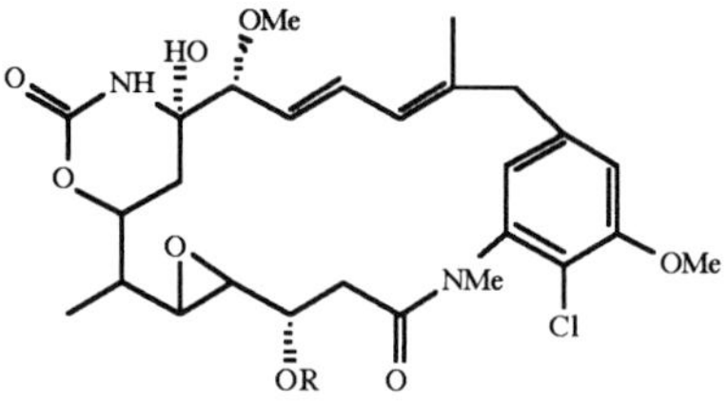

Maytansine

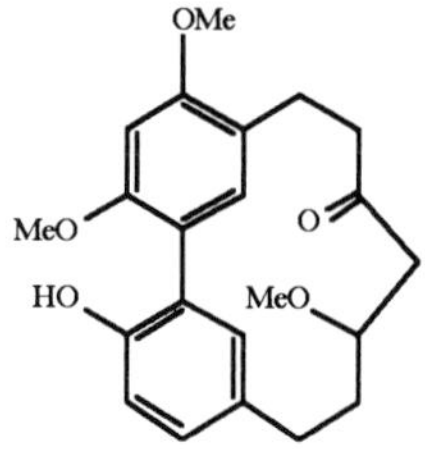

Garuganin 2

The first [7,7[paracyclophanes, called cylindro
cyclophanes have been isolated   (R. E. Moore et al,
J. Am. Chem. Soc., 1990, 112, 4061). These compounds
occur in species of blue-green algae and show
moderate cytotoxic activity (R. E. Moore et al,
Tetrahedron, 1992, 48, 3001). Their likely
biosynthesis is via a dimerisation of acetate
derived nonaketides.

A series of enediynes, calicheamicin, esperamicin,
dynemicin and neocarzinostatin are naturally
occurring, but not directly related in

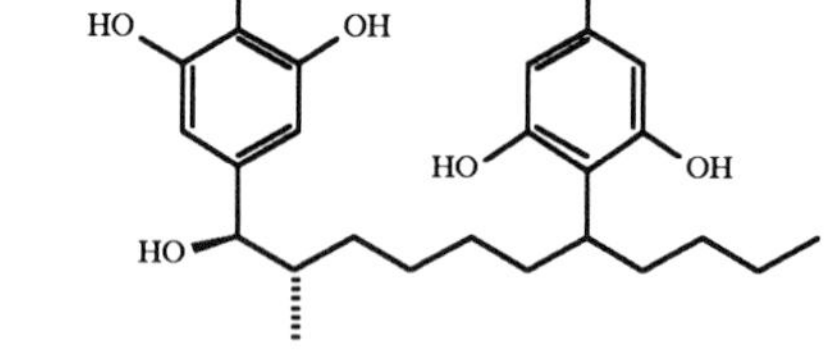

A cylindrocyclophane

structure to the subject of this chapter. However
their ability to cleave DNA and their unusual
capacity to generate benzenoid 1,4-diradicals has
attracted great interest (K. C. Nicolaou and W-M.
Oai, Angew. Chem. Int. Ed. Engl., 1991, 30, 1387;
and K. C. Nicolaou, ibid 32, 1377). Of relevance to
this section model compounds, e.g. golfomycin A have
the ability to cleave DNA and show antitumor
activity.

Golfomycin A

Second Supplements to the 2nd Edition of Rodd's Chemistry of Carbon Compounds, Vol. III F(Partial), G and H, by M. Sainsbury

33

Chapter 25

MONOCARBOXYLIC ACIDS OF THE BENZENE SERIES

MALCOLM SAINSBURY

1. Benzoic acids

(a) *Methods of preparation*

(i) *From arenes*

Certain arenes (ArX; X = H, Me, Et, ^{i}Bu, OMe, NMe$_2$, Cl, or Br) react with trichloroacetaldehyde in the presence of aluminium trichloride at -10 °C to yield the corresponding 4-(2,2,2-trichloro-1-hydroxyeth-1-yl)arene derivatives [4-X-ArCH(OH)CCl$_3$]. When these products are oxidised and hydrolysed through treatment with alkaline hydrogen peroxide they afford the appropriate benzoic acids (4-X-ArCO$_2$H) (P.Menegheli, M.C.Rezende and C.Zucco, *Synth. Commun.*, 1987, **17**, 457).

A one step oxidative carboxylation of benzene and other arenes with carbon monoxide is catalysed by palladium(II) acetate, ^{t}butyl hydroperoxide, and allyl chlorides. Some biphenyl and phenol are formed as by-products (Y.Fujiwara *et al.*, *J. Organomet. Chem.*, 1983, **256**, C35).

(ii) *From toluenes*

Optimum conditions for the oxidation of alkylarenes to benzoic acids, avoiding decarboxylation, have been determined [T.P.Kenigsberg, *Zh. Prikl. Khim.* (Leningrad), 1988, **61**, 2285; *C.A.*, 1989, **110**, 212228v].

Monoalkylbenzenes are oxidised to benzoic acids by oxygen in the presence of a cobalt/manganese catalyst [M.Komatu, T.Tanaka and H.Fujita, *Eup. P. App.*, 36,233 (1980); *C.A.* 1982, **86**, 64075s].

The oxidation of toluenes (especially nitrotoluenes) with potassium permanganate at 95 °C in nitrobenzene/water under phase-transfer conditions gives good yields of benzoic acids (J.Kulic *et al.*, *Zh. Obshch. Khim.*, 1990, **60**, 2370).

Another well established approach involves the photochlorination of toluenes, to give benzotrichlorides, followed by hydrolysis [see, for example, *F.R.*, 2,463,116 (1981); *C.A.*, 1982, **96**, 19836f].

2,3-Dibromotoluene can be converted into 2,3-dibromobenzoic acid. The first step is monobromination to afford 2,3-dibromobenzyl bromide and then reaction with pyridine to produce the corresponding N-benzylpyridinium bromide. In a modified Krönhnke procedure treatment of this salt with 4-(N,N-dimethylamino)nitrosobenzene then yields 2,3-dibromobenzaldehyde which, upon oxidation with potassium permanganate, finally produces the desired acid (M.Tashiro and K.Nakayama, *Org. Prep. Proced. Int.*, 1984, **16**, 379).

(iii) *From aryl halides by carbonylation*

The catalytic oxidative-carbonylation of iodobenzene to benzoic acid occurs under mild conditions. Thus it is sufficient to treat the iodide in aqueous sodium hydroxide/ethanol/benzene with the phase-transfer agent tetrabutylammonium bromide and to maintain the agitated mixture at 60 °C under an atmosphere of carbon monoxide in the presence of iron pentacarbonyl and dicobalt octacarbonyl as catalysts. Bromobenzene does not react in this way unless a small amount of iodobenzene is added as an initiator (J.J.Brunet and M.Taillefer, *J. Organomet. Chem.*, 1989, 361).

The double oxidative-carbonylation of 1,2-dihaloarenes is also possible using dicobalt octacarbonyl as the catalyst. In this case photo-irradiation of the reactants in aqueous sodium hydroxide at 65 °C under a pressure of carbon monoxide is recommended, although the yields are only moderate and the products are contaminated with monocarboxylated haloarenes (T.Kashimura *et al., Chem. Letters,* 1986, 483). The reaction is tolerant of carboxyl groups, thus halobenzoic acids are converted into phthalic acids through photo-stimulated carbonylation by carbon monoxide in the presence of various cobalt(II) salts acting as catalysts (K.Kudo *et al., Chem. Letters,* 1987, 577).

Palladium(II) acetate can also be used as a catalyst to carboxylate 4-substituted iodobenzenes with carbon monoxide in aqueous N,N-dimethylformamide containing potassium carbonate at 40-45 °C. Yields are highest when electron withdrawing substituents are present in the aromatic nucleus (N.A.Bumagin, K.V.Nikitin and I.P.Beletskaya, *Ivz. Akad. Nauk. SSSR, Ser. Khim.*, 1988, 1450).

Aromatic acids are also available from aryl bromides in much the same way. In one such procedure 4-bromophenol is converted into 4-hydroxybenzoic acid by heating it under an atmosphere of carbon monoxide in the presence of palladium(II) acetate, butanol and sodium carbonate [*J.P.* 04,193,847 (1992); *C.A.*, 1992, **117**, 191489h].

(iv) *From arynes*

Benzyne can be generated from either bromobenzene, or chlorobenzene, by treatment with sodamide. The aryne is then reacted *in situ* with *O*-silyl enolates of carboxylic esters to form bicyclic adducts. These adducts, after hydrolysis, yield 2-alkylbenzoic acids (S.M.Ali and S.Tanimoto, *J. Chem. Soc., Chem. Commun.*, 1988, 1465).

(v) *From benzyl and phenacyl halides*

Selective electrochemical oxidation of benzyl chlorides (4-RC$_6$H$_4$CH$_2$Cl, R = H, Me, Cl, or CH$_2$Cl) affords the corresponding benzoic acids in yields ranging from 78-96% (G.P.Borsotti, M.Foa, and N.Gatti, *Synthesis*, 1990, 207). Tetrabutylammonium periodate in dioxane can be employed to oxidise phenacyl bromides to benzoic acids (E.Santaniello, A.Manzocchi and C.Farachi, *Synthesis*, 1980, 563).

(vi) *From dienes by cycloaddition reactions*

(N.B. This section includes examples of the syntheses of benzoate esters; these, of course, yield the correponding acids on hydrolysis).

2-Methoxybenzoic acids and benzoates are formed by the Alder-Rickert reactions of 1-methoxy-1,3-butadienes and 3-alkylpropynoates, followed by thermal elimination of ethene from the initially formed adducts (C.C.Kanalem *et al.*, *J. Chem. Soc., Perkin Trans. 1*, 1989, 1907; also see R.K.Olsen and X.Q.Feng, *Tetrahedron Letters*, 1991, **32**, 5721).

R = H, Me, OMe; ^{1}R = Me, Et; ^{2}R = Me, Bu, Pentyl

Phthalic acids are synthesised by the cycloaddition of dialkynoic acids and 1,3-butadienes (H.Yamanaka *et al., Chem. Express,* 1989, **2**, 21). Polyfluoroalkylterephthalic acids are obtained in a similar manner by the cycloaddition of 3-polyfluoroalkylpropynoic acids with 2-methyl-1,3-butadiene, followed by aromatisation (bromination/dehydrobromination) and oxidation of the intermediate 4-methylbenzoic acids with potassium permanganate (M.Kuwabara *et al., J. Fluorine Chem.,* 1989, **42**, 105).

R = CF$_3$, CHF$_2$, (CF$_3$)$_2$CHCF$_2$,, CHF$_2$(CF$_2$)$_4$

An intra-molecular Wittig reaction is used to synthesise dimethyl 4-methyl-6-perfluroalkylisophthalates. Thus methyl 5-oxo-2-(triphenylphosphoranylidene)hexa-3-enolate when treated with methyl perfluoroalkynoates yields adducts, which on heating cyclise to isophthalate esters (D.Weiyu *et al., J. Chem. Soc., Perkin Trans. 1,* 1993, 855).

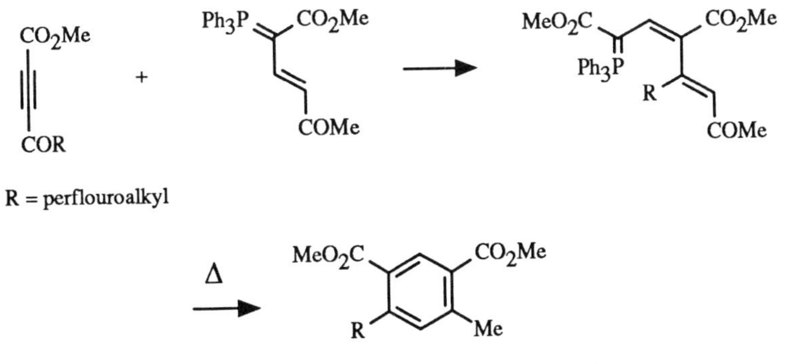

R = perflouroalkyl

N-*t*Butylvinylketenimine is formed by reacting allyl chloride with *t*butylisonitrile, this product undergoes cycloadditions with methyl acetylenecarboxylates to yield methyl 2-(*N*-*t*butylamino)benzoates (Y.Ito *et al.*, *Synth. Commun.*, 1980, **10**, 233).

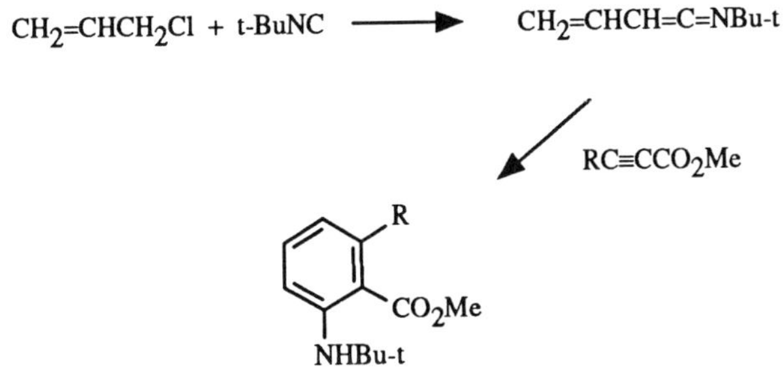

CH₂=CHCH₂Cl + t-BuNC ⟶ CH₂=CHCH=C=NBu-t

RC≡CCO₂Me

Birch reduction of arenes affords 1,4-cyclohexadienes, which readily undergo isomerisation to their conjugated forms on treatment with certain catalysts. When the conjugated dienes are reacted with activated ethynes at *ca* 100 °C cycloaddition occurs, followed by expulsion of ethene and the production of benzoate esters. 1,4-Dimethoxybenzene, for example, on reduction and subsequent treatment with dichloromaleic anhydride, yields 1,4-dimethoxy-1,3-cyclohexadiene. This product forms an adduct with dimethyl acetylenedicarboxylate, which on heating ultimately gives dimethyl 3,6-dimethoxyphthalate (P.A.Harland and P.Hodge, *Synthesis*, 1982, 223).

Similar cycloaddition reactions of dimethyl acetylenedicarboxylate and cyclohexadienes leading to dimethyl phthalates have been reported previously by (G.S.R.Subbarao and V.P.Sashikumar, *Indian J. Chem. Sect. B.,* 1979, **18**, 543); additionally, diethyl 3-methoxyphthalate is available from the cycloaddition of 1-methoxy-1,4-cyclohexadiene and diethyl acetylenedicarboxylate in the presence of dichloromaleic anhydride (M.S.Newman and K.Kanakarajan, *J. Org. Chem.,* 1980, **45**, 3523).

Another route to phthalates utilises cyclopentadienones as the diene component in cycloaddition reactions with dimethyl acetylene-dicarboxylate. Heat is required to initiate the additions and under these conditions the carbonyl group of the the [4+2] adduct is expelled as carbon monoxide thereby generating the required aromatic esters (S.Miki, M.Yoshida and Z.Yoshida, *Bull. Chem. Soc. Japan,* 1992, **65**, 932).

Effectively the same procedure is used in the cycloaddition of cyclopentadienone oximes and dimethyl acetylenedicarboxylate to give dimethyl phthalates, although in this case the adducts are decomposed under milder conditions by catalytic hydrogen-transfer with palladium on carbon and cyclohexadiene. The oxime starting materials are formed from their dimers by thermolysis (D.Mackay, D.Papadopoulos and N.J.Taylor, *J. Chem. Soc., Chem. Commun.*, 1992, 325).

(vii) *From benzaldehydes and related compounds*

There are numerous oxidants for the conversion of aldehydes into acids. Some modern recommendations include the following reagents and conditions.

Yields of benzoic acids in excess of 90% are claimed for the oxidation of araldehydes by iodine in potassium hydroxide (S.Yamada, D.Morizono and K.Yamamoto, *Tetrahedron Letters,* 1992, **33,** 4329). Potassium bromate is another suitable oxidant (H.Samaddar and A.Banerjee, *J. Indian Chem. Soc.,* 1982, **59,** 905). Alternatively, calcium hypochlorite in aqueous acetonitrile/acetic acid at room temperature can be used (S.O.Nawauka and P.M.Keen, *Tetrahedron Letters,* 1982, **23,** 3131).

30% Hydrogen peroxide in tetrahydrofuran, containing benzeneselenic acid as a catalyst, acts to oxidise benzaldehydes to benzoic acids, although hydrogen peroxide alone is ineffective (J.K.Choi, K.Young and S.Y.Hong, *Tetrahedron Letters,* 1988, **29,** 1967). Sodium chlorite - HClO and chlorine dioxide is also an effective oxidant combination for the conversion of araldehydes into aromatic acids (Z.Jaing, *Zhongguo, Yiayao, Gongye, Zazhi,* 1991, **22,** 1; *C.A.,* 1991, **115,** 8205j).

Cobalt(II) chloride functions as a catalyst for the oxidation of benzaldehydes by dioxygen and butanal/acetic anhydride (T.Punniya-murthy, S.J.S.Kalra and J.Iqbal, *Tetrahedron Letters,* 194, **35,** 2959).

Yet another efficient method utilises aldoximes, or their *O*-methyl ethers, as the substrates. These are converted into the corresponding acids by treatment with an alcoholic solution of 30% hydrogen peroxide in the presence of a catalytic amount of 2-nitrobenzeneselenic acid (S.B.Said, J.Skarzewski and J.Mlochowski, *Synth. Commun.,* 1992, **22,** 1851).

The propensity of cobalt to bind to molecular oxygen is utilised in the cobalt(II) ion mediated oxidations of araldehydes to aroic acids in the presence of acetic anhydride. Phenolic groups present in the substrates are *O*-acetylated under these conditions, but this does not affect the efficency of the oxidation step. Indeed, this is an advantage since it protects the phenols against alternative oxidative reactions (B.Bhatia and J.Iqbal, *Tetrahedron Letters,* 1992, **33,** 7961).

Sodium perborate in acetic acid is used for the selective oxidation of benzaldehydes to benzoic acids in the presence of nuclear alkylthio, alkoxy, alkyl, or halogeno substituents. Thus when 4-chlorobenzaldehyde is reacted with this oxidant at 45-50 °C 4-chlorobenzoic acid is obtained in 94% yield. The reagent is formed by treating a solution of sodium borate in acetic acid with hydrogen peroxide. [A.McKillop and D.Kemp, *PCT Internat. App. W.O.* 90 09,975 (1990); *C.A.,* 1990, **114,** 1013553v].

(viii) *By lithiation-carboxylation*

1,3,5-Tris(trifluoromethyl)benzene is carboxylated by deprotonation with butyl lithium and treatment of the corresponding anion with carbon dioxide. The product 2,4,6-trifluoromethylbenzoic acid cannot be esterified through reaction with an alcohol, such as ethanol, because of steric problems associated with the development of the required tetrahedral intermediate. However, as well as causing the deprotonation of aromatic acids, butyl lithium may act as a nucleophile. Thus benzoic acid forms small amounts of butyl phenyl ketone and 5-hydroxy-5-phenylnonane when treated with this reagent. The formation of butyl phenyl ketone can be rationalised through at least two pathways: *e.g.* one involving initial deprotonation of the acid, and the other direct nucleophilic attack (C.Einhorn, J.Einhorn and J.L.-Luche, *Tetrahedron Letters*, 1991, **32**, 2771) (see p.14).

A super basic mixture of butyl lithium and potassium *t*butoxide, in tetrahydrofuran at low temperature, deprotonates fluorobenzene, difluoro-, fluoro(trifluoromethyl)-, and bis(trifluoromethyl)-benzenes. The anions so formed may then be quenched with carbon dioxide to afford the corresponding benzoate salts (M.Schlosser, G.Katsoulos and S.Takagishi, *Synlett.*, 1990, 747). Similar reactions occur with fluorotoluenes giving methylfluorobenzoic acids, after aqueous work-up. However, if a mixture of lithium *i*propylamide and potassium *t*butoxide are employed then deprotonation in the nucleus does not occur. Now a proton is abstracted from the benzylic position and treatment with carbon dioxide *etc.*, leads to fluorophenylacetic acids (S.Takagishi and M.Schlosser, *ibid.*, p.119).

Much work has shown the utility of the 4,4-dimethyl-2-oxazoline unit as a latent carboxyl group and as a protecting group for acids (see 1st

Supplement Vol. IIIG p. 127; A.I.Meyers and W.B.Avola, *Tetrahedron Letters,* 1980, **21**, 3335). A good example is the generation of a benzyne from 2-(3-chlorophenyl)oxazoline by *ortho*-lithiation and a $E1_CB$) elimination of chloride ion. The reactions of the benzyne with a variety of reagents can then be exploited to afford, after deprotection, a range polysubstituted benzoic acids (P.D.Panesegrau, W.F.Riecker and A.I.Meyers, *J. Amer. Chem. Soc.,* 1988, **110**, 7178).

$Y = Me, CH_2OH,$
$CHO, CH(OH)Ph, H,$ etc.

Direct *ortho*-lithiation of 1,3-dialkoxy-, 1-alkoxy-3-(alkylthio)-, and 1,3-bis(alkylthio)-benzenes is possible, and further treatment of the lithiated compounds with carbon dioxide affords benzoic acids. Unfortunately the reactions are not selective and 2-, 4-, and 5-carboxylated products are formed. Some degree of steric control is observed, however, and as the size of the alkyl group in the substituent is increased the extent of lithiation *ortho* to the substituent decreases (S.Cabiddu *et al., Gazz.,* 1981, **111**, 123).

Lithiation of salicylic acid can be effected by halogen-metal exchange, the salts can then be used in reactions with a range of electrophiles to form derivatives substituted at the position of the original halogen atom (M.C.Rotger, A.Costa and J.M.Saá, *J. Org. Chem.,* 1993, **58**, 4083).

$$E = CH_2Ph, \ EtCO_2, \ CH(OH)Ph, \ MeS \ etc.$$

(ix) *Miscellaneous methods*

Industrial sources of phthalic acids and their allies have been surveyed (A.G.Bemis *et al., Kirk-Othmer Encycl. Chem. Technol. 3rd edn.,* 1982, **17**, 732, ed. M.Grayson and D.Eckroth, J.Wiley, N.Y.), as have methods for the manufacture and the use of salicylic acid, particularly in the synthesis of asprin (S.H.Erikson, *ibid.,* 1982, **20**, 500). The chemistry and synthesis of iodobenzoic acids and their derivatives has been reviewed (E.B.Merkushev *et al., Synthesis,* 1988, 923).

The preparation and reactions of mercaptobenzoic acids have been surveyed (J.Mirek, *Zesz. Nauk. Uniw. Jagiellon Pr. Chem.,* 1985, **29**, 57; *C.A.,* 1985, **103**, 104616s). For electron rich substrates such as 3,5-di-, or 3,4,5-tri- methoxybenzoic acids, or their esters, a methylthio group can be introduced at C-2 by reactions with methylthiomethyl chloride and zinc(II) chloride. Indeed with two equivalents of the regents 2,6-di(thiomethyl) derivatives are formed (J.B.Henderickson and P.M.DeCapite, *J. Org. Chem.,* 1985, **50**, 2112).

The cathodic reduction of aryl halides in *N,N*-dimethylformamide, saturated with carbon dioxide and containing tetraethylammonium 4-methylbenzenesulphonate, dichlorobis(triphenylphosphine)palladium and triphenylphosphine, leads to the formation of the corresponding benzoic acids in good yields. It is considered that the first step is palladation, followed by electron transfer from the (lead) cathode to give a radical anion. This species reacts with carbon dioxide and eventually affords the acid (S.Torii *et al., Chem. Letters,* 1986, 169).

Benzoic acid itself can be carboxylated by treatment with excess carbon tetrachloride in warm sodium hydroxide solution in the presence of β-cyclodextrin. The major product is terephthalic acid which is formed in 75% yield (H.Hirai and H.Mihori, *Chem. Letters,* 1992, 1523). The procedure is not confined to the carboxylation of acids, thus 4-hydroxybenzoic acid is prepared in 96% yield by reacting phenol with these reagents (M.Komiyama and H.Hirai, *Makromol. Chem. Rapid Commun.,* 1981, **2**, 661; M.Komiyama, I.Sugiura and H.Hirai, *J. Inclusion Phenom.,* 1984, **2**, 837).

Boveault-Blanc reduction of 3,4,5-trimethoxybenzoic acid leads to 3,5-dimethoxybenzoic acid. Presumably a radical anion, or its equivalent, participates in this reaction, and after protonation, the dihydro derivative which is formed expels methanol to form the dimethoxylated product (R.Sharda and H.G.Krishnamurty, *Indian J.Chem. Sect. B.,* 1980, **19B**, 410).

3,5-Dimethoxy-4-methylbenzoic acid can be prepared by the nucleophilic displacement of the bromine atoms of methyl 3,5-dibromo-4-methylbenz-ote by sodium methoxide in the presence of copper(II) chloride, followed by hydrolysis. The starting compound is obtained by dibromination of methyl 4-methylbenzoate with bromine and aluminium(III) chloride (P.S.Marchand *et al., Synthesis*, 1980, 409).

Various syntheses of orsellinic and isoanacardic acids have been reviewed (T.H.P.Tyman and N.Visani, *FECS Internat. Confer. Chem. Biotechnol. Biol. Act. Nat. Prod. [Proc.], 3rd,* 1985 (pub. 1987), **2**, 33).

3,5-Dichloro-6-ethyl-2,4-dihydroxybenzoic acid is the principal aromatic constituent of the antibiotic lipiarmycin A3. This acid has been synthesised in five steps from ethyl 2-oxobutanoate and ethyl pent-2-enoate (M.Alexy and H.D.Scharf, *Liebig's Annalen*, 1991, 1363). Similarly, lunularic acid a metabolite of certain *Bryophytes* has been synthesised by two routes (T.Eicher, K.Tiefensee and R.Pick, *Synthesis*, 1988, 525).

Lunularic acid

(x) *Isotopically labelled benzoic acids*

All of the possible isomers of benzoic acid bearing a single deuterium atom in the ring have been synthesised. Except for 2,3,4,5-*D*-benzoic acid which was obtained from tetrabromophthalic acid by deuteration and decarboxylation, all the others were prepared from the appropriate halogenobenzene by dehalogenation with Raney copper-aluminium alloy in 10% deuterated sodium hydroxide in deuterium oxide (M.Tashiro, K.Nakayama and G.Fukata, *J. Chem. Soc., Perkin Trans. 1,* 1983, 2315).

2-Deuteriobenzoic acids and the corresponding amides can be formed by halogen-deuterium exchange with deuterium oxide in the presence of group

VIII metal salts such as rhodium(III) chloride (W.J.S.Lockley, *J. Labelled Compd. Radiopharm.*, 1984, **21**, 45). The same methodology can be extended to furnish tritium labelled analogues (W.J.S.Lockley, *J. Chem. Res. Synop.*, 1985, 178).

(b) *Reactions*

(i) *Decarboxylation*

Benzoic acids are readily decarboxylated by treatment with the perfluorinated sulphonic acid resin, Nafion-H (G.A.Olah, K.Laali and A.K.Mehrota, *J. Org. Chem.*, 1983, **48**, 3360). In a more traditional procedure 2,4,5-trifluorobenzoic acid is decarboxylated to 1,3,4-trifluoro-benzene by heating it in quinoline and copper(II) oxide at 200 °C [L.B.Fertel and H.C.Lin, *Statutory Invent. Regist. U.S.* 992 (1991); *C.A.*, 1992, **116**, 105810f].

An improved methodology for the oxidative decarboxylation of toluic acids to cresols using copper salts and cupric oxide has been documented (M.P.Sharma and J.N.Chatterjea, *J. Chem. Technol. Biotechnol. Chem. Technol.*, 1983, **33A**, 328). Phenols are also available through the oxidative decarboxylation of aromatic acids in water containing copper and catalytic amounts of copper(I) and magnesium. The initial products are aryl arylcarboxylates, which undergo hydrolysis to a mixture of the appropriate acid and phenol [M.R.J.Offermanns, *Eur. P. App.*, 434,140 (1991); *C.A.*, 1991, **115**, 91841g]. An alternative procedure utilises $NiO/Fe_2O_3/Na_2O$ in a vapour-phase oxidation of benzoic acid to phenol (J.Miki *et al.*, *J. Chem. Soc., Chem. Commun.*, 1994, 1685).

Lithium benzoate when sonicated (38 kHz) with lithium and butyl chloride generates butyl phenyl ketone. However, if butyl iodide is used in a similar reaction benzil and benzoic acid are formed. At greater frequency (400 kHz) the latter reaction mixture affords benzoic acid and octane. the ketone forming processes are regarded as sonochemical variants of the Barbier reaction (Y.Danhui *et al.*, *J. Chem. Soc., Chem. Commun.*, 1994, 1815 (see p.9).

(ii) *Reduction to benzaldehydes and benzyl alcohols*

Benzoic acids are reduced to the corresponding benzyl alcohols by treatment with samarium diiodide in aqueous tetrahydrofuran. Benzoate esters and anhydrides can also be used as starting materials. Benzamides similarly yield alcohols under basic conditions. However, in such cases if an acid is added the reduction proceeds only as far as the appropriate benzaldehydes.

Benzonitriles afford benzylamines (Y.Kamochi and T.Kado, *Tetrahedron,* 1992, **48**, 4301).

Another method for the reduction of benzoic acids to benzaldehydes, depends upon the initial formation of pentacoordinated silyl carboxylates. Thus reaction of an acid with 8-(*N,N*-dimethlamino)-1-phenylsilylnaphthalene generates a complex in which the lone-pair electrons of the amino group interact with the silicon atom orientated *peri* to it. On heating at temperatures ranging from 110-170 °C, depending upon the acid, the complex decomposes to a cyclic siloxane and an aldehyde (R.J.P.Corriu, G.F.Lanneau and M.Perrot, *Tetrahedron Letters,* 1987, **28**, 3941).

$ArCO_2H$ + [silylnaphthalene complex] $\longrightarrow$ [ArCO₂-Si complex]

$\Delta \longrightarrow ArCHO$ + [cyclic siloxane]

$R =$ [methylnaphthalene with N(CH₃)₂ group]

Enzymic methods for the reduction of benzoic acids to benzaldehydes are also available, these utilise NAD-dependent reductases (I.Jezo and J.Zemek, *Chem. Pap.,* 1986, **40**, 279). Aldehydes are also formed by the reduction of benzoate esters by lithium tris(diethylamino)aluminium hydride in tetrahydrofuran solution at -78 °C. Yields range from 55-80% (J.S.Cha *et al., Tetrahedron Letters,* 1991, **32**, 6903; *Org. Prep. Proced. Int.,* 1992, **24**, 712). Sodium and lithium carboxylates are readily reduced to aldehydes by 9-borabicyclo[3.3.1]nonane (J.S.Cha *et al., Heterocycles,* 1988, **27**, 1595).

The addition of iodine to a solution of a benzoic acid in tetrahydrofuran containing sodium borohydride results in the formation of the corresponding benzyl alcohol. Alkenyl substituents, or ester groups, bonded to the aromatic ring of the acid are not reduced (J.V.B.Kanth and M.Periasamy, *J. Org. Chem.*, 1991, **56**, 5964).

Normally aromatic carboxylic acids are reduced by boranes to give benzyl alcohols, however, reactions with borane dimethyl sulphide lead to toluenes instead. Should boron trifluoride be present the products are benzyl dimethylsulphonium salts [(ArCH$_2$S$^+$Me$_2$)$_3$ B$_3$O$_6$$^{3-}$] (H.LeDeit, S.Cron and M.LeCorre, *Tetrahedron Letters,* 1991, **32**, 2759).

Electrochemical reduction of aromatic acids dissolved in diglyme containing sodium borohydride leads to the corresponding alcohols in good yields (R.Shundo *et al., Bull. Chem. Soc. (Japan),* 1992, **65**, 530). However, electrochemical methods may sometimes afford 1,2-dihydroaroic acids (see below).

(iii) *Reductive dehalogenation*

The optimal conditions for the reductive dechlorination of 2,4,6-trichlorophthalic acid have been defined. At 105 °C in 10% sodium hydroxide and with three equivalents of zinc 3,6-dichlorophthalic acid is obtained in 76% yield. However, if the concentration of the alkali is raised to 20% 3-chlorophthalic acid is the major product (L.B.Fertel, N.J.O'Reilly and K.Callaghan, *J. Org. Chem.,* 1993, **58**, 261).

(iv) *Birch reduction and the reactions of dihydrobenzoic acids*

Birch reduction and alkylation of the product has been applied to 3,5-dimethoxybenzoic acid. When reacted with excess sodium in ammonia and excess ethyl bromide this compound affords 3,6-diethyl-1,5-dimethoxy-1,4-cyclohexadiene (J.A.Guzman, R.Castanedo and L.A.Maldonato, *Synth. Commun.,* 1991, **21**, 1001).

It has now been confirmed that the Birch reduction of biphenyl-4-carboxylic acid, followed by rapid quenching with ammonium chloride, gives only 1,4- and 1,4′-dihydrobiphenyl-4-carboxylic acids. This is in contrast to earlier reports which implied that the products were benzyl alcohols and arenes (P.W.Rabideau *et al., J. Chem. Soc., Chem. Commun.*, 1980, 210).

1,4-Dihydrobenzoic acids when reacted under Vilsmeier conditions (*N,N*-dimethylformamide\phosphorus oxychloride) are formylated and decarboxylated giving mono- di- and triformyl-arenes. Benzoic acid itself affords 1,3,5-triformylbenzene. The initial step is considered to be the formation of the appropriate acid chloride, followed by multiple attack upon the product by the iminium salt $[H(Cl)=N^+Me_2 \; ^-OPOCl_2]$ obtained from the reaction of phosphorus oxychloride and *N,N*-dimethylformamide (B.Raju and G.S.K.Rao, *Synthesis*, 1987, 197).

Phthalic acids are reduced to 1,2-dihydrophthalic acids by sodium amalgam and also at a lead cathode with 5% aqueous sulphuric acid in dioxane as the electrolyte. These dienic products undergo cycloaddition reactions with a range of alkynes and at the temperature normally employed, 190-250 °C, the initially formed adducts eliminate fumaric acid and furnish the appropriate 1,2-disubstituted arenes (T.Ohno *et al.*, *Tetrahedron Letters*, 1993, **34**, 2629).

X and Y = CO_2R ; or X = H, or Ar, Y = CO_2R; or X = Ar, Y = CF_3

In addition, the ring can be variously substituted

(v) *Formation of ketones*

Symmetrical aromatic ketones are synthesised from benzoic acids under mild conditions through initial conversion of the acids into their *S*-(2-pyridyl) thiobenzoates and treatment of these products with bis(1,5-cyclooctadiene)nickel [Ni(COD)$_2$] (T.Goto, M.Onaka and T.Mukaiyama, *Chem. Letters*, 1980, 51).

(vi) *Nuclear bromination*

N-Bromosuccinimide, or dibromodimethylhydantoin, brominates certain

alkoxybenzoic acids in aqueous sodium hydroxide in high yields
(J.Auerbach *et al., Tetrahedron Letters,* 1993, **34**, 931):

Benzoic acid	Derivative	% Yield
2-methoxy-	5-bromo-	95
2,6-dimethoxy-	3-bromo-	94
2,3-dimethoxy-	6-bromo-	70
3,4,5-trimethoxy-	2-bromo-	83

(vii) *Nuclear hydroxylation*

Benzoic acid is monohydroxylated in an indiscriminate manner by
hydrogen peroxide in contact with a silica-supported palladium catalyst.
Dihydroxylation also occurs. However, under these conditions phthalic acid
only affords 4-hydroxyphthalic acid (A.Itoh *et al., J. Mol. Catal.,* 1991, **69**,
215).

Mutants of *Pseudomonas testosteroni* are capable of hydroxylating certain
activated benzoic acids. 3-Hydroxybenzoic acid, for example, undergoes
site-specific hydroxylation to afford 2,3-dihydroxybenzoic acid
(G.O.Daumy, A.S.McColl and G.C.Andrews, *J. Chem. Soc. Chem.,
Commun.,* 1979, 1073).

2. Benzoate esters

(a) *Preparation*

Some methods for the preparation of aryl esters have already been
discussed in Section 1 above.

(i) *From sodium benzoates and benzoic acids*

Sodium benzoates react under phase-transfer conditions with alkyl halides
(bromides, or iodides) to afford the corresponding esters (S.Pavlov *et al.,
Arh. Farm.,* 1986, **36**, 161; *C.A.,* 1987, **107**, 154023z).

Direct methods of converting aroic acids into their alkyl esters include
reacting them with trialkyl borates (C.I.Chiriac, *Rev. Roum. Chim.,* 1982,
27, 533). Methyl benzoates are conveniently prepared from benzoic acids
using *O*-methylcaprolactim as the 'methyl group carrier' at reaction
temperatures of 80-85 °C (E.Mohacsi, *Synth. Commun.,* 1982, **12**, 453).

More generally, however, acids and alcohols may be converted into esters in the presence of either 2-chloro-3,5-dinitropyridine (L.P.Prikazchikoya, B.M.Khutova and V.M.Cherasov, *Ukr. Khim. Zh.* (Russ. Edn.), 1982, **48**, 1302; *C.A.*, 1983, **98**, 125547p), 2-chloro-1,3,5-trinitrobenzene (M.Yamaguchi *et al., Bull. Chem. Soc. Japan*, 1981, **54**, 1470), or 8-quinolinesulphonyltetrazolide, as coupling agents (H.Takaku, N.Saoaki and K.Morita, *Chem. Pharm. Bull.*, 1982, **30**, 2633).

Phenols react directly with benzoic acids in the presence of dichloro-phenylphosphine oxide and pyridine. For example, benzoic acid and phenol give phenyl benzoate in 80% yield when heated together at 135-145 °C with these reagents (C.I.Chiriac, *Rev. Roum. Chim.*, 1982, **27**, 831).

A mixture of triphenylphosphine and triethylamine mediates the formation of phenyl carboxylic esters from 4-bromophenol and acids. No solvent is required, but a temperature of 200 °C is needed. An unusual mechanism is postulated in which bromide ion is expelled and a phosphonium salt, or its equivalent, is produced. This then fragments to the phenate anion which is esterified with loss of triphenylphosphine oxide (S.Hashimoto and I.Furukawa, *Bull. Chem. Soc. Japan*, 1981, **54**, 2839).

A highly selective method for the mono-methyl esterification of terephthalic acid requires that the dicarboxylic acid be chemisorbed on alumina, prior to treatment with diazomethane (T.Chihara, *J. Chem. Soc., Chem. Commun.*, 1980, 1215).

Cyclic esters such as phthalides can be synthesised by palladium promoted coupling reactions between 2-iodobenzoic acid and alkynes, or alkenes. 3-Hydroxy-2-methylbutyne, for example, yields the methylene derivative (1) (N.G.Kundu and M.Pal, *J. Chem. Soc. Chem. Commun.*, 1993, 86).

(ii) *From araldehydes*

Methyl benzoates are prepared by the electrochemical oxidation of araldehydes at a platinum anode. The starting compounds are dissolved in methanol containing flavin and *N*-benzyl-4-methylthiazolium bromide as catalysts and electrolytes (S.W.Tam, L.Jimenz and F.Diederich, *J. Amer. Chem. Soc.*, 1992, **114**, 1503).

(iii) *From aromatic nitriles*

An alternative approach to methyl benzoates is to hydrolyse benzonitriles with polyphosphoric ester containing methanol (F.D.Mills and R.T.Brown, *Synth. Commun.*, 1990, **20**, 3131).

(iv) *From aryl halides and aryl trifates*

Chlorobenzenes are carbonylated by carbon monoxide to afford methyl esters in the presence of methanol at 200 °C and over a palladium catalyst. Potassium dichromate enhances the activity of the catalyst (V.Dufaud, J.T.-Cazat and J.M.Basset, *J. Chem. Soc., Chem. Commun.*, 1990, 426).

$$ArCl + MeOH + CO \xrightarrow[NaOAc]{Pd \backslash C} ArCO_2Me$$

Methoxycarbonylation of aryl trifluoromethanesulphonates with carbon monoxide and methanol in contact with palladium(II) acetate and 1,3-bis(diphenylphosphino)propane gives methyl benzoates in good yields (R.E.Dolle, S.J.Schmidt and L.I.Kruse, *J. Chem. Soc., Chem. Commun.*, 1987, 904). Similar work has been published by S.Cacchi *et al.* (*Tetrahedron Letters*, 1986, **27**, 3931) which refers to the use of other alcohols than methanol, so that a range of esters is available by this route.

Should primary amines replace alcohols then amides are formed.

It is also possible to form 2,6-di-^{t}butylphenyl benzoates from lithium 2,6-di-^{t}butylphenates and aryl bromides in contact with carbon monoxide and palladium(II) chloride (Y.Kubota *et al.*, *J. Chem. Soc., Chem. Commun.*, 1994, 1553).

The catalyst need not contain palladium and the results of cobalt mediated carbonylation reactions of aryl halides have been disclosed. In these cases alkylcobaltcarbonyls are employed and these are either preformed, or generated *in situ*. The reactions proceed in the presence of metal alkoxides, sodium hydroxide, or potassium carbonate, and an aliphatic alcohol which supplies the alkyl unit of the product alkyl benzoates (M.Foa *et al.*, *J. Organomet. Chem.*, 1985, **285**, 293).

(v) *From coupling reactions*

Cross coupling mediated by palladium occurs when aryltributyltins (ArSnBu$_3$) are reacted with chloroformate esters (ClCO$_2$R). This leads to benzoate esters (ArCO$_2$R), but should the chloroformates be replaced by carbamoyl chlorides (ClCONR$_2$), then benzamides (ArCONR$_2$) are produced (L.Balas, *Organometallics*, 1991, **10**, 366).

A general method for the synthesis of alkyl benzoates requires reactions between arylmagnesium bromides and 2-pyridyl alkyl carbonates in tetrahydrofuran at 0 °C. The co-product is 2-pyridone (J.I.Lee and S.Kim, *Bull. Korean Chem. Soc.*, 1989, **10**, 611).

(vi) *From α,β-diketones*

1,3-Alkanediones react with ethyl (Z)-3-bromopropenoate under basic conditions to form the enolates of diketo esters, which then cyclise and aromatise during work-up to afford ethyl 4-hydroxybenzoates (A.B.Smith and S.N.Kiléyi, *Tetrahedron Letters*, 1985, **26**, 4419).

R' = H or Me, R = Me, pentyl, Ph

(vii) *Miscellaneous methods*

2,4,6-Triphenylpyrylium perchlorate reacts with α-ketoalkanoate esters in the presence of an equivalent of triethylamine/acetic acid to afford 3,5-diphenylbenzoates (2), whereas in the presence of two equivalents of the reagents alkyl 2-benzoyl-3,5-diphenylbenzoates (3) are produced (T.Zimmerman and G.W.Fischer, *J. Prakt. Chem.*, 1987, **329**, 499).

Ethyl 4,6-diaryl-3-cyano-2-hydroxybenzoates (5) can prepared by the cycloaddition of ethyl acrylates and ethyl α-cyanoacrylates in the presence of sodium hydride. The first formed products are cyclohexenones (4), which yield enolates with excess sodium hydride and then readily aromatise (C.Ivanov and T.Tcholakova, *Synthesis*, 1982, 731).

Methanolysis of coumalic acid (6) in methanol, dimethoxymethane and sulphuric acid leads to 1,3-dimethoxycarbonyl-4-methoxy-1,3-butadiene (7). This product affords dibenzoate esters through subsequent reactions with various types of alkyl ketones and esters (M.H.Nautz and P.L.Fuch, *Synth. Commun.*, 1987, **17**, 761).

Dihydroorsellinates (8) are oxidised to orsellinate esters (9) by the action of bromine in acetic acid and acetic anhydride (H.J.Dyke *et al.*, *Austral. J. Chem.*, 1987, **40**, 431).

Diethyl 2,5-dioxocyclohexane-1,4-dicarboxylate (10) reacts with primary alkylamines to give diethyl 2,5-bis(*N*-alkylamino)cyclohexa-1,4-dienoates (12), which are easily oxidised by treatment with either bromine, or chloroanil, to form diethyl 2,5-bis(*N*-alkylamino)terephthalates (H.Ulbricht, G.Loder and L.Kittler, *J. Prakt. Chem.*, 1979, **321**, 905; see also J.Sinnreich, *Synthesis*, 1980, 578).

R = SO$_2$Ph, CO$_2$Me

R = Ac, CO$_2$Me, CO$_2^t$Bu, SO$_2$Ph

R = Me, Pr, C$_5$H$_{11}$, C$_7$H$_{15}$

A route to trimethyl 3-amino-4-cyano-5-methylbenzene-1,2,3-tricarboxy-late involves the cycloaddition of dimethyl acetylenedicarboxylate to methyl 4,4-dicyano-3-methylbut-3-enoate in the presence of potassium carbonate (K.Gewald, U.Hain and M.Gruner, *Z. Chem.*, 1987, **27**, 32).

A Diels-Alder cycloaddition of 2,3-diphenyl-1,3-butadiene and methyl acrylate affords methyl 3,4-diphenylcyclohex-3-enoate, which can be easily oxidised to yield methyl 3,4-diphenylbenzoic acid (A.Kawamoto, H.Uda and N.Harada, *Bull. Chem. Soc. Japan,* 1980, **53**, 3279).

[2.2]Paracyclophanes bearing four ester groups are available through the cycloddition of dialkyl acetylenedicarboxylates and two mol. equivalents of 1,2,4,5-hexatetraene (G.Weber, K.Menke and H.Hopf, *Chem. Ber.,* 1980, **113**, 531).

Enol esters can be obtained from 4-methoxybenzoyl chloride through its reactions with tetramethylammonium pentacarbonyl(1-oxalkylidene)-

chromate(0) complexes. If the reactions are carried out at -40 °C and then allowed to warm slowly to room temperature the products have exclusively, or predominantly, the (Z)-configuration (B.C.Soderberg and M.J.Turbeville, *Organometallics*, 1991, **10**, 3951).

(b) *Physical properties and reactions*

(i) *Mass spectra*

The mass spectral fragmentation patterns of phthalic, isophthalic, and terephthalic esters have been documented, in order to assist the early recognition of these ubiquitous contaminants of biological extracts (M.P.Friocourt, D.Picart and H.H.Foch, *Biomed. Mass Spectrom.*, 1980, **7**, 193).

An ^{18}O label in the carboxyl group of certain esters may be scrambled during mass spectrometry. For example, this occurs in the molecular ions of 4-methylphenyl benzoate and 4-methylphenyl nitrobenzoate (J.B.Rowe, *Org. Mass Spectrom.*, 1979, **14**, 624).

(ii) *Reduction*

Benzoate esters are reduced to the corresponding alcohols by sodium borohydride and zinc chloride in the presence of *N,N*-dimethylaniline (T.Yamakawa, M.Masaki and H.Nohira, *Bull. Chem. Soc. Japan,* 1991, **64**, 2730). Another method, which selectively reduces aliphatic esters in the presence of aromatic ester functions, uses zinc borohydride in 1,2-dimethoxyethane with sonication of the reaction mixture. If *N,N*-dimethylaniline is added both types of esters are reduced (B.C.Ranu and M.K.Basu, *Tetrahedron Letters,* 1991, **32**, 3243).

It was claimed that lithium aluminium hydride in the presence of *N,N*-diethylamine is an extremely effective reductant for esters giving aldehydes. Although the reagents do react in this way, experience has shown that the yields of aldehydes are more modest (45-85%) than originally stated (see J.S.Cha and S.S.Kwou, *J. Org. Chem.,* 1990, **55**, 1692).

γ-Rays induce the reduction of benzoate esters dissolved in methanol, or 2-propanol. The products include benzaldehydes, and benzyl alcohols as well as 1,2-glycols and α-ketoalcohols. The reductions are caused by solvated electrons generated during the irradiation, but coupling reactions *via* solvent-derived hydroxyalkyl radicals lead to the formation of the glycols and the ketoalcohols (A.Sugimora and T.Yashima, *Chem. Letters,* 1980, 483).

X = CH_2 or Me_2C

Benzils are formed by the electrochemical reduction of aromatic esters in an undivided cell with a sacrificial magnesium anode freshly coated with cadmium (M.Heintz *et al., Tetrahedron,* 1993, **49**, 2249).

In common with many other aromatic compounds, methyl 2-methoxybenzoates are reduced under Birch conditions to their 1,4-dihydro derivatives. Should an alkyl halide be added to the reaction mixture, rather than a proton source, then the products are 1-alkyl-2-methoxy-1-methoxycarbonyl-1,4-dihydrobenzenes (J.M.Hook, L.N.Mander and M.Woolias, *Tetrahedron Letters*, 1982, **23**, 1095).

The benzyl esters of both aliphatic and benzoic acids are efficiently and chemoselectively cleaved in the presence of benzyl and benzyloxy ethers and also of *N*-benzyl groups by hydrogenolysis under hydrogen-transfer conditions. The catalyst is 10% palladium on carbon and cyclohexadiene is the hydrogen donor (J.S.Bajwa, *Tetrahedron Letters,* 1992, **33**, 2299).

(iii) *Alkylation*

Methyl 4-bromobenzoate can be alkylated by 2-methyl-3-butyn-2-ol in the presence of palladium(0) generated *in situ*. Hydrolysis and deprotection of the product in butanol at reflux containing potassium (or sodium) hydroxide gives 4-ethynylbenzoic acid in almost quantitative yield, after acidification. This compound slowly polymerises at room temperature, but its potassium and sodium salts are much more stable (A.P.Melissaris and M.H.Litt, *J. Org. Chem.*, 1992, **57**, 6998).

(iv) *Rearrangements*

Benzyl benzoate undergoes an intramolecular rearrangement to 2-benzylbenzoic acid (1) when it is heated to *ca* 350 °C over acidic alumina. At this temperature, however, the product may either decarboxylate to diphenylmethane, or cyclodehydrate to form anthrone (2). Even the latter product does not survive the reaction conditions, but degrades to anthracene (3) (K.Ganesan and C.N.Pillai, *Tetrahedron,* 1989, **45**, 877).

(v) *Formation of biaryls*

2-Methoxybenzoates (4) react with 1,3,5-trialkylarylmagnesium bromides (5) to afford 1,1′-biphenyl-2-carboxylates (6) (T.Hattori, T.Suzuki and S.Miyano, *J. Chem. Soc., Chem. Commun.*, 1991, 1375).

Biphenyls are also available through the reactions of 'electron-poor' aroyl chlorides and disilanes in the presence of a palladium catalyst. For example, trimellitic acid anhydride acid chloride (7) combines with 1,2-dichlorotetramethyldisilane in boiling mesitylene to afford, 4,4′-bis(phthalic anhydride (8) in 85% yield (T.E.Krafft, R.D.Rich and P.J.McDermott, *J. Org. Chem.*, 1990, **55**, 5430).

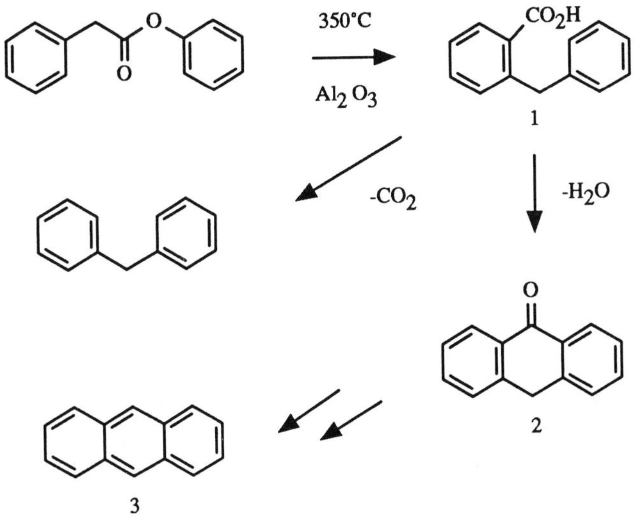
350°C
Al_2O_3
CO_2H
1
$-CO_2$
$-H_2O$
O
2
3

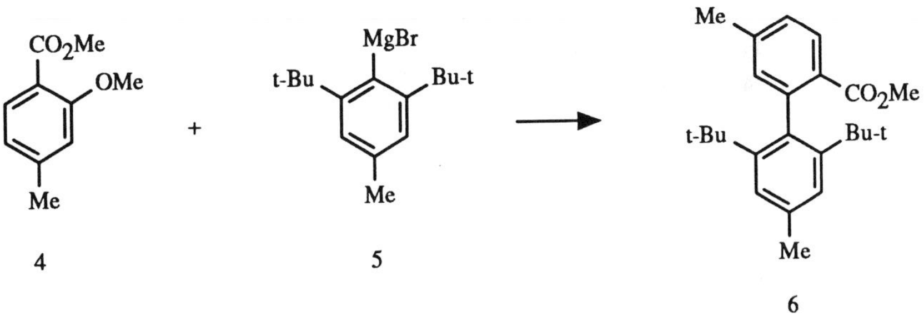
CO_2Me
OMe
Me
4
+
MgBr
t-Bu
Bu-t
Me
5
Me
CO_2Me
t-Bu
Bu-t
Me
6

2,2′-Disubstituted biphenyls can be obtained by intramolecular Ullmann couplings of di-(2-iodobenzoates) of α,ω-alkanediols, and also by the cyclisation of the di-(2-iodobenzoates) of salicyl alcohol. Optimum yields for the first group of reactions are noted when the ring of the cyclised products comprises eleven atoms (M.Takahashi *et al.*, *Tetrahedron Letters*, 1991, **32**, 6919; 1992, **33**, 5103).

n = 2-6, R = alkoxyl groups

When 4-aminobenzoic acid and 4-chloro-3,5-dinitrobenzoic acid are ground together at room temperature co-crystals are formed. Within the matrix aromatic nucleophilic substitution occurs and 4-(4-carboxyphenyl)-3,5-dinitrobenzoic acid is formed (M.C.Etter, G.M.Frankenback and

J.Bernstein, *Tetrahedron Letters*, 1989, **30**, 3617).

(vi) *Oxetane formation*

Alkyl benzoates (PhCO$_2$R, R = Me, Et, or ^{i}Pr) form three types of product when reacted under photochemical conditions with 2,3-dimethyl-2-butene. The first are oxetanes, formed by direct [2+2] cycloaddition, the second are alkoxyalkenes arising from the oxetanes by [2+2] reverse cycloaddition, and the third are β,γ-unsaturated ketones. The last type of compound is formed by hydride abstraction from the allylic position of the butene by the carbonyl oxygen atom of a photoexcited ester (T.C.Cantrell and A.C.Allen, *J. Org. Chem.*, 1989, **54**, 135).

(vii) *Nucleophilic displacement of nuclear substituents*

Examples of this type of reaction have already been cited, see section (v) above.

Benzylamine reacts with methyl 3-chloro-2-nitrobenzoate in *N,N*-dimethylformamide at 120 °C to give methyl 2-*N*-(benzyl-amino)-3-chlorobenzoate, through a nucleophilic displacement of the nitro substituent. However, 3-chloro-2-nitro-*N,N*-dimethylbenzamide reacts with the same reagent, but in the presence of copper(II) chloride and water, to cause the ejection of the chlorine atom thereby forming 2-*N*-(benzyl-amino)-2-nitro-*N,N*-dimethylbenzamide (T.L.Ho and A.Bergmanis, *Chem. Ind.*, 1988, 529).

3. Aroyl halides

(a) *Synthesis*

(i) *From aryl halides*

Aryl bromides can be carbonylated under a pressure of carbon monoxide in *N,N*-dimethylformamide containing potassium fluoride. The products are the corresponding aroyl fluorides. Such reactions are promoted by the addition of a phase-transfer reagent, but the yields are reduced by the presence of water (T.Okano, N.Harada and J.Kiji, *Bull. Chem. Soc. Japan,* 1992, **65**, 1741). This work postdates earlier studies which showed that palladium, platinum, cobalt and rhodium poly(triphenylphosphines) catalyse the carbonylation of aryl halides in contact with ceasium fluoride in acetone, or acetonitrile (T.Sakakura *et al.*, *J. Organomet. Chem.,* 1987, **334**, 205).

(ii) *From benzoic acids*

Benzoyl bromides are synthesised by reacting aromatic acids with dibromotriphenylphosphine in dichloromethane at room temperature, or with dibromotriphenylphosphine, *N*-(trimethylsilyl)oxazolidin-2-one and triethylamine in dichloromethane (J.M.Aizpurua and C.Palomo, *Synthesis*, 1982, 684).

(iii) *From miscellaneous sources*

Benzoylchlorobis(triphenylphosphine)platinum(II) [PhCOPt(PPh$_3$)$_2$Cl], or the analogous palladium(II) complex, reacts with chlorine, bromine, or iodine to afford the corresponding benzoyl halides (M.Kubota *et al.*, *Organometallics*, 1989, **9**, 1616).

Potassium benzoate when heated with benzotrichloride yields benzoyl chloride and potassium chloride. The reaction also occurs with other alkali metal benzoates giving mixed aroyl and benzoyl chlorides (E.G.Friedlin *et al., Zh. Org. Khim.*, 1980, **16**, 1040).

$$ArCO_2K + PhCCl_3 \longrightarrow ArCOCl + PhCOCl + KCl$$

Contary to earlier reports it now seems that the half ester acid chlorides of 3-nitro- and 3-chloro-phthalic acids do not exist in tautomeric equilibrium, but can be isolated as individual compounds (M.V.Rao and M.V.Bhatt, *Indian J. Chem. Sect. B.,* 1979, **18B**, 532: *ibid.*, 1980, **19B**, 323). However, it is known that the isomeric acid chlorides, obtained by treating the half-esters of 3-methoxyphthalic acids with thionyl chloride, interconvert so readily that it is impossible to isolate one free from the other. An oxonium ion acts as an intermediate in the equilibration process (D.M.Gupta, P.Hodge and P.N.Hurley, *J. Chem. Soc., Perkin Trans. 1,* 1989, 391).

(b) *Reactions*

(i) *Reduction*

Tributylgermanium hydride can be used to reduce aroyl chlorides to the corresponding benzaldehydes in the presence of tetrakis(triphenyl-phosphine)palladium. The reaction conditions are mild and the yields are generally in the range 80-93% (L.Geng and X.Lu, *J. Organomet. Chem.,* 1989, **376**, 41).

Aroyl chlorides are reduced to the corresponding aldehydes by triethylsilane in the presence of a transition metal enolate of dimethyl malonate. The metal ion may be trivalent Cr, Mn, Fe, or Co, or divalent Co, or Ni (M.G.Voronkov *et al., Obshch. Khim.,* 1990, **60**, 1584). The reduction proceeds further to the corresponding alcohol if the reagent is sodium borohydride and a drop of methanol is added (S.K.Kang and D.H.Lee, *Bull. Korean Chem. Soc.,* 1988, **9**, 402), however, if methanol is omitted and pyridine is added, to scavenge borane, the reduction only proceeds as far as the benzaldehydes (J.H.Babler, *Synth. Commun.,* 1982, **12**, 839).

μ-Bis(cyanotrihydroboronato)tetrakis(triphenylphosphine)dicopper(I) {[(Ph$_3$P)$_2$CuBH$_3$CN]$_2$} is a selective pH dependant reducing agent which under neutral conditions can be used to reduce benzoyl chlorides to benzaldehydes (R.O.Hutchins, and M.Markowitz, *Tetrahedron Letters,* 1980, **21**, 813).

It was claimed that exposure of aroyl fluorides to Wilkinson's catalyst

[(Ph$_3$P)$_3$RhCl] causes decarbonylation to give the corresponding aryl fluorides. This is now disputed and it seems that the true products are simply arenes (R.E.Ehrenkaufer, R.R.MacGregor and A.P.Wolf, *J. Org. Chem.*, 1982, **47**, 2489).

Reduction of benzoyl chlorides to benzyl chlorides is effected by treatment with sodium borohydride under phase-transfer conditions (A.Costello and D.J.Milner, *Synth. Commun.*, 1986, **16**, 1173).

(ii) *Coupling reactions*

Aroyl chlorides are susceptible to the conditions of the Heck reaction and, for example, they react with butoxyethene in contact with a palladium catalyst and triethylamine in boiling toluene to afford β-arylvinyl butyl ethers (C.-M.Anderson and A.Hallberg, *J. Org. Chem.*, 1988, **53**, 235).

Benzoyl chloride undergoes reductive self-coupling with copper at -78 °C in tetrahydrofuran, or in toluene, to give predominantly the (Z)-isomer of benzil. At room temperature and with nickel instead of copper the (E)-isomer is preferred. Other aroyl chlorides behave similarly affording diaryl diketones. In these reactions it is concluded that the initial step is a one electron-transfer to the the aroyl chloride yielding an aroyl radical. This which may either dimerise directly, or alternatively undergo further reduction to an aroyl anion, prior to reaction with a second molecule of the aroyl chloride (T.C.Wu and R.D.Rieke, *J. Org. Chem.*, 1988, **53**, 2381).

If iron tricarbonyl is employed as the reductant both isomers of benzil are formed (T.-Y.Luh, K.S.Lee and S.W.Tan, *J. Organomet. Chem.*, 1983, **248**, 221).

Benzoyl chloride reacts with alkylmanganese(II) chlorides to give alkyl phenyl ketones. Benzophenone is obtained if phenylmanganese chloride is used. The addition of copper(II) chloride sometimes improves the yield of product (G.Cahiez and B.Laboue, *Tetrahedron Letters,* 1989, **30**, 7369).

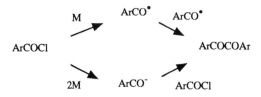

(iii) *Insertion reactions*

Aroyl fluorides insert sulphur trioxide at low temperature to form thermolabile arylcarbonic fluorosulphonic anhydrides. On warming these products decompose reversing the reactions by which they were formed (F.Effenburger, K.Huthmacher and M.Keil, *Chem. Ber.,* 1981, **114**, 1967).

$$ArCOF + SO_3 \rightleftharpoons ArCOOSO_2F$$

(iv) *Formation of aroylsilanes*

Benzoyltrimethyl silanes can be obtained from benzoyl chlorides through reactions with hexamethyldisilane in the presence of a palladium(II) catalyst (K.Yamamoto, S.Suzuki and J.Tsuji, *Tetrahedron Letters,* 1980, **21**, 1653).

R = H, Me, OMe, Cl, Br, CO_2Me, NO_2

4. Aromatic acid anhydrides

Synthesis

(i) *From benzoyl halides*

Aroyl chlorides form the corresponding acid anhydrides when they are reacted under phase-transfer conditions with aqueous sodium, or potassium hydrogensulphate and benzene (or acetonitrile) (J.Wang, Y.Hu and W.Cui, *J. Chem. Research (Synop.)*, 1990, 84).

(ii) *From benzoic acids*

Benzoic acids ($ArCO_2H$) can be used to form anhydrides in two steps: through initial reactions with chlorosulphonyl isocyanate ($ClSO_2NCO$) and triethylamine, followed by treatment of the products ($ArCO_2CON^-SO_2Cl$ - Et_3NH^+) with a second equivalent of the acid. Should alcohols be used in the second step, instead of acids, then esters are produced. Amides are obtained if amines are employed (K.S.Keshavamurty, Y.D.Vankar and D.N.Dhar, *Synthesis*, 1982, 506).

A more direct and traditional approach to symmetrical anhydrides is to react the acid with a suitable dehydrating reagent (see, C.I.Chiriac, *Rev. Chim. Bucharest*, 1984, **35**, 593). A new variant utilises Mitsunobu reagents: triphenylphosphine and diethyl diazodicarboxylate, in tetrahydrofuran at room temperature (E.Grochowski, H.Stepowski and C.J.Michejda, *Bull. Pol. Acad. Sci. Chem.*, 1984, **32**, 129).

(iii) *From arenes*

Arenes also serve as starting materials for anhydrides, thus benzene when heated with 1,2-dibromoethane and carbon monoxide at 100 °C in contact with palladium(II) acetate yields benzoic acid anhydride in 31% yield. Chlorobenzene similarly affords a mixture of the anhydrides of *o-*, *m-*, and *p*-chlorobenzoic acids (Y.Fujiwara *et al.*, *J. Chem. Soc., Chem. Commun.*, 1982, 132).

5. Benzamides

(a) *Synthesis*

(i) *From azides*

Aryl azides are converted into the corresponding benzamides through reactions with hydrogen sulphide in the presence of pyridine (H.S.P.Rao and S.D.Doss, *Sulphur Letters* 1992, **14**, 61).

(ii) *From benzonitriles*

The hydrolysis of benzonitriles to benzamides is accomplished by exposing them to hydrated alumina catalysts coated with potassium fluoride. These catalysts can also be used to convert aliphatic nitriles to amides (C.G.Rao, *Synth. Commun.*, 1982, **12**, 177). Immobilised whole cell *Rhodococcus* sp. may also utilise benzonitriles as substrates and hydrolyse them to benzamides. In the case of dinitriles, hydrolysis of only one of the cyanide groups may occur and it is also possible to effect hydrolysis at a more crowded site with some degree of selectivity. By combining the first step with a follow-up amidase promoted hydrolysis it is possible to convert the intermediate amides, without isolation, directly into the corresponding aroic acids (J.Crosby *et al.,J. Chem. Soc., Perkin Trans. 1*, 1994, 1679).

(iii) *From benzaldehydes*

Aromatic aldehydes also act as starting materials for amides. The reagents needed are a secondary amine, potassium cyanide and air. Presumably an iminium salt is formed initially and this reacts immediately with cyanide ion to form a pseudocyanide. This product undergoes autoxidation with final loss of hydrogen cyanide (T.Chuang *et al.*, *Synlett.*, 1990, 733).

(iv) *From benzoyl halides*

Aroyl chlorides can be used as a source of primary benzamides when they are reacted with bis(trimethylsilyl)amide, followed by hydrolysis of the intermediate *N,N*-bis(trimethylsilylbenzamides with acidified methanol (R.Pellegata *et al., Synthesis*, 1985, 517).

(v) *From benzamides and imides*

The *N*-alkylation of benzamides by primary alcohols occurs in the presence of dichlorotris(triphenylphosphine)ruthenium as catalyst. For example, benzamide affords 70% *N*-octylbenzamide when treated with 1-octanol at 180 °C (Y.Watanabe, T.Ohta and Y.Tsuji, *Bull. Soc. Chem. Japan,* 1983, **56**, 2647).

$$\text{ArCONH}_2 + \text{RCH}_2\text{OH} \xrightarrow{[\text{Ru}]} \text{ArCONHCH}_2\text{R}$$

Secondary and tertiary 2-aminobenzamides are synthesised from 2-aminobenzoyloxyimides through their reactions with primary, or secondary amines, respectively (C.Hinman and K.Vaughan, *Synthesis*, 1980, 719).

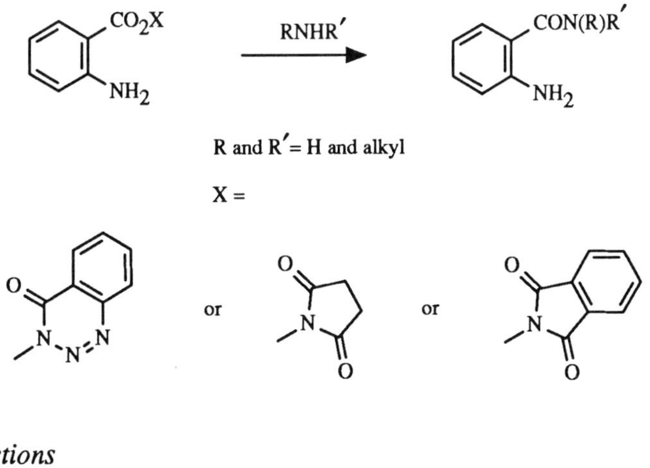

R and R′ = H and alkyl

X =

or

or

(b) *Reactions*

(i) *Ortho-lithiation and related reactions*

An important application in synthesis is the *ortho*-lithiation of benzamides. The products formed can be reacted with a variety of electrophiles to yield the corresponding 2-substituted benzamides. The subject has been reviewed (see, for example, P.Beak and V.Snieckus, *Acc. Chem. Research,* 1982, **15**, 306; V.Snieckus, *Chem. Reviews,* 1990, **90**, 879).

An illustration of the use of this procedure is in the synthesis of substituted saccharins. Here *N,N*-diethylbenzamides are deprotonated with *s*butyl lithium and the lithiated derivatives are then reacted in turn with sulphur dioxide, and hydroxylamine-*O*-sulphonic acid and sodium hydroxide. This affords the corresponding sulphonamides which on treatment with acetic acid cyclise to the heterocycles (D.J.Hlasta, J.J.Court and R.C.Desai, *Tetrahedron Letters,* 1991, **32**, 7179).

The regioselectivity of metalation is dependent upon the ring substitution of the benzamide and also upon the base used. When bulky groups are present which shield attack at the *ortho* position(s) to the amide unit the addition of TMEDA is recommended. This changes polymeric *s*butyl lithium into its dimeric form which is less sterically demanding (M.Khaldi, F.Chrétien and Y.Chapleur, *Tetrahedron Letters,* 1994, **35**, 401).

Chiral amides, obtained from benzoyl chloride and (*S*)-(-)-*N*-benzoyl-*O*-methylphenylalaniol, or (*S*)-(-)-*N*-(3-methoxybenzoyl)-*O*-methylphenyl-alaniol, are deprotonated by the action of 2.2 equiv. of *s*butyl lithium in TMEDA\THF at 0 °C. The resulting lithiated species may then be reacted with aldehydes to afford *ortho*-substituted products with a high degree of diastereoselectivity. When hydrolysed under acidic conditions phthalides are produced (S.Matsui *et al., Chem. Soc., Perkin Trans. 1*, 1993, 701).

The deprotonation of benzanilides is often accomplished in two steps the first utilising butyl lithium and the second methyl lithium as bases. The dilithio salts thus formed react with a range of electrophiles including alkyl halides and carbon dioxide (A.R.Katritzky and W.Fan, *Org. Prep. Proced. Int.,* 1987, **19**, 263).

$$E = D, \text{alkyl}, CO_2^-$$

However, benzanilides are also deprotonated at both the NH site and also at the *ortho*-position by two equivalents of butyl lithium. If the dilithio salts are reacted with *N,N*-dimethylformamide then isoindolinones are generated (J.Epsztajn, A.Józiak and A.K.Szcześniak, *Tetrahedron,* 1993, **49**, 929).

N-Allylbenzamides are isomerised by lithiation with LDA to afford *N*-propenylbenzamides and the further addition of two equivalents of either *s*butyl litium, or *t*butyl lithium, at low temperature generates their *O*,2-dilithiated derivatives. These dianions react easily with a range of electrophiles (E = Me, SMe, SPh, CONHPh *etc*.) to give adducts which upon hydrolysis with 50% aqueous acetic acid release 2-substituted primary benzamides in high yields. It follows that the procedure is equivalent to the *ortho*-lithiation of primary benzamides (L.E.Fischer, J.M.Muchowski and R.D.Clarke, *J. Org. Chem.*, 1992, **57**, 2700).

If 3-methoxy-*N*-(*t*butyl)benzamides are converted into their η⁶-chromium tricarbonyl complexes, prior to successive lithiation and reactions with electrophiles, both 2- and 4-substituted products are formed (M.Uemura, N.Nishikawa and Y.Hayashi, *Tetrahedron Letters*, 1980, **21**, 2069).

3,6-Dibromo-1,2-(*N,N*-diethylamido)benzenes undergo metal-halogen exchange when treated with butyl lithium and the dilithio derivatives formed can then be reacted with numerous electrophiles to afford other 3,6-disubstituted 1,2-(*N,N*-diethylamido)benzenes (R.J.Mills *et al., Tetrahedron Letters*, 1985, **26**, 1145).

R = D, SMe, I, CONEt₂

Ortho-lithiation of suitable benzamides, followed by reactions with methyl borate affords 2-boric acid derivatives (1), which undergo palladium mediated cross-coupling with bromobenzenes to give biphenyl-2-carboxyamides (B.I.Alo *et al., J. Org. Chem.*, 1991, **56**, 3763).

Ortho-lithiation and copper transmetalation of *N,N*-dimethylbenzamides can be carried out as a 'one pot' operation if copper(I) cyanide-lithium chloride is used as the reagent. The arylcyanocuprates which are formed can be coupled directly with aliphatic halides (*e.g.* allyl bromide) to yield 2-alkylbenzamides (2) (P.Pini, S.Superchi and P.Salvadori, *J. Organomet. Chem.*, 1993, 452).

The formation of 2-(pyridinoyl)benzoic acids is possible through the *ortho*-lithiation of anilides, followed by the addition of a pyridine carboxamide. The reactions proceed by the intermediacy of hydroxyisoindolinones which ring open under hydrolytic conditions (J.Epztajn, A.Jóźwiak and J.A.Krysiak, *Tetrahedron*, 1994, **50**, 2907).

(ii) *N-Halogenation*

N-Bromination of benzamides is achieved by reacting them with benzyltrimethylammonium tribromide in ice-cold aqueous sodium hydroxide (S.Kajiaeshi *et al.*, *J. Chem. Soc.*, *Perkin Trans. 1,* 1989, 1702) and *N*-nitration occurs in reactions between amides and tetrabutyl-ammonium nitrate and trifluoroacetic anhydride in dichloromethane at -30 °C [E.Carvalho *et al.*, *J. Chem. Res. (Synop.),* 1989, 260].

(iii) *Reactions with phosphines and the synthesis of enamides and dienamides*

N-Methylbenzamide reacts with ethoxydiphenylphosphine in the presence of paraformaldehyde and trimethylsilyl choride to yield *N*-methyl-*N*-(di-phenylphosphinoyl)methylbenzamide. This reagent can then be treated with arylaldehydes to give enamides, or with cinnamaldehydes to give dienamides (A.Couture, E.Deniau and P.Grandclaudon, *Tetrahedron Letters*, 1993, **34**, 1479).

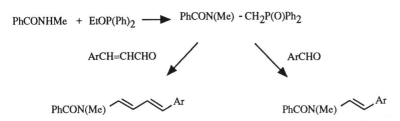

Another route to styrylamides involves the reactions of benzamides with 2,2-dimethoxylphenylethane in benzene containing trifluoroacetic acid (J.E.Estevez *et al.*, *Synth. Commun.,* 1993, **23**, 1081).

$$ArCONH_2 \quad + \quad PhCH_2CH(OMe)_2 \quad \longrightarrow \quad ArCONHCH=CHPh$$

(iv) *Alkylation and the formation of ketones*

N,N-Diethylbenzamides react with alkyl and aryl lanthanum triflates to afford ketones. The reagents are prepared by reacting anhydrous lanthanum tritriflate with the appropriate alkyl, or aryl, lithiums (S.Collins and Y.Hong, *Tetrahedron Letters,* 1987, **28**, 4391).

$$\text{ArCONEt}_2 \xrightarrow{\text{RLa(CF}_3\text{SO}_3)_2} \text{ArCOR}$$

(v) *Reduction*

Lithium aminoborohydrides are powerful reductants formed by treating amine-borane complexes with butyl lithium. Thus when *N,N*-di-^{i}propylbenzamide is reacted with lithium *N,N*-di-^{i}propylaminoborohydride *N,N*-di-^{i}propylbenzylamine is obtained. However, with lithium pyrrolidinoborohydride benzyl alcohol is produced (G.B.Fischer *et al., Tetrahedron Letters,* 1993, **34**, 1091).

(vi) *Hydroxylation*

N-Benzoyl-2-methylalanines, obtained from aroyl chlorides and 2-methylalanine, are *ortho*-hydroxylated by treatment with molecular oxygen in the presence of copper and trimethylamine *N*-oxide to give the corresponding salicylamides. These products can be hydrolysed to the corresponding benzoic acids through hydrolysis with 15% sulphuric acid (O.Reinaud, P.Capdevielle and M.Maumy, *Synthesis,* 1990, 612).

(vii) *Coupling reactions*

4'-Substituted 2-chlorobenzanilides photocylise to phenanthridones when irradiated in cyclohexane solution with light of λ_{max} 254 nm. The reactions conditions are very sensitive and neither bromo- nor iodo-benzanilides can be used as substrates (J.Grimshaw and A.P.DeSilva, *J. Chem. Soc., Chem. Commun.*, 1980, 302).

However, both 2-iodo-*N*-methylbenzanilides and their 2'-iodo analogues are cyclised to *N*-methylphenanthridones by tributyltin hydride in the presence of the radical initiator AIBN in boiling toluene exposed to light (W.R.Bowman, H.Heaney and B.M.Jordan, *Tetrahedron*, 1991, **47**, 10119).

(viii) *Benzamides as nucleophiles*

When benzamide is reacted with tetramethylammonium hydroxide (TMA.OH) and nitrobenzene under anaerobic conditions *N*-(4-nitrophen-yl)benzamide and azoxybenzene are formed. Nucleophilic attack by benzamide upon nitrobenzene is regarded as the first step in this reaction sequence and the adduct so formed reacts oxidatively with more nitrobenzene, perhaps as shown below, to generate nitrosobenzene and thence azoxybenzene. The other product is a nitronate anion which when treated with water protonates and aromatises to form N-(4-nitrophenyl)benzamide. This last compound is easily hydrolysed and with water at 70 °C it affords tetramethylammonium benzoate and 4-nitroaniline (M.K.Stern and B.K.Cheng, *J. Org. Chem.*, 1993, **58**, 6883).

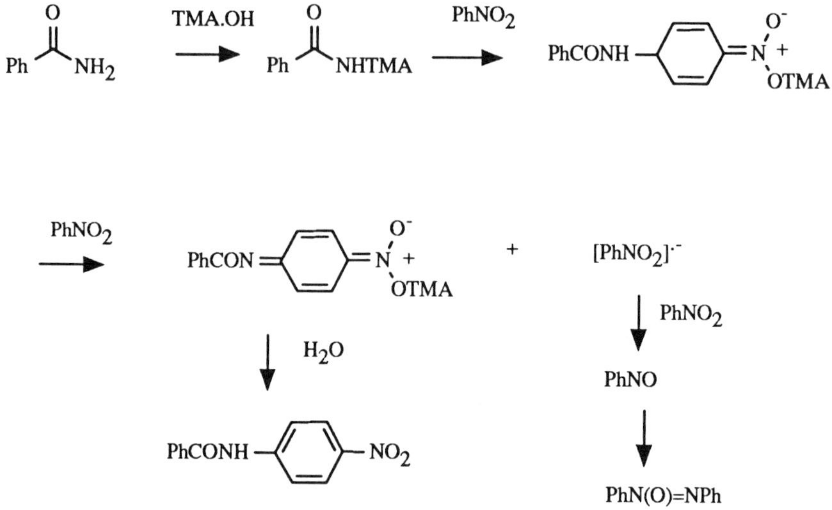

(ix) *Pyrolysis*

O-Allylsalicylic alkylamides are converted into isoindolones by vacuum-flash pyrolysis (M.Black, J.I.C.Cadogan and H.McNab, *J. Chem. Soc., Perkin Trans. 1*, 1994, 155). The first step is the formation of a phenoxy radical which rearranges by the transfer of a hydrogen atom from an *N*-alkyl group. The radical then formed attacks the aromatic ring at the *meta*-position to the hydroxy group to give the corresponding 3-hydroxyisoindolin-2-one.

(x) *Spectroscopy*

The effects of 3- and 4-substiuents in the aromatic rings of benzamides upon the chemical shift of the carbonyl carbon resonance have been analysed (M.DeRosa *et al. J. Chem. Soc., Perkin Trans.* 2, 1993, 1787). The results indicate that π-polarisation plays an important role in determining the ^{13}C chemical shift of the carbonyl carbon resonance (*c.f.* C.Dell'Erba *et al. J. Chem. Soc., Perkin Trans.* 2, 1990, 2055) and, moreover, that the carbonyl group is more sensitive to the nature of certain substituents than is the carbonyl group of esters (D.J.Craik and R.T.C.Brownlee, *Prog. Phys. Org. Chem.*, 1983, **14**, 1).

6. Benzonitriles

(a) *Synthesis*

Methods for the syntheses of aromatic nitriles have been reviewed (N.R.Butkeikhanov and B.V.Suvorov, *Tr. Inst. Khim. Nauk. Kaz. SSR.*, 1980, **51**, 3).

(i) *From benzamides*

A commonly used method for the synthesis of nitriles is the dehydration of amides. Acetic anhydride in contact with either finely powered copper or nickel is recommended as a reagent for this type of reaction [J.F.Stenberg, *Chem. Ind. (Dekker),* 1985, **22** (*Catal. Org. React*), 373]. Other reagent systems are: trichloroacetyl chloride and triethylamine (A.Saednya, *Synthesis*, 1985, 184) and silver oxide and ethyl iodide (M.L.Sznaidman, C.Crasto and S.M.Hecht, *Tetrahedron Letters,* 1993, **34**, 1581).

Thionyl chloride in dimethylformamide can also be used, as in the conversion of 1,2-diamido-4,5-dichlorobenzene into 1,2-dicyano-4,5-di-chlorobenzene (D.Wöhrle *et al., Synthesis*, 193, 194).

(ii) *From alkylarenes*

Toluenes undergo vapour phase ammoxidation over a vanadium-titanium oxide catalyst to give benzonitriles. The slow step of the reaction is the homolytic fission of a carbon-hydrogen bond of the methyl group. This homolysis is impeded when the ring to which the methyl group is attached bears electron donating substituents (F.Cavani, F.Perrenello and F. Trifiro, *J. Mol. Cat.,* 1987, **43**, 117). The relative efficiences of copper-bearing zeolite ZSM-5 catalysts for the ammoxidation of toluene have been analysed [J.C.Oudejans, F.J.Van derGraag and H.van Bekkum, *Proc. Int. Zeolite Conf. 6th,* 1983 (publ. 1984), 536].

Dimethylarenes form dicyanoarenes when they are reacted with ammonia over chromium and vanadium oxides, with, or without, boron oxide, on alumina at 300-500 °C (C.S.Ray *et al., Proc. Catsympo. 80, Natl. Catal. Symp. 5th,* 1980 (published 1983) 303; *C.A.* 1983, **99**, 104918r). A review on this subject is available [S.C.Ray *et al., Adv. Catal.* (*Proc. Natl. Sympos. Catal.*), 7th, 1985, 413].

The electrochemical cyanation of ^{t}butylated anisoles can be acheived through electrolysis in methanol solution containing sodium cyanide at a platinum anode. Two types of reaction are noted, one leading to substitution of an aromatic hydrogen and the other replacement of the methoxy group. The latter process is favoured if there are ^{t}butyl groups at all *ortho* and *para* positions to the methoxy group (K.Yoshida, K.Takeda and T.Fueno, *J. Chem. Soc., Perkin Trans. 1,* 1993, 3095).

(iii) *From araldehydes, aldoximes and hydrazones*

Aromatic nitriles can be synthesised from benzaldehydes in a 'one pot' procedure using hydroxylamine hydrochloride and potassium fluoride adsorbed upon an aluminate support in the absence of solvent. The reactions are initiated by microwave irradiation and oximates are formed. These become absorbed on the support, but they are released and converted into the nitriles by treatment with carbon disulphide (D.Villemin, M.Lalaoui and A.B.Alloum, *Chem. Ind.,* 1991, 176).

A similar reaction, although one not requiring microwave initiation, involves the treatment of benzaldehydes with hydroxylamine hydrochloride in acetic acid containing polyphosphoric acid (I.Ganboa and C.Palomo, *Synth. Commun.,* 1983, **13**, 999).

Yet another procedure involves heating benzaldehydes with nitroethane in pyridine containing hydrochloric acid (D.Dauzonne, P.Demepseman and R.Royer, *Synthesis,* 1981, 737).

Araldehydes (4-RC_6H_4CHO, R = H, Me, MeO, or Cl) generate the corresponding benzonitriles when they are treated with trimethylsilyl azide and a catalytic amount of zinc(II) chloride. While this procedure works reasonably well for the examples indicated above, the reaction is inefficient for nitrosubstituted araldehydes, unless a stoichiometric amount of zinc(II) chloride is present (K.Nishiyama and A.Watanabe, *Chem. Letters,* 1984, 773).

Similarly, benzaldehydes form benzonitriles when they are reacted in succession with 1,1-dimethylhydrazine, methyl iodide, and sodium methoxide. It is also reported that araldehyde oximes can be dehydrated to the corresponding nitriles using a mixture of formic acid and sodium formate as the reagents. (X.Shi, *Huaxue Shiji,* 1990, **12**, 312, 293; *C.A.,* 1991, **114**, 206700j).

Dehydrations of aldoximes are also catalysed by copper(II) carboxylates (Z.Zhang *et al., Wuji Huaxue,* 1988, **4**, 112; *C.A.,* 1989, **111**, 133736b).

N,N-Dimethyldichloromethaniminium chloride ($Cl_2C=N^+Me_2$ Cl^-) can be used as yet another dehydrating agent for aldoximes. It also converts phenylhydrazones into benzonitriles (B.Kokel, G.Menichi and M.H.-Hab-art, *Synthesis*, 1985, 201).

$$ArCH=NNHPh \quad + \quad \begin{array}{c} Cl \\ Cl \end{array}\!\!\!\bigg\rangle\!\!=\!\!N^+\!\!\begin{array}{c} Me \\ Me \end{array} \quad \xrightarrow{-HCl}$$

Cl⁻

Ar

$$H\!\!-\!\!\!\bigg\langle\!\!\begin{array}{c} Ph \\ N\!-\!N \end{array}\!\!\!\bigg\rangle\!\!=\!\!N^+\!\!\begin{array}{c} Me \\ Me \end{array}$$
Cl

Cl⁻

$$\xrightarrow{-HCl} \quad ArCN + PhN=C(Cl)NMe_2$$

2-Methylbenzaldehyde gives 2-methylbenzonitrile in 88% yield when it is treated with cupric nitrate and potassium persulphate in ammonium hydroxide solution (P.K.Arora and L.M.Sayre, *Tetrahedron Letters,* 1991, **32**, 1007). Araldehydes can also be converted into the corresponding nitriles by reacting them, under pressure, with aqueous ammonia containing potassium persulphate and nickel(II) sulphate as the catalyst. Similar treatment of benzyl alcohol gives benzonitrile in 82% yield (S.Yamazaki and Y.Yamazaki, *Chem. Letters*, 1990, 571).

4-Nitro-*N*-alkoxybenzaldoximes are deprotonated by the action of sodium hydride in *N,N*-dimethylformamide and this ultimately affords 4-nitrobenzonitrile, hydrogen and a sodium alkoxide. In a follow-up reaction the alkoxide anion reacts with the benzonitrile to displace the nitro group and to yield 4-alkoxybenzonitriles (D.Mauléon, R.Granados and C.Minguillón, *J. Org. Chem.,* 1983, **48**, 3105).

(iv) *From aryl iodides*

Benzonitriles can be obtained by reacting aryl iodides (but not chlorides, or bromides) with trimethylsilyl cyanide in boiling triethylamine containing a catalytic amount of tetrakis(triphenylphosphine)palladium (N.Chatani and T.Hanafusa, *J. Org. Chem.*, 1986, **51**, 4714).

(v) *From benzoic acids and benzoyl halides*

Benzoic acids form the appropriate nitriles when they are heated with urea and phosphoric acid (W.Huang *et al.*, *Beijing Gongye Daxue Xuebao*, 1989, **15**, 57 ; *C.A.*, 1990, **112**, 76552h), or when they are heated with tosylamine and phosphorus pentachloride (Q.Lu and G. Lu, *Xuaxue Tongbao*, 1989, 31; *C.A.*, 1990, **112**, 234923m).

3,4-Difluorobenzoic acid affords 3,4-difluorobenzonitrile when it is reacted with 1,4-dicyanobutane (R.G.Pews, J.A.Gall and J.C.Little, *J. Fluorine Chemistry*, 1990, **50**, 365).

Another method for the production of nitriles requires the initial conversion of aroyl chlorides into the corresponding *N*-alkoxyamides (ArCONHOR) through reactions with *N*-alkoxyamines. These products, on treatment with carbon tetrabromide and triphenylphosphine in acetonitrile solution, afford *N*-alkoxyimidoyl bromides [ArC(Br)=NOR]. Alternatively, the *N*-alkoxyamides can be reacted with phosphorus pentachloride/phosphorus oxychloride to give *N*-alkoxyimidoyl chorides. Both bromides and chlorides can be reduced with zinc in acetic acid/*N*,*N*-dimethylformamide to obtain the desired nitriles (T.Sakamoto *et al.*, *Synthesis*, 1991, 750).

(vi) *From alkyl thiobenzoates and thiobenzamides*

When thiobenzoic acid esters are heated at 380-400 °C with ammonia and boron phosphate as a catalyst the corresponding benzonitriles are produced [B.V.Suvorov *et al.*, *Zh. Prikl. Khim. (Leningrad)*, 1987, **60**, 677]. Thiobenzamides are also a source of benzonitriles when they are reacted under phase-transfer conditions with 30% aqueous sodium hydroxide and benzyl chloride in benzene at 30 °C (Y.Funakoshi, T.Takido and K. Itabashi, *Synth. Commun.*, 1985, 15, 1299).

Nitriles are obtained when thiobenzamides are reacted with elemental sulphur and sodium nitrite in liquid ammonia. This method represents a new version of the Willgerodt-Kindler process (R.Sato *et al.*, *Chem. Letters*, 1984, 1913).

(vii) *By cycloaddition reactions*

Isophthalonitriles are prepared by reacting 1,1-dicyano-2-phenyl-1,3-butadienes with substituted acetonitriles (A.M.El Torgoman *et al.*, *Chem.*

Industry, 1987, 60). The cycloaddition of ethyl 4-phenylacetoacetate and 1,1-dicyano-2-ethoxyethene gives ethyl 4-amino-5-cyano-2-hydroxy-3-phenylbenzoate (H.W.Schmidt and M.Kores, *Monatsh. Chem.*, 1988, **119**, 91).

X = CSNH$_2$, CN, CO$_2$Et, COPh

(viii) *Miscellaneous*

Alkyl-2,4,5-tricyanobenzenes can be formed from 1,2,4,5-tetracyanobenzene when a solution in acetonitrile containing an alkanoic acid is irradiated with light. The reactions may proceed through initial electron-transfer from the acids to the tetracyanobenzene forming its radical anion, together with a proton and an alkanoate radical. In the next phase of the reaction the alkanoate radical loses carbon dioxide and forms an alkyl radical. This is then captured by the tetracyanobenzene radical anion with concomitant loss of a cyanide ion (K.Tsujimoto, N.Nakao and M.Ohashi, *J. Chem. Soc., Chem. Commun.*, 1992, 366).

$$\text{(tetracyanobenzene)} + RCO_2H \xrightarrow{-H^+} \text{(radical anion)} + RCO_2\bullet$$

$$RCO_2\bullet \longrightarrow R\bullet + CO_2$$

$$\text{(radical anion)} \xrightarrow{R\bullet} \text{(R-adduct anion)} \xrightarrow{-CN^-} \text{(R-substituted tricyanobenzene)}$$

(b) Reactions

(i) Hydrolysis

Rhodococcus rhodochrous in suspension catalyses the partial hydrolysis of dinitriles to cyanocarboxylic acids under very mild conditions. The reactions are efficient, thus 1,3-dicyanobenzene affords 3-cyanobenzoic acid in 95% yield (C.B.-Gabier and A.L.Gutman, *Tetrahedron Letters,* 1988, **29**, 2589).

Reductive hydrolysis of benzonitriles gives araldehydes. One recommended method uses a metal, aluminium, or iron, covered with nickel as a promoter in contact with an acid (Z.Bodnar, T.Mallat and J.Petro, *J. Mol. Catal.,* 1991, **70**, 53).

(ii) Reduction

The direct hydrogenation of a benzonitrile to a toluene is possible, if the reduction is carried out with hydrogen at 1 atmosphere pressure and 30% nickel on alumina is used as the catalyst. A temperature of 150 °C is required (J.G.Andrade *et al., Synthesis*, 1980, 802).

Benzonitriles can be reduced to benzaldehydes by initial *N*-alkylation with Meerwein's reagent $[(Et)_3O^+BF_4^-]$ and then treatment with triethylsilane. This affords benzaldimines which are readily hydrolysed to the corresponding benzaldehydes. Selective reductions of cyanide groups in polysubstituted aromatic compounds is possible by this technique which does not lead to over reduction and the formation of anilines (J.L.Fry and R.A.Ott, *J. Org. Chem.*, 1981, **46**, 602).

$$ArCN + Et_3O^+BF_4^- \longrightarrow ArC^+{=}NEt\ BF_4^- \xrightarrow{Et_3SiH}$$

$$\xrightarrow{H_2O}$$

$$ArCH{=}N(Et) \longrightarrow ArCHO + EtNH_2$$

Birch reduction of benzonitrile with potassium in a 4:1 mixture of ammonia and tetrahydrofuran, followed by quenching of the reaction mixture with methyl iodide leads to 6-cyano-6-methyl-1,4-cyclohexadiene. Similar treatment of 1-naphthonitrile affords 1-cyano-1,4-dihydro-1-methylnaphthalene (T.A.Vaganova, I.I.Bil'kis and V.D.Shteingarts, *Z. Org. Khim.*, 1986, **22**, 2239; A.G.Schultz and M.Macielag, *J. Org. Chem.*, 1986, **51**, 4983).

In the presence of excess lithium powder (1:14 molar equivalents) and a catalytic amount of 4,4-di-*t*butylbiphenyl (TBBP) benzonitrile reacts with benzaldehyde, or alkanones, with elimination of the cyanide group and the formation of diphenylmethanol, or alkylphenylmethanols, respectively (D.Guijarro and M.Yus, *Tetrahedron*, 1994, **50**, 3447).

(iii) *Reactions with azides*

Benzonitriles react with azides to form tetrazoles, recent illustrations of this type of reaction include the synthesis of 5-(2-fluorophenyl)tetrazole from 2-fluorobenzonitrile and sodium azide in the presence of acetic acid (R.K.Russell and W.V.Murray, *J. Org. Chem.*, 1993, **58**, 5023), and the preparation of biarylated tetrazoles from 2-cyanobiaryls and tributyltin azide (J.V.Duncia, M.E.Pierce and J.B.Santelle (III), *J. Org. Chem.*, 1991, **56**, 2395).

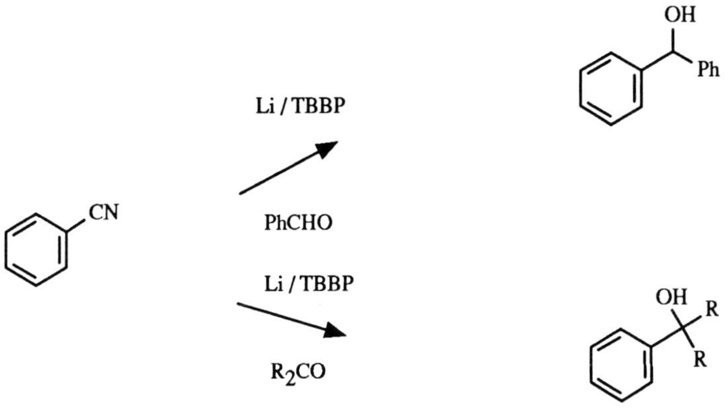

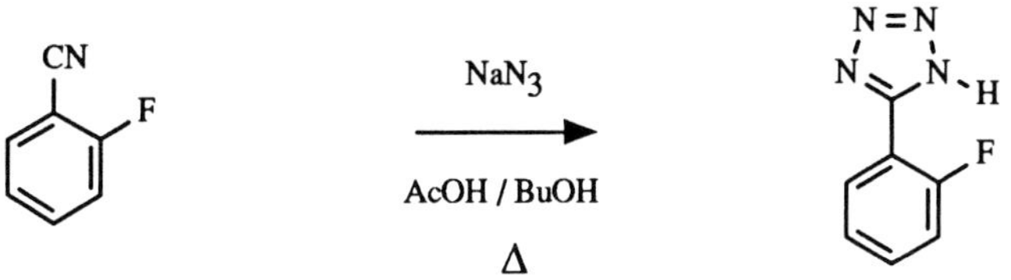

OH
Ph

Li / TBBP

PhCHO

Li / TBBP

R₂CO

C_6H_5CN (benzonitrile)

OH R
R

CN, F

NaN₃

AcOH / BuOH

Δ

N=N
N—N—H
F

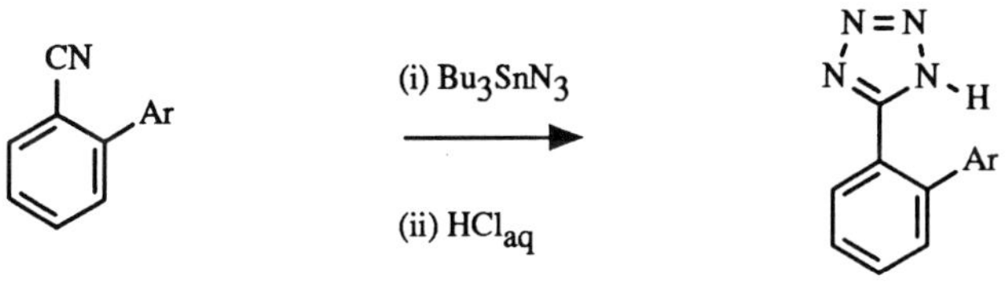

CN, Ar

(i) Bu₃SnN₃

(ii) HCl$_{aq}$

N=N
N—N—H
Ar

A further reagent which is recommended for the conversion of nitriles into 5-aryltetrazoles is trimethylsilyl azide in contact with a catalytic amount of dibutyl, or dimethylstannane, in toluene at 93-100 °C (S.J.Wittenberger and B.G.Donner, *J. Org. Chem.*, 1993, **58**, 4139).

(iv) *Formation of N-(ᵗbutoxycarbonyl)arylamines*

Primary aromatic amides react and rearrange when treated with lithium *t*butoxide in THF containing copper(II) bromide to afford *N*-(*t*butoxycarb-onyl)arylamines (ArNHCO$_2$*t*Bu) (J.Yamaguchi, K.Hoshi and T.Takeda, *Chem. Letters,* 1993, 1273).

7. Benzonitrile oxides

Arene carbonitrile oxides continue to have much value as 1,3-dipolarophiles (see S.Auricchio *et al., Tetrahedron Letters,* 1993, **34,** 4363) and are commonly used in heterocyclic synthesis. They can be prepared by reacting trichloromethylarenes with hydroxylamine in ethanol to form α-chlorooximes [ArC(Cl)=NOH]. Both (*E*)- and (*Z*)-isomers are formed; the latter eliminate hydrogen chloride to afford the required products (D.B.Brokhovetskii, L.I.Belen'kii and M.M.Krayushkin, *Izv. Akad. Nauk. SSSR, Ser. Khim.,* 1990, 1692; *C.A.,* 1991, **115,** 279557y].

Lewis acids promote the electrophilicity of aromatic nitrile oxides and allow them to be used as hydroxylnitrilium equivalents. For example, 2,6-dichlorobenzonitrile oxide and aluminium(III) chloride react with benzene to afford the oxime of 2,6-dichlorobenzophenone in 70% yield (J.N.Kim and E.K.Ryu, *Tetrahedron Letters,* 1993, **34,** 3567).

8. Aroyl cyanides

(a) *Synthesis*

(i) *From aroyl halides*

A common method for the synthesis of aroyl cyanides from aroyl chlorides uses copper(I) cyanide as the source of the cyanide unit, however, when 2-nitrobenzoyl chorides are treated in this way a vigorous and uncontrollable reaction may ensue (B.Goodwin, *Chem. Britain,* 1987, **23,** 118).

Aroyl cyanides are also prepared by reacting aroyl chlorides with potassium cyanide in acetonitrile. It is not necessary for the acetonitrile to be dry, but yields are lowered if tributyltin chloride is added, or if 18-crown-6 is present (M.Tanaka and M.Koyanagi, *Synthesis*, 1981, 973). Sodium, or potassium, cyanide impregnated on Amberlite XAD-2, or XAD-4 resins, are recommended reagents for the conversion of aroyl chlorides into aroyl cyanides [K.Sukata, *Bull. Chem. Soc. (Japan)*, 1987, **60**, 1085].

(ii) *From araldehydes*

Aroyl cyanides (ArCOCN) are synthesised by reacting benzaldehydes with potassium cyanide and trimethylsilyl chloride, followed by the addition of chromium trioxide to the reaction mixtures. Comparable yields are achieved by an alternative procedure in which the aldehyde is treated successively with potassium cyanide, acetic anhydride, and pyridinium chlorochromate/pyridinium 4-toluenesulphonate (S.K.Kang, H.K.Sohn and S.G.Kim., *Org. Prep. Proced. Int.*, 1989, **21**, 383).

(iii) *From arylglyoxals and phenacyl halides*

Arylglyoxals also serve as starting materials for the synthesis of aroyl cyanides. Such compounds can be reacted with 1-amino-4,6-diphenyl-2-pyridone to produce intermediate imines which on heating fragment into the corresponding aroyl cyanide and 4,6-diphenyl-2-pyridone (M.Alajarin, P.M.Fresneda and P.Molina, *Synthesis*, 1980, 844).

(b) *Reactions*

Sodium bromide is an effective promoter for the hydrolysis of aroyl cyanides while still retaining the carbon atoms of the side chain. Thus treatment of benzoyl cyanide with 85% aqueous sulphuric acid containing sodium bromide gives benzoylformamide ($PhCOCONH_2$). If methanol and sulphuric acid are used the product is methyl benzoylformate ($PhCOCO_2Me$) (J.M.Photis, *Tetrahedron Letters,* 1980, **21**, 3539).

9. Aroyl azides

(a) *Synthesis*

(i) *From benzoic acids and benzoyl halides*

Aroyl azides can be synthesised from aromatic acids by reactions with phenyl dichlorophosphate and sodium azide at room temperature in dichlorobenzene containing either tetrabutylammonium bromide or pyridine (J.M.Lago, A.Arrieta and C.Palomo, *Synth. Commun.,* 1983, **13**, 289). Another efficient method involves reactions of aroyl chlorides with trimethylsilyl azide using zinc(II) iodide as catalyst. Under these conditions subsequent Curtius rearrangements of the azides are not observed (G.K.S.Prakash *et al., J. Org. Chem.,* 1983, **48**, 3358).

(ii) *From phenacyl dibromides*

Phenacyl dibromides also afford aroyl azides when they are reacted with sodium azide (G.Weber *et al., Synthesis,* 1983, 191).

(iii) *From aroylhydrazines*

Benzoyl azide is obtained from *N*-benzoylsemicarbazide (PhCONHNH-$CONH_2$) by reacting it with nitrous acid at 100 °C (Yu I.Smushkevich and

M.I.Usorov, *Zhur. Org. Khim.*, 1988, **24**, 2450).

(b) *Reactions*

(i) *Reduction*

The reduction of benzoyl azides by sodium borohydride gives the corresponding benzyl alcohols as the principal products, plus smaller amounts of benzamides (H.S.P.Rao, *Synth. Commun.*, 1990, **20**, 45).

(ii) *Curtius rearrangements to isocyanates*

Mellitic acid can be converted into benzenehexacarbonyl azide and then subjected to the conditions of the Curtius rearrangement reaction thus forming benzenehexaisocyanate (L.Pozolotina *et al., Ivz. Akad. Nauk. Kirg. SSR.*, 1983, 45; *C.A.*, 1983, **99**, 53278x). Aryl isocyanates can also be synthesised *in situ* from aroyl chlorides through the action of sodium azide in a mixed solvent system of toluene/dimethylformamide/acetonitrile (V.A.Zlobin and A.K.Tarasov, *Izv. Vssh. Uchebn. Zaved. Khim. Khim. Teknol.*, 1989, **32**, 35).

10. Aroyl cyanates

Benzoyl cyanate (PHCOCNO) is formed by reacting benzoyl chloride with sodium cyanate in the presence of either tin tetrachloride or zinc dichloride as a catalyst (M.Z.Dang *et al., Tetrahedron Letters,* 1988, **44**, 6079).

11. *N*-Aroylhydrazines and aroylazoalkenes

Aroylhydrazines react with alkyl 2-chloroacetoacetates to form semicarbazides. These products are dehydrochlorinated by treatment with aqueous base to yield alkyl 3-[(*N*-aroyl)azo]-2-butenoates (O.Attanasi, M.Grossi and F.S.-Zanetti, *Org. Prep. Proc. Int.*, 1985, **17**, 385).

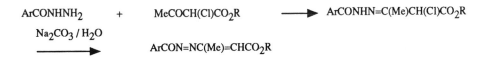

$ArCONHNH_2$ + $MeCOCH(Cl)CO_2R$ $\longrightarrow$ $ArCONHN=C(Me)CH(Cl)CO_2R$

Na_2CO_3 / H_2O

$\longrightarrow$ $ArCON=NC(Me)=CHCO_2R$

A series of arylhydrazonates is synthesised from arylhydrazonyl bromides and phenols when they are reacted together under phase-transfer conditions. 1,2-Diaroylhydrazines are similarly formed if the phenols are replaced by benzoic acids (A.S.Shawali *et al., Bull. Chem. Soc. Japan,* 1981, **54**, 2545).

12. Benzimidoyl halides and related compounds

Aryl imidates are readily formed when benzimidoyl chlorides are reacted with phenols under phase-transfer conditions (J.E.Rowe, *Synthesis*, 1980, 114).

$$PhC(Cl)=NR + ArOH \quad \xrightarrow[\text{H}_2\text{O / DCM}]{\text{NaOH / Bu}_4\text{NBr}} \quad PhC(OAr)=NR$$

N-Arylbenzimidoyl chlorides in which the the *N*-aryl group is unsubstituted at the 2- and 4-positions are nitrated by treatment with silver nitrate to form *N*-(nitroaryl)benzamides. However, if these sites are blocked the products are *N*-aryl-*N*-nitrobenzamides. Both types of product may arise through the initial formation of *O*-nitroimidates, which undergo homolysis to nitro and benzamide radicals. The nitro radicals may then attack either the aromatic nucleus or the *N*-atom of the amide.

N-Alkylbenzimidoyl chlorides react with silver nitrate to form *N*-alkyl-*N*-nitrobenzamides, as well as the *N*-nitroso analogues. Whereas the origins of the *N*-nitro compounds are probably the same as those of their aromatic counterparts, the formation of the nitrosoamides requires a different reaction pathway (J.Iley *et al.*, *J. Chem. Soc.*, *Perkin Trans. 2*, 1992, 281).

N-Phenylbenzimidoyl chlorides (ArC(Cl)=NPh) are benzoylated by reaction with benzaldehyde in the presence of a catalytic amount of 1,3-dimethylimidazolium iodide. In this way *N*-(α-benzoylbenzylid-ene)anilines (ArC(COPh)=NPh) are formed, which when reacted with dilute hydrochoric acid afford benzils (ArCOCOPh) in excellent yields (A.Miyashita, H.Matsuda and T.Higashima (*Chem. Pharm. Bull.*, 1992, **40**, 2627).

The chlorination of araldehyde oximes (RC$_6$H$_4$CH=NOH, R = H, 2- and 4-NO$_2$, 4-Cl, 4-Br, 4-MeO *etc.*) with hydrogen chloride and a slight excess of sodium hypochlorite yields the corresponding hydroximic acids (RC$_6$H$_4$C(Cl)=NOH). The aromatic ring is not attacked during such reactions (A.C.Coda and G.Tacconi, *Gazz.*, 1984, **114**, 131).

13. Aminobenzoic acids and related compounds

Previous it was claimed that anthranilic acid reacts with thionyl chloride in benzene, or toluene, under an atmosphere of nitrogen to give 3,2,1-benoxathiazin-4(1*H*)-one 2-oxide. However, this is not the case and the correct product is a benzoyl chloride bearing an *ortho* sulphinylamino (thiazte NSO) group (J.Garín *et al.*, *Tetrahedron Letters*, 1991, **32**, 3263).

One application of anthranilic acid is its use, through diazotisation, in the synthesis of benzyne. Recently the structure of the highly unstable precursor 2-diazoniobenzenecarboxylate has been determined by single crystal X-ray structure analysis (C.J.Horan, C.L.Barnes and R.Glaser, *Chem. Ber.*, 1993, **126**, 243).

14. Thiobenzoic acids, their derivatives, and selenium and tellurium analogues

(a) *Thiobenzoic acids*

The preparation and reactions of thiobenzoic acid anhydrides have been surveyed (A.A.Martin and G.Bainikov, *Z. Chem.*, 1987, **27**, 90). Thiobenzoic acid can be prepared from benzoyl chloride in two steps: first reaction with *N,N*-dimethylthioformamide and then decomposition of the adduct (PhCOSCH=N$^+$Me$_2$ Cl$^-$) which is formed by treatment with methanol. Other products formed in the reaction are benzoic acid and methyl benzoate (S.Hibino *et al., Nippon Kagaku Kaishi*, 1985, 898; *C.A.,* 1986, **104**, 68525w).

(b) *Thiobenozate esters*

(i) *Synthesis*

Thiobenzoate esters (ArCOSR) are obtained from benzoic acids by reacting them with thiols (RSH) in the presence of polyphosphoric ester (T.Imamoto, M.Kodera and M.Yokoyama, *Synthesis*, 1982, 134). Another coupling agent for this type of reaction is 1-fluoro-2,4,6-trinitrobenzene in the presence of 4-(*N,N*-dimethylamino)pyridine (S.Kim and S.Yang, *Chem. Letters*, 1981, 133).

S-Methyl benzothiolates (ArCOSMe) are available from benzoic acids by reacting them with methyl thiocyanate and trifluoroacetic acid, smaller amounts of benzonitriles are formed as by-products (Yu Polivin *et al.*, *Izv. Akad. Nauk. Ser. Khim.*, 1992, 412).

Aryl thiobenzoates (ArCOSAr') are formed by treating aromatic anhydrides with thiophenols (Ar'SH) in acetonitrile solution containing cobalt(II) chloride (S.Amad and J.Iqbal, *Tetrahedron Letters,* 1986, **27**, 3791).

N-Nitrosobenzamide, or *N*-nitrobenzamide, reacts with sodium thiolates (RSNa, R = Et, *t*Bu, or Ph) in tetrahydrofuran to yield the corresponding thiobenzoates (PhC(O)SR) (R.Bevenguer, J.Garcia and J.Vilarrasa, *Synthesis*, 1989, 305).

(ii) *Mass spectra*

The mass spectra of pentyl and hexyl thiobenzoates have been recorded. Among the more important processes observed are *O* to *S* -alkyl shifts and subsequent fragmentation to yield aroyl ions. These features are less important in the spectra of the analogous methyl and ethyl analogues (G.W.Wood and B.T.Kiremire, *Org. Mass Spectrom.*, 1979, **14**, 596).

(iii) *Natural product*

The structure of the antibiotic resorthiomycin has been revised to (1) (M.Tahara *et al., J. Antibiot.,* 1991, **44**, 255).

(c) *Thiobenzamides*

(i) *Synthesis*

N-Phenylthiobenzamides (ArC(S)NHPh) are available through the

1

reactions of arenes with phenyl thiocyanate in the presence of aluminium trichloride in nitromethane (T.Jagodzinski, *Synthesis*, 1988, 717). *N,N*-Disubstituted benzothioamides (ArCSNR$_2$) can be prepared by reacting phosphonate carbanions [ArCH⁻P(O)(OEt)$_2$] with elemental sulphur in the presence of secondary amines. The selenium analogues are made similarly using selenium rather than sulphur (K.Okuma, K.Ikari and H.Ohta, *Chem. Letters*, 1992, 131).

A synthesis of *N,N*-disubstituted thiobenzamides [PhCSN(Me)R; R = Me, or R = Ph] requires the reactions of thiuram monosulphides [R(Me)NCSSSCN(Me)R] with phenyllithium in tetrahydrofuran at -78 °C. If tetramethylthiuram disulphide (Me$_2$NCSSSSCNMe$_2$) is reacted in this way phenyl *N,N*-dimethyldithiocarbamate (PhSCSNMe$_2$) is produced (S.Gronowitz, A.-B.Hörnfeldt and M.Temcine, *Synthesis*, 1993, 483).

(ii) *Physical properties and reactions*

The ESR spectra of radical anions of thiobenzamides have been studied and it is noted that the CSNR$_2$ group withdraws as much spin density from the aromatic nucleus as does a nitro group. The selenoamide group has similar properties (J.Voss and F.R.Brun, *Liebig's Annalen*, 1979, 1931).

When oxidised with ceric ammonium nitrate thiobenzamide gives rise to 3,5-diphenyl-1,2,4-thiadiazole (D.N.Dhar and A.K.Bag, *Indian J. Chem. Sect. B*, 1985, **24B**, 445).

Primary thiobenzamides can be alkylated at the sulphur atom by reaction with Meerwein's salts. Thus thiobenzamide itself yields *S*-methylbenzo-thioimidate [PhC(SMe)=NH] when treated with trimethyloxonium tetra-fluoroborate (M.A.Cascade, B.Di Rienzo and F.M.Moracci, *Synth. Commun.*, 1983, **13**, 753).

Thiobenzamide reacts with furan during irradiation with ultraviolet light from a high pressure mercury lamp in a atmosphere of nitrogen to form 2-phenylpyrrole-4-carbaldehyde. The reaction is considered to proceed *via* isomerisation to the thioimine and a [2+2] cycloaddition of this compound with furan. The adduct then undergoes ring expansion to 4-phenyloxepine, followed by ring contraction to 5-phenyl-2,3-epoxidopyridine. Finally this compound isomerises in two steps to the ultimate product (K.Oda and M.Machida, *J. Chem. Soc., Chem. Commun.*, 1993, 437).

Photolyses of thiobenzanilides (*e.g.* 1) in the solid state give 3-arylpropanoic acids. Initially it was concluded that this type of reaction involved a Norrish II type γ-hydrogen abstraction (M.Sakamoto *et al., J. Chem. Soc., Perkin Trans. 1*, 1991, 347); however, new evidence points to the formation of a cyclic acylium species (2), which when treated with water hydrates and cleaves into a propanoic acid and a *N*-arylthiobenzamide (T.Y.Fu J.R.Scheffer and J.Trotter, *Tetrahedron Letters*, 1994, **35**, 3235).

(d) *Aroylsulphenyl halides*

Aroylsulphenyl iodides (ArCOSI) are obtained when phenylmercuric thiocarboxylates (ArCOSHgPh) are reacted with 0.1M iodine in chloroform. These compounds have limited stability and upon heating to 40 °C, or exposure to sunlight, they decompose evolving iodine (S.Kato *et al.*, *Angew. Chem.*, 1982, **94,** 148).

(e) *Diaroyl disulphides and related compounds*

Diacyl disulphides (ArCOSSCOAr) are conveniently formed by adding aroyl chlorides to a solution of sulphur in sodium hydroxide (J.Wang *et al.*, *Chin. Chem. Letters*, 1990, **1**, 193). Trithio analogues (ArCOSS(S)CPh) are available by reacting aroylsulphenyl halides (ArCOSX, X = Cl or Br) with dithiobenzoic acid (T.Murai *et al.*, *Tetrahedron Letters*, 1986, **27**, 4593).

Symmetrical bis(aroyl) disulphides are synthesised through the reactions of aroyl chlorides with disodium disulphide under phase-transfer conditions (M.Kodomari, M.Fukuda, and S.Yoshitomi, *Synthesis*, 1981, 637), whereas their unsymmetrical counterparts can be obtained from potassium thiobenzoates by first reacting them with *N*-chlorosuccinimide, to form the corresponding *N*-(benzoylthio)succinimide, and then treatment of these products with a thioaroic acid (M.Mizuta *et al., Synthesis,* 1980, 721).

$$ArCOSK \quad + \quad \overset{O}{\underset{O}{\bigvee}}N{-}Cl \quad \longrightarrow \quad ArCOS{-}N\overset{O}{\underset{O}{\bigvee}}$$

$$\xrightarrow{Ar'COSH} \quad ArCOSSOCAr'$$

(f) *Dithiobenzoic acids and esters*

Dithiobenzoate esters ($PhCS_2R$) can be prepared from sodium phenyldithiocarboxylate and alkylthioformylchoride (RSCOCl) (P.Beslin, A.Dlubala and G.Levesque, *Synthesis,* 1987, 835).

Dithiobenzoic acid reacts with ketene to produce acetyl thiobenzoyl sulphide [PhC(S)SCOMe] (G.Barnikow and A.A.Martin, *J. Prakt. Chem.,* 1983, **325**, 337).

Dithiobenzoic acids yield adducts with imines preformed from the reactions of anilines and araldehydes. When treated with phenacyl bromide the adducts afford phenacyl dithiocarboxylates (M.Ishida, S.Kato and M.Mizuta, *Z. Naturforsch.,* 1981, **36B**, 1047).

$$ArCS_2H + PhN{=}CHR \quad \longrightarrow \quad ArCS_2CH(Ph)NHR$$

$$PhCOCH_2Br \diagdown$$

$$ArCS_2CH_2COPh$$

(g) *Seleno- and telluro-benzoic acids*

Selenobenzoic acids (ArCOSeH) on heating form selenoacid anhydrides (ArCOSeOCAr) (Y.Hirabayashi, H.Ishihara and M.Echigo, *Nippon Kagaku Kaishi,* 1987, 1430; *C.A.,* 1988, **108**, 55604d). The tellurium analogues arise when aroyl chlorides are treated with sodium telluride (S.Kato, T.Kakigano and M.Ishida, *Z. Chem.,* 1986, **26**, 179).

(h) *Seleno- and telluro-benzoate esters*

Seleno and telluro esters (ArCOXPh, X = Se, or Te) are obtained from diaryldiselenides, or ditellurides (Ar$_2$X$_2$) respectively, by treatment with carbon monoxide under pressure and in the presence of octacarbonyl-dicobalt. Such reactions can be made catalytic in the presence of phosphines (H.Takahashi *et al.*, *J. Organomet. Chem.*, 1987, **334**, C43). Methyl benzothionoselenate [PhC(S)SeMe] can be prepared by reacting methyl benzothionate with dimethylaluminium methylselenolate (Me$_2$AlSeMe) (M.Khalid, J.L.Ripoll and Y.Vallee, *J. Chem. Soc., Chem. Commun.*, 1991, 964).

(i) *Arylcarbonyldiselenides and selenium, or tellurium bis(dithio-carboxylates*

Arylcarbonyldiselenides (ArCOSeSeAr') are synthesised from potassium arylselenocarboxylates (ArCOSeK) and arylselenyl bromides (Ar'SeBr) (H.Ishihara, N.Matsunami and Y.Yamada, *Synthesis*, 1987, 371).

Piperidium, or sodium aryldithiocarboxylates, react with Na$_2$ZS$_4$O$_6$; Z = Se, or Te) to form the appropriate selenium, or tellurium bis(dithiocarboxylates [(ArCS$_2$)$_2$Z] (S.Kato *et al.*, *Chem. Ber.*, 1985, **118**, 1695).

Bis(benzoyl)ditelluride [(PhCOTe)$_2$] is obtained from potassium tellurobenzote through oxidation with either iodine or phenylsulphenyl chloride (T.Kakigano *et al.*, *Chem. Letters*, 1987, 475).

Second Supplements to the 2nd Edition of Rodd's Chemistry
of Carbon Compounds, Vol. III F(Partial), G and H, by M. Sainsbury
© 1995 Elsevier Science B.V. All rights reserved.

Chapter 26

BENZOCYCLOPROPENE, BENZOCYCLOBUTENE AND INDENE, AND THEIR
DERIVATIVES

M. M. COOMBS

It should be noted that Chemical Abstracts now indexes
benzocyclopropene and benzocyclobutene as bicyclo[4.1.0]hepta-
1,3,5-triene and bicyclo[4.2.0]octa-1,3,5-triene, respectivley. The
former is also frequently named cyclopropabenzene, and hence
cyclopropanaphthalene and cyclopropaarenes in general, in the
American chemical literature.

Recent *ab initio* calculations of ring strain (among other
parameters) in benzocyclopropene, benzocyclobutene, and
indene indicate that these are 52.6, 28.4, and 5.9 kcal/mol,
respectively (R.Benassi *et al., J. Chem. Soc. Perkin Trans.II,* 1991,
1381). Thus while ring strain in the two former hydrocarbons
leads to characteristic reactions involving opening of the small
fused ring, this is not in general evident for indene and its
derivatives.

1. Benzocyclopropene and its derivatives

Since the last review (Rodd, supplement to vol. IIIF , pp.163-174)
there has been intense interest in this hydrocarbon. Because of
its high strain and associated high energy, it possesses unusual
chemical properties which are of theoretical as well as synthetic
importance. An X-ray study at room temperature of
benzocyclopropene (**1**) and its 7,7-bis(trimethylsilyl) derivative
(**2**) (Neidlein *et al., Angew. Chem. Int. Ed. Engl.,* 1988, 27, 294)
indicates fairly normal aromatic bond lengths although C(1)-C(6)
is short (1.334Å), shorter than benzene (1.395Å) but longer than
cyclopropene (1.296Å). However substantial deformations of the
bond angles from the regular 120^0 of benzene are apparent:-

C(3)-C(4)-C(5) 122.4^0
C(4)-C(5)-C(6) 113.2^0
C(1)-C(6)-C(5) 124.5^0

1   **2**

The silyl derivative (**2**) has similar dimensions. For a detailed discussion of molecular distortion and ring strain as calculated by molecular orbital theory see Y.Apeloig and D.Arad (*J. Amer. Chem. Soc.*, 1986, 108, 3241). The chemistry of benzocyclopropene is rationalised by frontier MO theory as being dominated by its highest occupied molecular orbital [localised between C(1)-C(2) and C(5)-C6)] and the possibility of relief of strain by cleavage of the cyclopropene ring. P.C.Hiberty *et al.* (*J. Amer. Chem. Soc.*,1985, 107, 3095) discuss the Mills-Nixon effect [the short double bond C(1)-C(6)], while C.Wentrup *et al.* (*Tetrahedron*, 1985, 41, 1601) apply SFC force field calculations to this problem. The strain energy of benzocyclopropene (**1**) is 68 kcal/mol, and its trimethylsilyl derivative is cleaved 64 times as rapidly as benzyltrimethylsilane from which it is estimated that its pKa ~ 36 (C.Eaborn *et al., J. Organomet. Chem.*, 1977,124,C-27) in agreement with calculation (pKa = 33) (C.Eaborn and J.G.Stamper, *J. Organomet. Chem.*,1980, 192,155). For the ^{13}C nmr spectrum of (**1**) see K.H.Albert and H.Duerr (*Org. Mag. Reson.*, 1979,12, 687); in its mass spectrum (12 or 70 eV) the carbon atoms in the molecular ion lose their positional identity before expulsion of acetylene.

(a) Syntheses of benzocyclopropenes

Benzocyclopropene is available from cyclohexadiene through the action of dichlorocarbene followed by dehydrochlorination (M.G.Banwell *et al., J. Chem. Soc. Perkin Trans.I*, 1977, 2165): -

67%   **1**

This dechlorination reaction has been studied in detail by A.R.Browne and B.Halton (*Tetrahedron*, 1977, 33, 345).

Benzocyclopropene is always accompanied by a small quantity of benzyl t-butyl ether, but this does not arise from solvolysis of the main product. In the case of its benzo homologue it is evident that the following pathway occurs leading, through *ortho* migration of the cyclopropene carbon, to the 1-butyl ether:-

In another synthesis, Diels-Alder condensation of butadiene with readily available 1,2-dichloro-3,3-difluorocycloprop-1-ene gives the 7,7-difluoro derivative (**3**) from which the hydrocarbon (**1**) can be obtained through reduction with lithium aluminium hydride (P.Muller *et al.*, *Helvetica Chim. Acta,* 1978,<u>61</u>,2081). Dissolution of the difluoro compound in cold fluorosulphonic acid yields the monofluoro ion which is stable enough for a nmr study; with water the solution gives benzoic acid.

(b)   Reactions involving opening of the cyclopropene ring

Many examples of opening the 3-membered ring to form a cycloheptatriene are known. Thus by treatment of benzocyclo-propene with iodine or thiocyanogen under irradiation with a high pressure mercury lamp this is the main reaction; in both cases cleavage of the cyclopropene ring to benzyl compounds occurs to only a minor extent (K.Okazaki *et al., Angew. Chem. Int. Ed. Engl.*, 1981, 20, 799):-

| (SCN)$_2$ | 61% | 27% |
| I$_2$ | 67% | 4% |

1,6-Diiodohepa-1,3,5-triene has been employed by E.Vogel *et al.* (*Angew. Chem.*, 1986, 98, 727) to construct the interesting bismethano-bridged phenanthrene (4). The nmr spectrum shows that this compound is aromatic, and its structure is confirmed by an X-ray study.

Straightforward Diels-Alder reactions of benzocyclopropene with electron-poor dienes have been reported by Y.Tanchuk *et al.* (*Zh. Org. Khim.*, 1989, 25,776) :-

$R^1=R^4=$Me,Et,Ph,CO$_2$Me
$R^2=R^3=$Ph or 2,2'-biphenyl

Anthracene adds to yield homotriptycenes (K.Sayo, *et al., Org. Prep. Proc. Inter.*, 1991, 23, 196):-

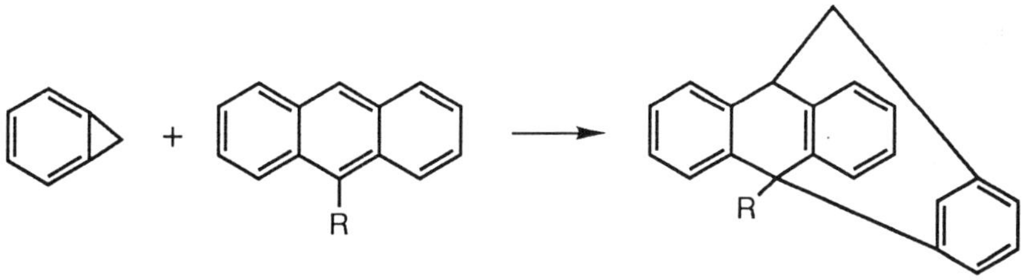

R = H, Me, or CH$_2$Ph

However most cycloadditions have involved electron-poor heterocyclic compounds, and there are numerous examples of this. With phenylcyanates in ethanol solution for one week at 0^{0}C benzocyclopropene gives mainly the expected heterocycles (Nitta *et al., Chem. Lett.*, 1979, 1431) :-

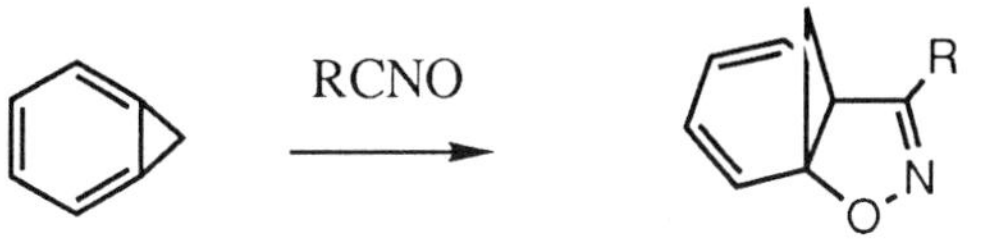

R = Ph or mesityl

Addition of diphenylisobenzofuran occurs readily in THF at 80^{0}C to give mainly the *endo*-adduct (**5**), together with the unsymmetrical adduct (**6**) with ring expansion to a central seven-membered ring (U.H.Brinker and H.Wuster, *Tetrahedron Lett.*, 1991, 32, 593).

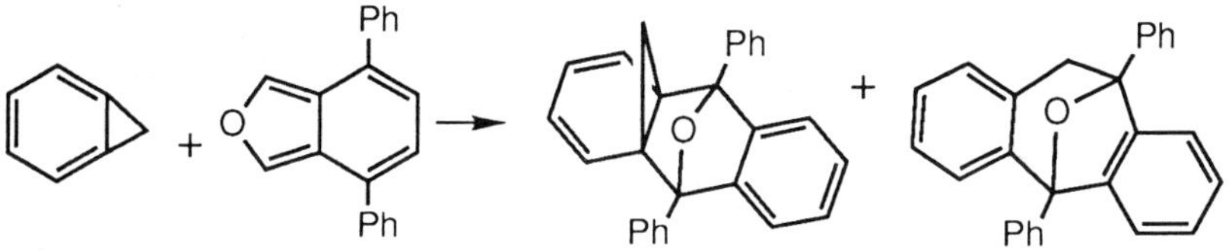

| **5** | 52% | **6** | 23% |

P.Muller and J-P.Schaller (*Helvetica Chim. Acta*, 1989, 72, 1608) have devised a new route to cycloproparenes by cycloaddition of benzoisofuran to 1-bromo-2-chloro-cyclopropene, followed by aromatisation of the product with low-valent titanium. The isoquinoline analogue (7) of benzocyclopropene has also been prepared by this route :-

7

Addition of electron-deficient triazines to benzocyclopropene yields metho-bridged isomers of this ring system by cycloaddition followed by loss of N_2 (J.C.Martin and J.M.Muchowski, *J. Org. Chem.*, 1984, _49_, 1040). An analogous reaction occurs with dimethyl 1,2,4,5-tetrazine-3,6-dicarboxylate (R. Neidlein and L.Tadesse, *Helvetica Chim. Acta*,1988, _71_, 249).

Benzocyclopropene also reacts readily in the same manner with a mesoionic oxathiazoliumate with the extrusion of CO_2 to yield methanothiazonines. With a mesoionic dithioliumolate it reacts with loss of SCO or of sulphur (in the presence of tributylphosphine) to give a variety of sulphur-containing heterocycles (H.Kato *et al.*, *J. Chem. Soc. Perkin Trans I*, 1990, 2035).

Ring expansion to a cycloheptatriene also occurs under the influence of zero-valent nickel, and in the presence of triphenylphosphine leads to a tetramer (R. Neidlein *et al.* (*Angew. Chem. Int. Ed. Engl.*, 1986, 25, 367).

Linear polymerisation occurs when benzocyclopropene is treated with benzoyl peroxide, boron trifluoride etherate, or titanium tetrachloride to yield poly(methylene-1,2-phenylene) (K.Lim and S.Choi, *J. Polymer. Sci., Part C: Polymer Lett.*, 1986, 24, 645).

(c)   Reactions leading to enlargement of the cyclopropene ring

Reactions leading to enlargement of the 3-membered ring with retention of the benzene ring are also known. Thus reactions of benzocyclopropene with α, β-unsaturated carbonyl compounds catalysed by Yb(fod)₃ give 1,3-dihydroisobenzfuran derivatives, whereas 2-substituted indans result from its reaction with α, β-unsaturated hydrazones (R.Neidlein and B.Kramer, *Chem. Ber.*, 1991, 124, 353).

When benzocyclopropene is treated with bromoform or chloroform in the presence of sodium hydroxide and a catalytic quantity of TEBA ring enlargement to dihalogeno-benzocyclobutenes occurs in 91-96% yield (S.Kagabu and K.Saito, *Tetrahedron Lett.*, 1988, 29, 675) :-

The 7,7-bis(trimethylsilyl) derivative (**2**) reacts with triethene nickel by insertion of the nickel atom into the 3-membered ring (C.Kruger *et al., Chem. Ber.*, 1987, 120, 471). The structure of the product has been confirmed by a X-ray study; the ligand L can be readily exchanged.

Finally, it was noted by H. Schwager *et al.* (*Angew. Chem. Int. Ed. Engl.*, 1987, 26, 67) that when benzocyclopropene and two of its 7,7-substituted derivatives were exposed to a ferrocenyl propene palladium complex in the presence of trimethylphosphine, three distinct outcomes including ring cleavage to a benzyl compound, cycloaddition, and ring enlargement to a 4-membered ring could be distinguished as follows :-

(d)   Reactions at the cyclopropene methylene group

As already mentioned, the cyclopropene carbon atom in benzocyclopropene is acidic and can be readily metallated with butyl lithium; quenching the anion with chlorotrimethylsilane yields the 7-silyl derivative, and this procedure can be repeated to give the 7,7-disilyl compound (**2**). The mono-silyl anion reacts with a variety of ketones with the formation of 7-alkylidene derivatives which combine both a fused cyclopropene ring and cross conjugation in the same molecule (B.Halton *et al., J. Amer. Chem. Soc.*, 1986, 108, 5949).

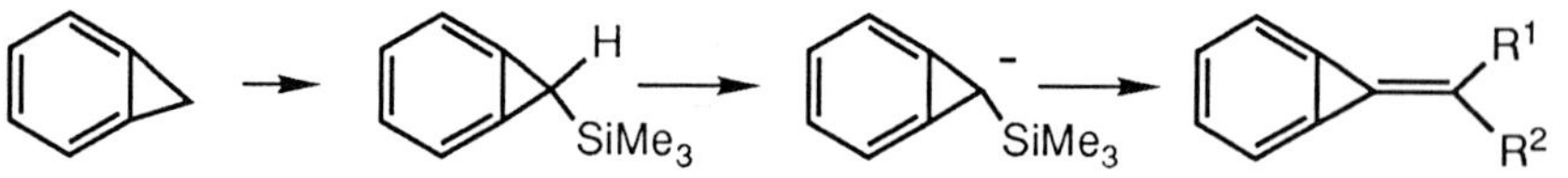

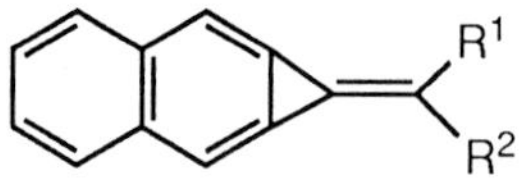

114

The corresponding naphthyl analogues are obtained both in better yield and are more stable than the benzo derivatives. Using this procedure B.Halton *et al.*(*Tetrahedron Lett.*, 1986, 27, 5159) has condensed the anions with fluorenone to give benzocalicenes (8) as stable, coloured crystalline solids. The first triahepta-fulvalenes (9) and (10) have also been obtained by reaction of the anion with benzotropolones. The measured dipole moments of these molecules are generally in agreement with the calculated values.

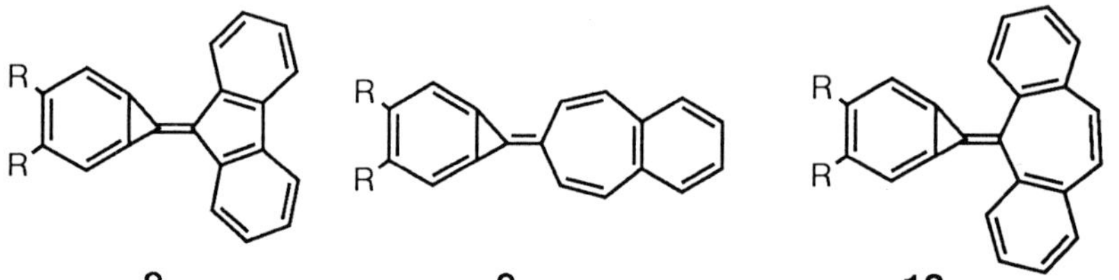

8   9   10

R = H
or R,R = benzo

Exposure of the benzocalicene (8, R = H) to concentrated hydrochloric acid causes cleavage of the small ring with the formation of 9-benzylfluorene; by contrast similar treatment of the diphenyltropilidene (10, R = H) leads to the chloroohepta-tetraene derivative (11) (B.Halton *et al.*, *J. Org. Chem.*, 1988, 53, 2418).

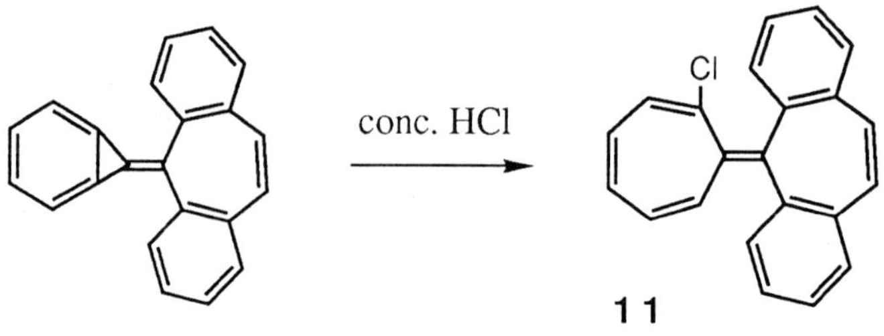

conc. HCl

11

Starting from 2,5-diphenyl-7,7-dichlorobenzocyclopropene treatment with butyl lithium causes dimerisation to the symmetrical compound doubly bonded at C(7) (R.Neidlein *et al., Angew. Chem. Int. Ed. Engl.,* 1986, 98, 735); hydrogenation opens both 3-membered rings :-

The preparation, reactions, dipole moments and fluorescence of alkylidenebenzocyclopropenes are the subject of a review by B. Halton in *Pure and Applied Chem.,* 1990, 63, 541.

(e)   Other reactions of benzocyclopropene

7,7-Bis(trimethylsilyl)naphthocyclopropene reacts with Cr[(CH3CN)3(CO)3] at the free aromatic ring to form a π complex which after desilylation is converted to the anion with butyl lithium at -100⁰C. Quenching with methyl iodide then provides a 1:1 mixture of the isomeric 7-methyl derivatives. The corresponding bis-silylated benzocyclopropene does not react under these conditions whereas the 3-membered ring in the hydrocarbon itself is enlarged by the insertion of chromium (P.Muller *et al., Helvetica Chim. Acta,* 1992, 75, 1995).

Nitration of 7,7-bis(isopropylsilyl)benzocyclopropene with 67% nitric acid gives the 3-nitro derivative without opening the 3-membered ring. The nitro compound may be reduced with zinc and hydrochloric acid to the 3-amino compound, which can be diazotised normally and and reacts with a second molecule to yield the *o*-aminoazo derivative (R.Neidlein and D.Christen, *Helvetica Chim. Acta,* 1986, 69, 1623). Reduction of the nitro compound with lithium aluminium hydride also leads to an azo compound, but reduction with zinc and sodium hydroxide results in ring cleavage to the benzyl derivative.

Evidence for the existence of highly strained benzyne derivatives of benzocyclopropene comes from treatment of both 2- and 3-bromobenzocyclopropene with the t-butoxide/ammonia (t-BuO⁻/NH₂⁻) complex in the presence of furan, when two adducts are isolated. Cleavage of the cyclopropene rings in these adducts leads to the same benzyl ether (P.Apeloig *et al.*, *J. Amer. Chem. Soc.*, 1986, 108, 4932).

There is less conclusive evidence for the existence of the highly strained naphtho-1,2:3,4-bis(cyclopropene) (12) (U.Brinker *et al.*, *Angew.Chem.*, 1987, 99, 585). Dehydrobromination of the dibromo compound already bearing two cyclopropane rings using potassium t-butoxide in THF/DMSO in the presence of 1,3-

diphenylisofuran leads to two bis-adducts apparently arising from (**12**) and its double bond isomer (shown below each adduct). However it is not certain whether the main adduct arises from (**12**) or in two consecutive steps.

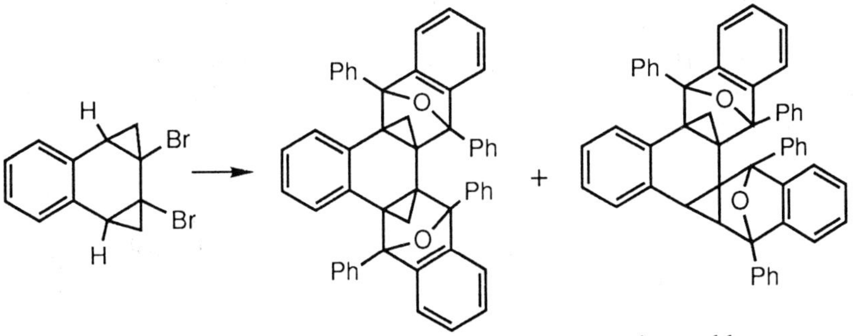

main adduct   minor adduct

1 2

The highly strained benzocyclopropen-7-one has been identified for the first time from photoactivation of benzocyclobutenedione by ultraviolet / visible and infrared spectroscopy of a specimen held in an argon matrix at $120K$ (J.Simon *et al.*, *Chem. Phys. Lett.*, 1992, 200, 631).

There are many other polycyclic aromatic and heterocyclic systems containing cyclopropene rings in the chemical literature, and some are included in a review of cycloproparene chemistry by B.Halton in *Chem. Rev.*, 1989, 89, 1161.

2. Benzocyclobutene and its derivatives

There continues to be much interest in this ring system and the name cyclobutabenzene is used to some extent. The systematic name for the hydrocarbon (**14**) is in fact 1,2-dihydrobenzocyclobutene, but this is seldom employed and will not be used here.

The ^{1}H nmr spectrum of benzocyclobutadiene (**13;** the systematic name is benzocyclobutene) has been obtained in a flow cell by mixing the 7-trimethylsilyl-8-mesylate with tetramethylammonium fluoride in acetonitrile-d$_3$ solution. Three signals are observed at δ 6.26 and 5.78 (6-membered ring protons) and 6.36 (cyclobutadiene protons) from which it is concluded that this hydrocarbon is polyolefinic, neither aromatic nor anti-aromatic (W.S.Trahanovsky and D.R.Fischer, *J.Amer.Chem Soc.*, 1990, 112, 4971).

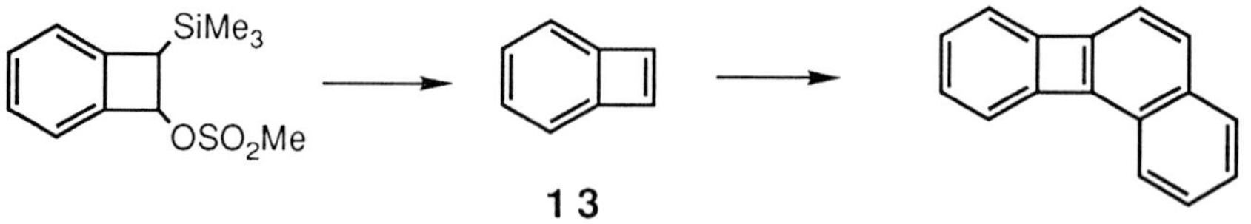

1 3

Within minutes at room temperature this highly strained molecule dimerises as shown above. The two six-membered ring signals in benzocyclobutadiene are close to those in the ^{1}H nmr spectrum of its 7,8-bis(trimethylsilyl) derivative, and it is suggested that these two compounds are best represented as quinodimethylenes (*o*-xylylenes):-

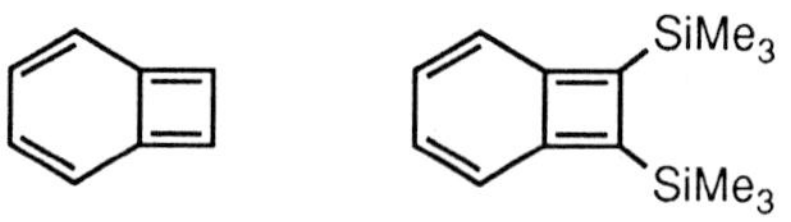

The X-ray structure of benzocyclobutene (**14**) and linear benzodicyclobutene (**15**) measured at -177^{0}C show that, as in benzocyclopropene, the bond lengths are not greatly affected although they are all shorter than the aromatic bonds in benzene (R.Boese and D. Blaser, *Angew. Chem.* 1988, 100, 293). However distortion of the bond angles from the 120^0 of benzene is evident, and is the greater for the highly strained benzodicyclobutene in which the 3'-1-2 bond angle of 112.1^0 is even less than that of the corresponding angle (113.2^0) in benzocyclopropene.

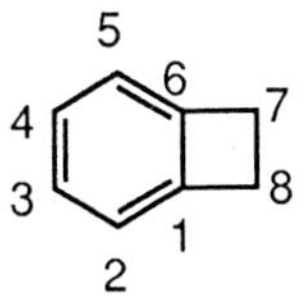

14

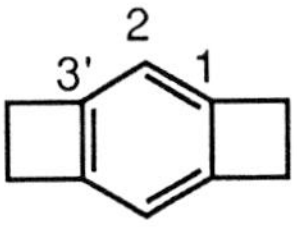

3 - 4 - 5	121.7°
5 - 6 -1	122.3°
4 - 5 - 6	116.0°

15

1 - 2 - 3	124.0°
3' - 1- 2	112.1°

(a)   Syntheses of benzocyclobutene

Several different methods for the synthesis of benzocyclobutene have appeared. Diels-Alder condensation of butadiene with methyl cyclobutene dicarboxylate at 100^0C in a sealed tube gives the adduct in 73% yield. After saponification, oxidation with lead tetra-acetate in pyridine leads to the hydrocarbon (**14**). In a similar manner both the linear and angular benzodicyclobutenes are also obtained, and in addition the two corresponding cyclobutenindenes are synthesised by this route (R.P.Thummel and W.Nutakul, *J. Org. Chem.*, 1977, 42, 300).

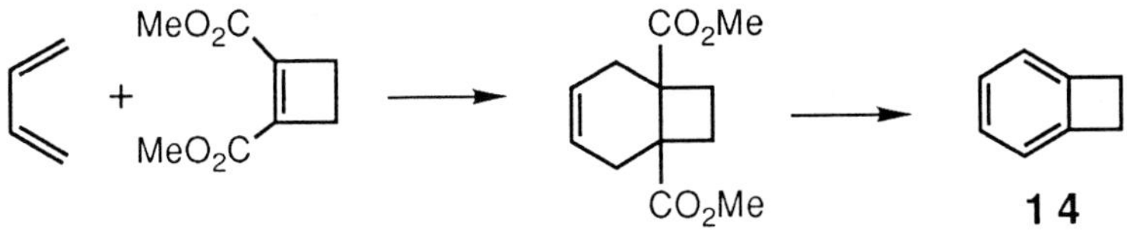

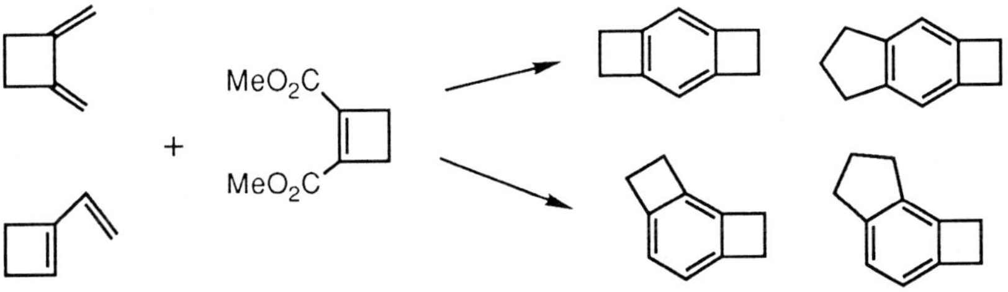

Co-oligomerisation of α,ω-diynes with monoacetylenes catalysed by cyclopentadiene cobalt dicarbonyl provides useful yields of benzocyclobutenes in one step (K.P.Vollhardt, *Acc. Chem. Res.* 1977, 10, 1).

$$
\begin{array}{c}
\text{C}\!\equiv\!\text{CR}_1 \\
\big| \\
\text{C}\!\equiv\!\text{CR}_2
\end{array}
\quad + \quad
\begin{array}{c}
\text{CR}_3 \\
\text{|||} \\
\text{CR}_4
\end{array}
\quad \longrightarrow
$$

R_1	R_2	R_3	R_4	yield%
H	H	CO_2Me	CO_2Me	14
H	H	Ph	H	17
H	H	Ph	Ph	48
H	H	C_6H_{13}	H	13
H	H	$SiMe_3$	$SiMe_3$	65
H	H	CH_2OMe	$SiMe_3$	55
H	H	CH_2OMe	H	14
Me	Me	Ph	H	20
Me	Me	CO_2Me	CO_2Me	28
CH_2OMe	CH_2OMe	$SiMe_3$	H	25
$SiMe_3$	$SiMe_3$	$SiMe_3$	H	2
$SiMe_3$	H	$SiMe_3$	CH_2OMe	16

The trisilylmethyl groups R_3 and R_4 in the resulting benzocyclobutenes undergo electrophilic displacement at different rates, R_3 leaving 36-42 times faster than R_4, thus allowing unsymmetrically disubstituted derivatives to be prepared in quantitative yield:-

room temp.
15 minutes

$X = H$
D
Br

room temp.
18 hours

$X = Y = H$
D
Br
$X = Br, Y = I$
$X = Br, Y = OAc$

Treatment of the bromo-iodo derivative with butyl lithium in the presence of furan yields the oxygen bridged dihydronaphthalene whereas in its absence a dimer and trimer result as follows:-

2.3:6,7-Dicyclobutabiphenylene is one of the most strained hydrocarbons known. It is unstable in air and is hydrogenated instantly at normal pressures to the totally saturated hydrocarbon without skeletal change, while hydrogenation of the trimer occurs to leave the central benzene ring intact.

A photolytic synthesis of cyclobutanones is described by M.Yoshioka *et al.* (*J. Amer. Chem. Soc.*, 1990, 112, 374). Irradiation of phenylpropandiones containing an *o*-methylene group with Pyrex-filtered light gives cyclobutanols in 44-60% yield. On being heated at 150-180°C these undergo retro-aldol cleavage with the formation of benzocyclobutanones.

R^1	R^2	R^3
H	H	H
H	Me	H
H	H	Me
Me	H	H

Treatment of 3-trimethylsilyl-4-methoxymethylbenzocyclo-butene with bromine followed by butyl lithium generates the linear cyclopropacyclobutabenzene (C.J. Saward and K.P. Vollhardt, *Tetrahedron Lett.*, 1975, 4539).

In a more recent and improved synthesis of this compound from acetylenes, catalysed by cyclopentadiene cobalt dicarbonyl, the overall yield is 32% (A.T.McNichols and P.L.Stang, *Synlett.*, 1992 971).

A mild, high yielding synthesis of benzocyclobutenes consists of treating 1-iodo- or 1-bromo-2-bromoethylbenzenes with butyl lithium at -100 to -78^0C (P.D.Brewer *et al.*, *Tetrahedron Lett.*, 1977, 4573):-

R_1	R_2	X	yield %
H	H	Br	87
OMe	H	I	93
OMe	OMe	I	91
-O-CH$_2$-O-		I	96

An alternative synthesis of benzocyclobutenes is by pyrolysing 3-isochromanones, obtainable from indan-2-ones by peracid oxidation in 70-80% yield, in a stream of nitrogen at 565-575^{0}C (R.J.Spangler *et al.*, *J. Org. Chem.*, 1977, **42**, 2988).

R$_1$	R$_2$	yield%
H	H	85
OMe	H	70
OMe	OMe	40
-O-CH$_2$-O-		90
OMe	OH	low

Also passing indanone in argon through a plasma formed by an inductively coupled 13.6 MHz radiofrequency generator gives benzocyclobutene in 40-50% yield (at 30% conversion of starting material) with only traces of styrene (M.Tokuda *et al.*, *J. Org. Chem.*, 1979, **44**, 4504).

Good to moderate yields of benzocyclobutenes result from vacuum pyrolysis of *o*-disubstituted benzenes (P.Schiess *et al.*, *Tetrahedron Lett.*,1982, **23**, 3669). 7-Chlorobenzocyclobutene is a useful intermediate and the following derivatives have been prepared from it by Grignard reactions in 51-74% yield :- 7-CO$_2$H, -CHO, -COCH$_3$, -SiMe$_3$, -CH$_3$,-C$_4$H$_9$, and -CH=CH$_2$.

X	Y	R$_1$	R$_2$	yield%
CH$_2$Cl	CH$_3$	H	H	77
CH$_2$Cl	CH$_2$Cl	Cl	H	53
CHCl$_2$	CH$_3$	Cl	H	80
CCl$_3$	CH$_3$	Cl	Cl	35
CHClCN	CH$_3$	CN	H	42
CCl$_2$CN	CH$_3$	CN	Cl	19

A patent describes pyrolysis of *o*-chloromethyl toluenes at 750^{0}C to give benzocyclobutenes in 30-40% yield (U.S.P.Appl. 612,164,

Chem. Abstr., 1987,<u>106</u>, 49806). The mechanism of this reaction has been studied by M.J.Morello and W.S.Trahanovsky (*Tetrahedron Lett.*, 1979, 4435). Flash vacuum pyrolysis of *o*-methyl benzylchloride-α,α-D$_2$ gives benzocyclobutene that retains its deuterium label quantitativly, thus suggesting δ-elimination of HCl or possibly a 1,3-sigmatropic chlorine shift followed by β-elimination of HCl.

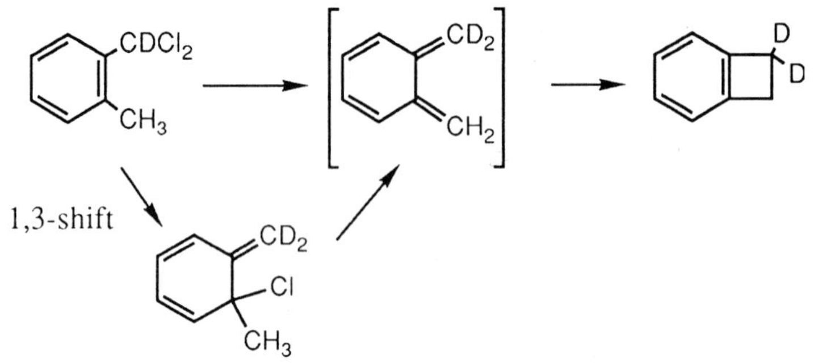

Decarbonylation of all three isomeric tolualdehydes with atomic carbon generated in an electric arc leads to benzocyclobutenes and styrenes in the ratios 2.35 : 0.8; 0.94 : 0.13; and 0.95 : 0.4 from the *o*-, *m*-, and *p*-isomers, respectively (M.Rahman and P.B. Shevlin, *Tetrahedron Lett.*, 1985, <u>26</u>, 2959). Heating the azoseleniumcyclo-octatriene heterocycle shown below with copper powder at 180^0C also leads to benzocyclobutene in 24% yield (N.Hanold and H.Meier, *Chem. Ber.*, 1985, <u>118</u>, 198):-

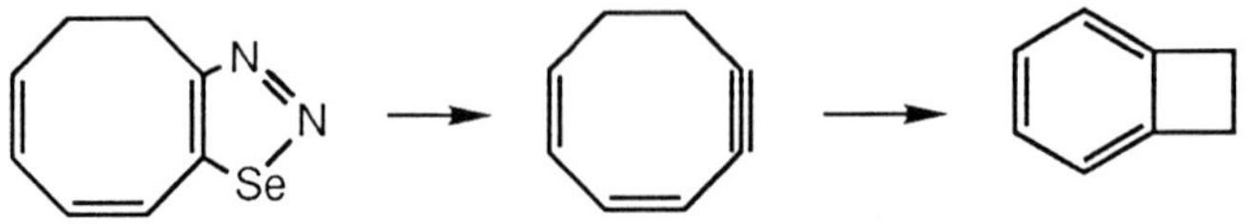

Finally, a practical route to 7-substituted benzocyclobutenes involves additionof dibromocarbene to cycloheptatriene in a one-pot reaction , giving 7-bromobenzocyclobutene in 26% yield (M.R.DcCamp and L.A.Viscogliosi, *J. Org. Chem.*, 1981, <u>46,</u> 3918). Reduction of this intermediate with tributyltin hydride or with tributytin chloride and lithium aluminium hydride gives the hydrocarbon itself, while numerous 7-substituted derivatives are available by appropriate treatments.

$$CHBr_3 + K_2CO_3$$

18-crown-6
10h, 140°C

(b) *Reactions involving opening of the four-membered ring*

Characteristic of benzocyclobutene is thermolytic opening of
its four-membered ring to the *o*-xylylene intermediate which is
of course reactive, thus providing the possibility of significant
synthetic and commercial applications. In the absence of an
acceptor the main product is styrene. In a study with [7-^{13}C]-
benzocyclobutene it is evident from ^{13}C-nmr spectroscopy that
the following mechanism operates, leading to β- and *o*-labelling
(O.L.Chapman and U.E.Tsou, *J. Amer. Chem. Soc.*, 1984, 106, 7974):-

This reaction has also been followed using [7-D$_2$] labelling
(W.S.Trahanovsky and M.E.Scibner, *J. Amer. Chem. Soc.*, 1984, 106,
7976). Scrambling of the deuterium in the side chain and to the *o*-
position is found in agreement with this cycloheptatriene
intermediate. The same mechanism is operative during pyrolysis
and photolysis (>470 nm) of *o*-tolyldiazomethane (O.L.Chapman *et*
al., J. Amer. Chem. Soc.,1988, 110, 501). The heat of formation of
benzocyclobutene, measured by combustion calorimetry, is 47.7
kcal/mol (W.R.Roth *et al., Chem. Ber.*, 1978, 111, 3892). From the
kinetics of its reaction with maleic anhydride the activation
parameters for *o*-xylylene formation are:- activation energy,
39.9kcal/mol and frequency factor 2.8×10^{14}/sec. The
corresponding values for the back reaction are 29.3 kcal/mol and
2.2×10^{13}/sec. The kinetics of o-xylylene formation from the

thermal decomposition of the bridged ketone shown below and its cyclisation to benzocyclobutene have been studied by a shock tube technique (W.R.Roth and B.Scholz, *Chem. Ber.*, 1981, 114, 3741). The enthalpy of formation of the intermediate is 60.8 kcal/mol and the activation energies are as shown:-

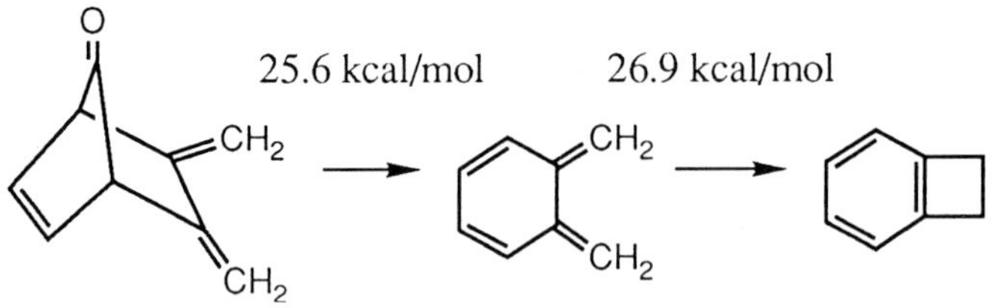

For the application of Huckel MO theory to the benzocyclobutene - *o*-xylylene question see C.F.Wilcox and B.K.Carpenter, (*J. Amer. Chem. Soc.* 1979, 101, 3897).

The direction in which ring opening occurs has been investigated both theoretically and also from a practical angle (C.W.Jefford *et al., J. Amer. Chem. Soc.*, 1992, 114, 1157) by trapping the reactive *o*-xylylenes with maleic anhydride or N-phenylmaleimide. 7-Cyanobenzocyclobutene gives two adducts in the ratio of major to minor of 19:1, from which it is concluded that conrotary ring opening occurs mainly with outward twisting of the cyano group.

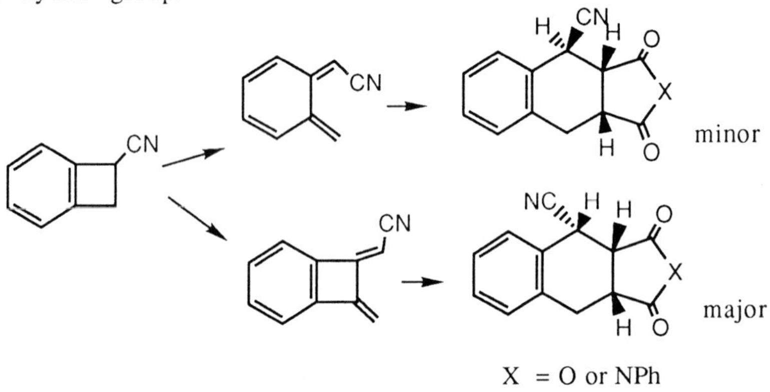

X = O or NPh

In a similar manner methyl benzocyclobutene-7-carboxylate gives major : minor adducts in the ratio 10:1. The 7-hydroxy-7-phenyl derivative on being heated with maleic anhydride yields a single lactone acid:-

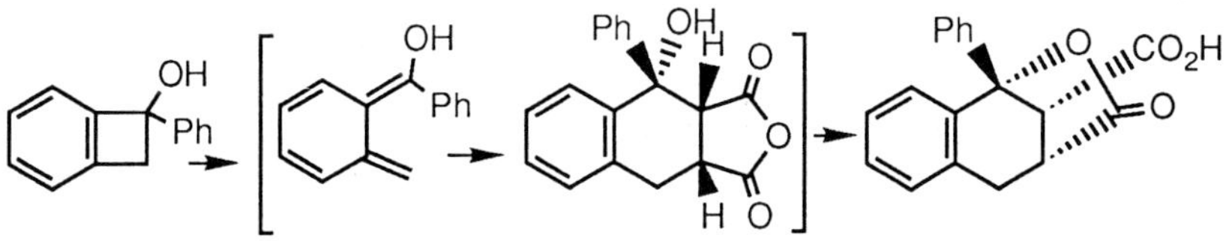

In contrast, N,N-dimethylbenzocyclobutene-7-carboxamide reacts with inward twisting of the 7-substituent to the extent of 75%. With the 7-methyl derivatives of both the 7-cyano and 7-carboxylic ester the methyl group shows outward movement. In general it seems that the proportion of [E] and[Z] xylylenes is dictated by electronic factors and determines the stereochemical outcome of the reaction.

Diels-Alder reactions occur in high yield and stereospecifically to give *cis*-fused adducts when carried out with dienophiles at 230^0C under pressure in toluene for 48 hours (S.V.D'Andrea *et al.*, *J.Org.Chem.*, 1990, 55, 4356).

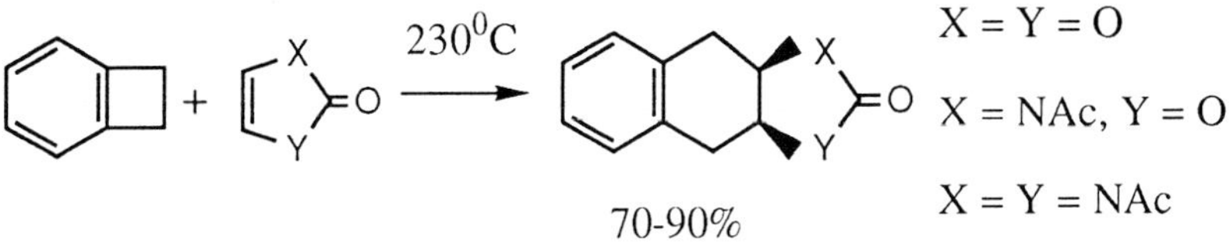

$$X = Y = O$$
$$X = NAc, Y = O$$
$$X = Y = NAc$$

70-90%

At the other extreme, in the presence of 7-hydroxy or 7-acetoxy groups α-oxy-o-xylylenes are generated below 0^0C by the use of butyl lithium, and react with dimethyl maleate, dimethy fumarate, or γ-crotonolactone to give Diels-Alder adducts in moderate to good yield (W.Choy and H.Yang, *J. Org. Chem.*, 1988, 53, 5796).

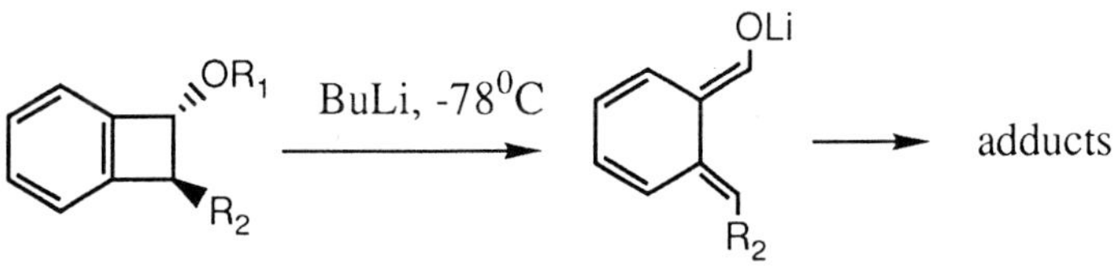

	R_1	R_2
adducts	H	H
	Ac	H
	H	Ph
	Ac	Ph

A good example of the practial usefulness of this thermolytic benzocyclobutene ring-opening reaction is its application to natural product synthesis. T. Kametani *et al.*, (*J. Chem. Soc. Perkin Trans.I*, 1985, 2151) have prepared the 1-benzocyclobutenyl-3,4-dihydroisoquinoline in several steps; on being heated at 180°C in dichlorobenzene it smoothly gives (±)-tetrahydropalmatine in 62% yield.

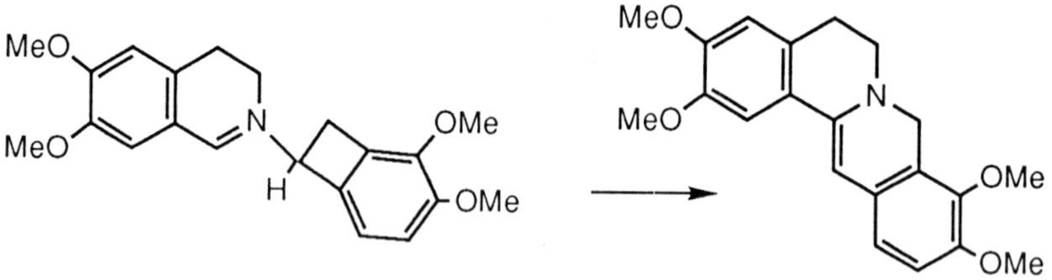

Benzocyclobutene has been employed with advantage in a new general synthesis of 3-amino-2-naphthalene carboxylates (K.Kobayashi *et al.*, *J. Chem. Soc. Chem. Commun.*, 1992, 780):-

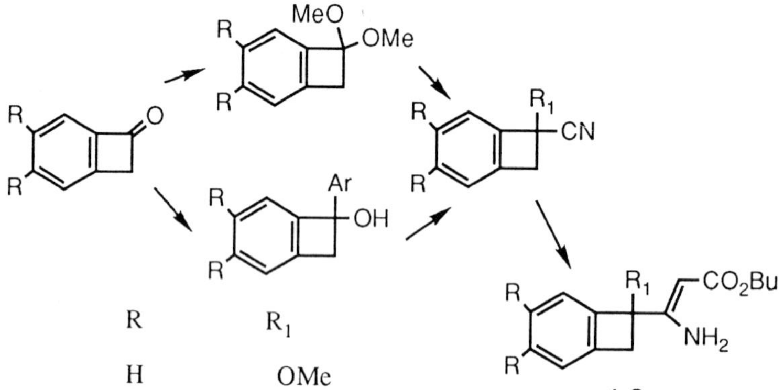

R	R$_1$
H	OMe
OMe	OMe
H	Ph
H	4-OMe-C$_6$H$_4$
H	3,4-OMe-C$_6$H$_3$

The intermediate benzocyclobutenes (**16**) on being heated in boiling dichlorobenzene in the presence or absence of air give naphthalenes as shown below.

when R_1 = OMe

when R_1 = OMe, Ph, or substituted Ph

It has also been used as a starting point in the synthesis of cyclopropa-anthracene (P.Muller and M.Ray, *Helvetica Chim. Acta*, 1982, 65, 1157) ; the Diels-Alder adduct formed with maleic anhydride in mesitylene at 210°C and obtained in 63% yield was the first step in this synthesis. Indoles are available from benzocyclobutene -7-azides (G.Adam *et al.*, *Tetrahedron*, 1985, 42, 399) through ring enlargement catalysed by concentrated sulphuric acid in chloroform:-

The thermolytic ring opening reaction makes benzocyclobutene and its derivatives useful as cross-linking agents in polymers, and many references to their application in this capacity have appeared over the last few years. For example K.A.Walker *et al.* (*Synthesis,* 1992, 1265) describe a practical route to benzocyclobutene-3,6-dicarboxylic acid, obtainable in 70% overall yield. Difficulty in converting the dibromo compound to the diacid is successfully overcome by the use of Pd(OAc)₂ and Ph₃P under carbon monoxide in butanol at 90°C for 45 hours to give the dibutyl ester (95%), followed by saponification (97%). After conversion to its acid dichloride it is condensed with the appropriate monoamines to give the two model compounds (17) and (18), both of which melt normally, but at higher temperatures undergo irreversible, exothermic reactions. By the use of this diacid together with terephthalic acid in polymerisations with diamines, it is possible to alter the physical properties of the resulting polymers in a controlable manner by subsequent heat treatment.

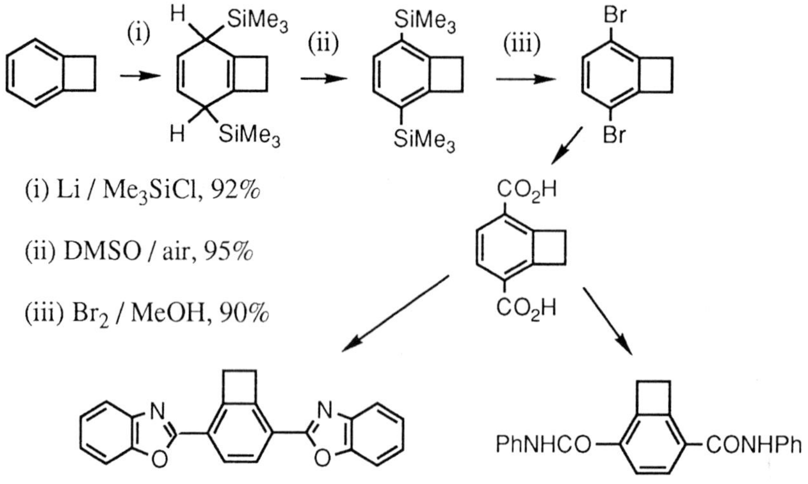

(i) Li / Me$_3$SiCl, 92%

(ii) DMSO / air, 95%

(iii) Br$_2$ / MeOH, 90%

17

18

In a recent patent (*Eur. Pat. Appl.* EP 410,039; *Chem. Abstr.*, 1991, 115, 50486) Friedel-Crafts acylation of benzocyclobutene with trimellitic anhydride acid chloride and Fe$_2$O$_3$ yields the 3-acyl derivative :-

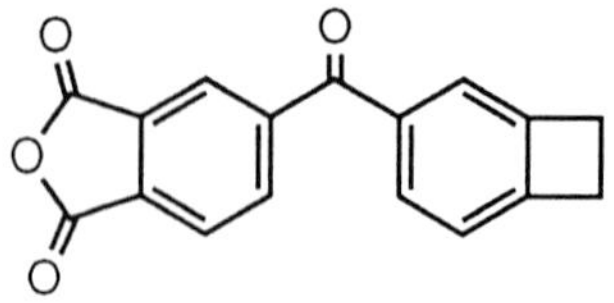

When this is heated with methylenedianiline it gives a product that undergoes ring-opening polymerisation to a cross-linked polymer with good resistance to aqueous potassium hydroxide. Polybenzocyclobutene itself possesses a lower dielectric constant and lower water adsorption characteristics than other polyolefinic polymers, and these render it a superior material for electrical circuit applications (*Chem. Abstr.*, 1991, 114, 248,773).

(c) *Other reactions of benzocyclobutene*

Electrophilic substitution of the aromatic ring in benzocyclobutene without alteration of the four-membered ring is possible, as in the above example of the Friedel-Crafts reaction. Another recent patent (*Eur. Pat. Appl.* EP 364,959; *Chem. Abstr.*, 1990, <u>113</u>, 71,043) describes direct bromination with bromine in an aqueous medium to give 3-bromobenzocyclobutene 77% yield. Nitration of the hydrocarbon with nitric acid in acetic acid/acetic anhydride at 0^0C gives the 3-nitro derivative, which after catalytic reduction is converted into a cyclobutaquinoline by a Skraup reaction (J.H.Markgraf *et al.*, *J. Org. Chem.*, 1979, <u>44</u>, 3261).

All these reactions occur with retention of the cyclobutene ring. However treatment of benzocyclobutene with the sodium-potassium alloy in tetrahydrofuran at room temperature leads to ring cleavage, the main product being 1-ethyl-9,10-dihydrophenanthrene (A.Maercker *et al.*, *Tetrahedron Lett.*, 1984, <u>25</u>, 1701).

(d) *Organometallic derivatives of benzocyclobutene*

Organometallic derivatives of benzocyclobutene have been the subject of much work during the last decade. The chromium tricarbonyl derivative is readily obtained in almost quantitative yield by chromium exchange with chromium tricarbonyl naphthalene in ether at 70^0C under pressure. It is more simply made in 80% yield by reaction of the hydrocarbon with $Cr(CO)_3(NH_3)_3$ in boiling dioxan (E.P.Kuendig *et al.*, *J. Organomet. Chem.*, 1985, <u>286</u>, 183); for its nmr spectrum see J.C.Boutonnet *et al.* (*J. Organomet. Chem.*, 1985, <u>290</u>,153). In a later paper the synthetic utility of this chromium complex is explored (E.P.Kuendig *et al., Helvetica Chim. Acta*, 1990, <u>73</u>, 1970). Lithiation occurs smoothly at C-3 with butyl lithium at -100^{0}C in tetrahydrofuran and the metal can be replaced by electrophilic groups as shown below in 50-92% yield. This treatment can be repeated leading to 3,6-disubstituted derivatives. Chromium can be removed by oxidation with Ce^{IV}(NH$_2$)$_2$(NO$_2$)$_6$, or in the case where R = CHO with air and light, in 76-92% yield. Benzocyclobutene itself, with lithium and trimethylsilyl chloride at 0-10^{0}C, yields first the 3,6-dihydro-3,6-bis(trimethylsilyl) derivative in 91% yield, and this is oxidised to 3,6-bis(trimethylsilyl)benzocyclobutene in air. Stereospecificity is complete, demonstrating the enhanced acidity of the 3- and 6-positions due to the ring strain. By contrast, with the similar chromium complex of indene in which there is very little ring strain, this reaction gives a mixture of *ortho* and *meta* trimethylsilyl derivatives in the ratio of 1 : 2.

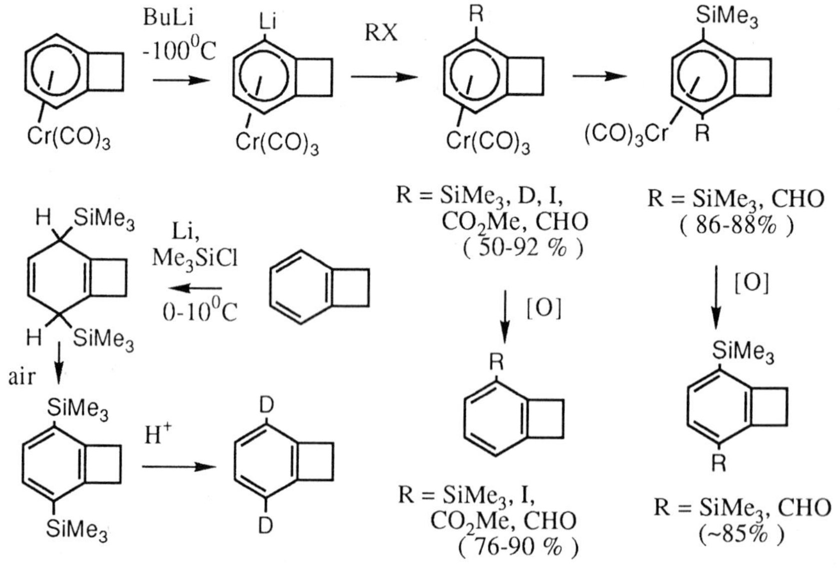

Complexation of chromium with 7-substituted benzocyclobutenes gives rise to asymmetry at C-7. H.G.Wey *et al..* (*Chem. Ber.*, 1991, <u>124</u>, 465) have determined the ratio of *endo* to *exo* forms with a number of complexes of this type using nuclear Overhauser enhancement as a guide, confirmed in one case by an X-ray structural study. In general only bulky groups such as $SiMe_3$ or $SnMe_3$ show a marked effect. The X-ray work indicates that the chromium atom is directly above (or below) the aromatic ring, as depicted in the structure below.

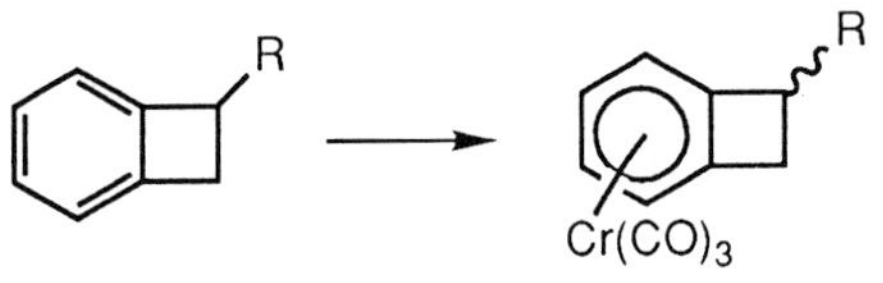

R	endo	exo
D	50	50
CH_3	40	60
$(CH_2)_3CH_3$	50	50
$(CH_2)CH=CH_2$	50	50
$Si(CH_3)_3$	12	88
$Sn(CH_3)_3$	24	76

Benzylideneacetone iron tricarbonyl reacts with cyclo-octatriene in benzene at 61^0C to yield a yellow oil shown by nmr to be bicyclo[4.2.0]octa-2,4-diene iron tricarbonyl (C.R.Graham *et al.*, *J. Amer. Chem. Soc.*, 1977, <u>99</u>, 1180).

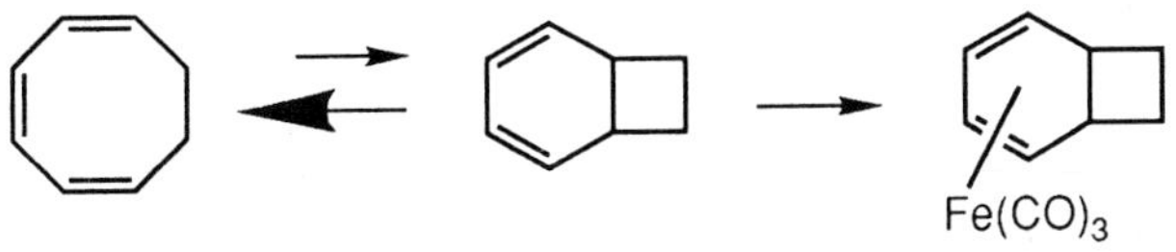

One or both the chloride ligands in [Pt(cod)Cl$_2$)] (cod = cyclo-octa-1,5-diene) may be replaced by benzocyclobuten-3-yl (R) to give [Pt(cod)RCl] or [Pt(cod)R$_2$] by reaction with benzocyclo-buten-3-yl trimethyltin (C.Eaborn *et al.*, *J. Chem. Soc., Dalton Trans.*, 1978, 357). Both cod and chloride ligands in the former complex can be replaced by cyanide in an organic solvent in the presence of 18-crown-6 to yield [Pt(CN)$_3$R] [K(18-crown-6)]$_2$ (J.F.Almeida and A.Pidcock, *J. Organomet. Chem.*, 1981, <u>208</u>, 273).

3. Indene and its derivatives

The name indene is reserved for the fully unsaturated cyclopentabenzene (**19**). The compound with the partially saturated five-membered ring (**20**) is usually known as indane (or indan), but it is indexed systematically in Chemical Abstracts as 2,3-dihydro-1H-indene. Both compounds will be considered in this review.

134

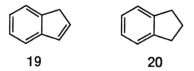

19   **20**

Very many molecules and associated polymers with this ring system are now listed, with a remarkable range of commercial uses and medical applications. For the present purpose inevitably a selection must be made, generally on the basis of novelty and chemical importance.

(a)   Syntheses of indenes

Several new syntheses of this ring system have appeared over the last 15 years. Thus in a manufacturing process, pyrolysis of *o*-ethyltoluene at 710^0C in a stream of H_2S/N_2 over a bed of alumina containing CoO, NiO and MoO$_3$ and/or WO$_3$ giving indene in 58% yield (at 86.4% conversion of ethyltoluene) is described in a patent (USP 4,613,711; *Chem. Abstr.*, 1987, <u>106</u>, P4,689n). In another process (USP 4,374,293; *Chem. Abstr.*, 1983, <u>98</u>, P178,995q) indenes are obtained by catalytic oxidation of allylbenzenes over WO$_3$; for allylbenzene itself a yield of 41.9% is obtained at 33.2% conversion. A high yielding synthesis of indane by decarbonylation of 2-allylbenzaldehyde by means of chlorotris(triphenylphosphine)rhodium is described by J.A.Kampmeier *et al.*(*J. Org. Chem.*,1984, <u>49</u>, 621). The internal rhodium hydride is thought to be an intermediate:-

Another high yielding synthesis involves hydroboronation of *o*-bromoallylbenzene and treatment of the product with dichloro[1,1'-bis(diphenylphosphino)ferrocene] palladium (II), when indane is obtained in 86% yield (N.Miyaura *et al.*, *J. Amer. Chem. Soc.*, 1989, <u>111</u>, 314).

Indane and 1-methylindane are produced in moderate yield when allyl magnesium bromide is allowed to react with benzyne (from *o*-bromofluorobenzene and magnesium in ether), followed by hydrolysis (J.Duboudin *et al., J. Chem. Soc. Chem. Commun.*, 1977, 454):-

when R = H	27%	65%	8%
R = Me	34%	57%	9%

Facile thermal rearrangement of a Dewar benzene leads to indane (J.Van Straten *et al., Rec. Trav. Chim. Pays-Bas*, 1985, 104, 89). Starting from 1,2-dimethylenecyclopentane, the dicyclopropene is synthesised in three steps. Treatment with silver fluoroborate in acetonitrile at 0^0C then causes rearrangement to the two Dewar benzenes, symmetrical and unsymmetrical in the ratio 2:3. The unsymmetrical isomer readily undergoes quantitative conversion to indan with a half-life of 60 minutes at 49^0C, whereas the other isomer requires much more vigorous conditions and gives mostly the corresponding cyclophane.

(b)   Syntheses of indanones

Several new syntheses of indan-1-ones and -1,3-diones have been reported. An efficient, selective method for indan-1-ones and 2-acetylindan-1-ones involves methylenation of 1-arylpropan-1,3-diones with methoxyacetyl chloride and aluminium chloride in nitromethane at 80^0C. Overall yields are in the range 62-74% (G.Sartori et al., Tetrahedron Lett., 1992, 33, 4771); the mechanism is as follows:-

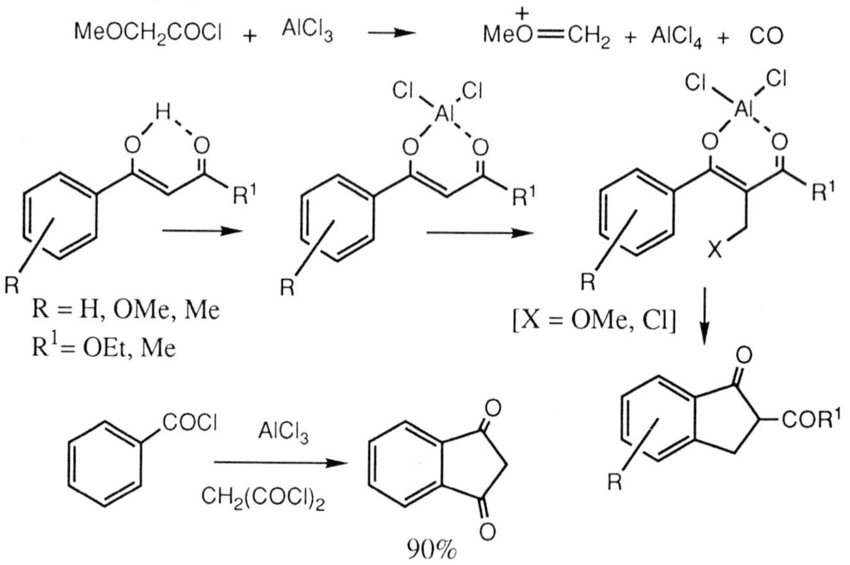

$$MeOCH_2COCl + AlCl_3 \longrightarrow Me\overset{+}{O}=CH_2 + AlCl_4 + CO$$

R = H, OMe, Me
R¹= OEt, Me

[X = OMe, Cl]

90%

In an associated process high yields of indane-1,3-diones are obtained by Friedel-Crafts acylation of benzoyl chloride with malonyl dichloride and aluminium chloride in nitrobenzene at 80^0C for 5 hours (G.Sartori et al., J. Chem. Soc. Perkin Trans. I, 1992, 2985). Carbonylative cyclisation of 2-(o-iodobenzyl)malonic esters with carbon monoxide under pressure in the presence of Li_2CuCl_4 or $NiBr_2$ catalysts leads to indan-1-one 2,2-dicarboxylates, also in high yield (E.Negishi et al., J. Amer. Chem. Soc., 1989, 111, 8018):-

In recent interesting syntheses of 3-hydroxyindan-1-ones and indane-1,3-diones (P.Dallemagne *et al., Bull. Soc. Chim. Fr.*, 1993, 121) condensation of benzaldehydes with malonic acid and ammonium acetate in boiling ethanol to give 3-amino-3-arylpropionic acids is followed by cyclisation with trifluoroacetic acid/anhydride. After hydrolysis the 3-aminoindan-1-ones are diazotised at 60^0C in dilute acid to give 3-hydroxyindan-1-ones which may be oxidised with the Jones reagent to the diones. The overall yields are 10-20%.

R^1 = H, H, H, H, H,
R^2 = OMe, OMe, OMe, O:CH$_2$ O:CH$_2$
R^3 = H, OMe, OMe, O:CH$_2$ O:CH$_2$
R^4 = H, H, OMe, H, H,

Chiral 2- or 3-substituted indan-1-ones are readily obtainable employing as a chiral auxilaries imindazolidinones prepared from urea and (+) or (-)-ephedrine. This causes 1,5-asymmetric induction during 1,4-addition of ArMgBr,Cu^I to the enone side chain. The auxilary is removed with lithium hydroxide and hydrogen peroxide. Cyclisation is achieved with the acid chloride and aluminium chloride in benzene for 3 hours at 25^0C. Optically active indanes result in all cases (H.Poras *et al., Chem. Ind., London*, 1993, 206).

X	R²	R¹
3-Me	H	Me
H	H	Me
H	H	Et
H	Me	H
H	H	CHMe₂

(c)   Reactions at the cyclopentene double bond in indene

The double bond in the five-membered ring of indene is reactive and readily takes part in cycloadditions. For example it reacts with diazoalkanes at room temperature to give diazocyclopentane derivatives which on being heated extrude nitrogen to yield indenocyclopropanes (A.Padwa and H.Ku, *J. Org. Chem.*, 45, 3756).

A German patent describes a Diels-Alder reaction of indene with acrolein, followed by hydrogenation of the product to the indenopyran with a grassy-like odour for use in perfumes (Ger.Offen. 2,900,421; *Chem. Abstr.*,1981, 94, P121,321e). A Diels-Alder reaction of indene with 1,2-dicyanocyclobut-1-ene gives 2,3-dicyanotetrahydrofluorene in 70% yield (D.Bellus *et al.*, *Org. Synth.* 1978, 58, 67).

(i) acrolein
(ii) Pd/C, H$_2$

Cycloaddition also occurs with 2-methoxy-p-benzoquinone in CH$_2$Cl$_2$ at -78^{0}C catalysed by titanium tetrachloride with or without titanium tetrapropoxide to yield different products (T.A.Engler *et al., J.Amer.Chem.Soc.*, 1988, 110, 7931). With the 2,6-disubstituted-p-benzoquinone under the latter conditions a third type of addition product is obtained exclusively (T.A.Engler *et al., J. Org. Chem.*, 1990, 55, 5810).

1H-indene undergoes a Diels-Alder reaction with dimethyl acetylenedicarboxylate without isomerisation to 2H-indene to give a 1:1 adduct with loss of aromaticity (W.E.Noland *et al., J. Org. Chem.*, 1980, 45, 4564). However on being heated above 140^{0}C with maleic anhydride or N-phenylmaleimide isomerisation occurs followed by addition and reformation of the aromatic ring.

α, α-Dibromodeoxybenzoin reacts with indene in the presence of samarium di-iodide to yield a dihydrofuran (S.Fukuzawa *et al.*, *J. Chem. Soc. Chem. Commun.*, 1987, 919).

Irradiation indene and vinyl acetate in cyclohexane with Vycor filtered light from a 450watt Hanovia lamp produces a complex mixture from which the cycloadduct (**21**) can be isolated in 20% yield. It can be converted in several steps into the sesquiterpene modhephene (**22**) in 8.2% overall yield (P.A.Wender and G.B.Dreyer, *J. Amer. Chem. Soc.*, 1982, <u>104</u>, 5805).

21   **22**

Other additions to the indene double bond also occur readily. Thus with xeon difluoride in dichloromethane at 20^0C fluorine adds to give mainly the *trans*-difluoride; the *cis*-difluoride is unstable (M.Zupan and A.Pollak, *J. Org. Chem.*, 1977, <u>42</u>, 1559). With dimethyl sulphoxide and phosphorus oxychloride or PhOPOCl$_2$ in dichloromethane at -20 to 20^0C indene is converted into its 1,2-dithioether in 81% yield (H-J. Liu and J.M.Nyangul, *Terahedron Lett.*, 1988, <u>29,</u> 5467).

With benzene tellurinyl trifluoracetate, generated *in situ* from benzenetelluric anhydride and trifluoroacetic acid, addition of ethyl carbamate to the double bond occurs in the presence of boron trifluoride etherate in chloroform. When this reaction is carried out at a higher temperature in boiling 1,2-dibromoethane cyclisation with elimination of the tellurium gives *cis*-indeno[1,2-d]oxazolidin-2-one in 79% yield (N.X.Hu *et al.*, *J. Org. Chem.*, 1989, <u>54</u>, 4398):-

(d)   Reactions at the benzylic methylene group in indene

The benzylic methylene group in indene is acidic ; for example it forms a magnesium iodide salt, used by E.S.Lazer *et al.* (*J. Med. Chem.*,1979, 22, 845) in the first step in the synthesis of a cocaine analogue; reaction of it with β-chloropropionaldhyde gives the desired starting material. In the presence of alumina and potassium fluoride, benzaldehyde condenses readily with indene at room temperature overnight; KF/Al_2O_3 is more basic than Al_2O_3 alone   (D.Villemin and M.Ricard, *Tetrahedron Lett.*, 1984, 25,1059).

Indane forms a complex with chromium tricarbonyl in which the Cr atom is bonded to the benzene ring; H.Lumbroso *et al.*(*J. Organomet. Chem.*, 1975, 165, 341) discuss its structure in relation to its dipole moment and infrared spectrum. Complexation enhances the acidity of the benzylic carbons so that in the presence of potassium t-butoxide it condenses with diethyl oxalate and benzaldehyde (M.C.Senechal-Tocquer *et al.*, *J. Organomet. Chem.*, 1985, 291, C5-C8):-

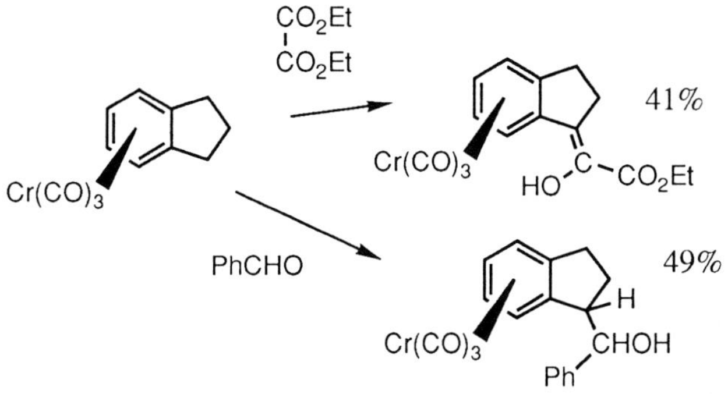

Expansion of the benzene ring in the Cr(CO)3 complex of indane to a heptatriene occurs when it is treated with benzylchloride and lithium diisopropylamine (G.Simonneaux *et al.*, *Tetrahedron*, 1980, <u>36</u>, 893). A similar ring expansion in indane itself is brought about by addition of methyl diazoacetate catalysed by tetrakis(perfluorocarboxylato)di-rhodium (A.J.Anciaux *et al.*, *J. Org. Chem.*, 1981, <u>46</u>, 873).

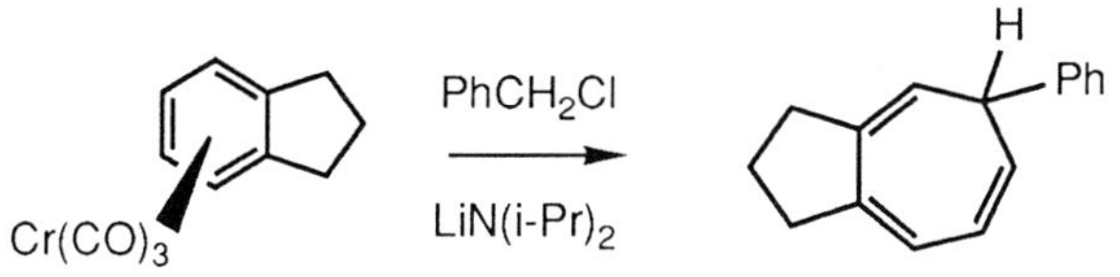

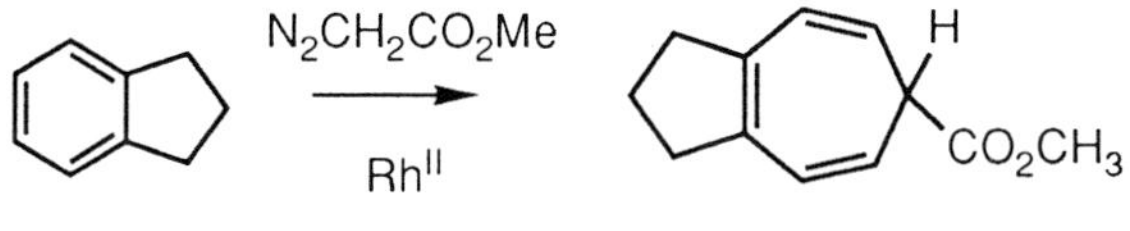

and *o*- and *m*-isomers

(e)   Spiro-indenes

Treatment of indene with phthaloyl dichloride in the presence of sodium hydride in the phase-transfer catalyst tetrabutyl ammonium bromide leads directly to the 1,1-spiro-indene (**22**) in 30% yield (S.N.Naik *et al.*, *Synth. Commun.*, 1988, <u>18</u>, 638).

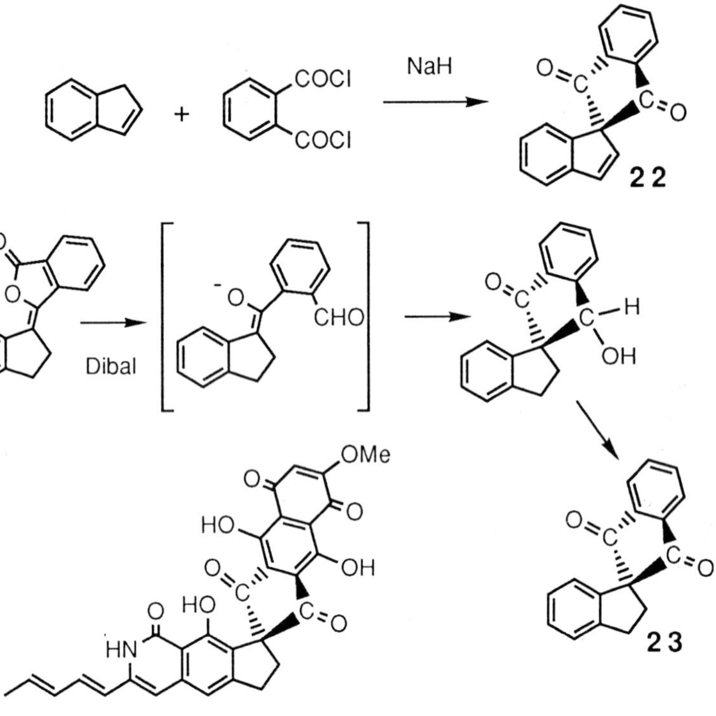

Fredricamycin A

The related 1,1-spiro-indane (23), an important intermediate in the total synthesis of the anti-cancer natural product fredricamycin A, is also obtained in several steps as shown. The full paper describing this work contains a wealth of indene chemistry (T.R.Kelly *et al., J. Amer. Chem. Soc.*, 1988, 110, 6471). Unlike the majority of anti-cancer drugs fredricamycin A does not show mutagenicity in the Ames test.

(f)   Oxidation of indenes

Factors influencing the Cr^{VI} benzylic oxidation of indane are examined by J.Muzart and A Ajjou (*J. Mol. Cat.*, 1991, 66, 165). 1-Indanone is obtained in 86% yield when indan is oxidised with t-butylhydroperoxide (4 eqiv.) and chromium trioxide (0.05 equiv.) in dichloromethane. Use of the latter with 30% hydrogen peroxide leads only to low yields. Asymmetric osmylation of indene can be achieved employing chiral N,N'-dialkyl or diaryl-2,2'-bipyrrolidines (M.Hirama *et al., J. Chem. Soc. Chem. Commun.*, 1989, 665). Thus in the presence of the diphenyl derivative at -78⁰C in toluene [1S,2R]indanediol is formed in 85% yield and 35% e.e.

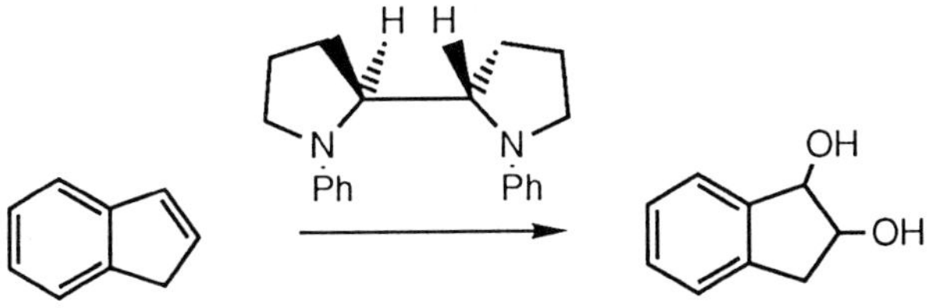

Several reports on the oxidation of indenes using metal porphyrins have appeared. Oxidation of indane with hydrogen peroxide over a porphyrin in the presence of imidazole gives indan-1-ol (48%) plus indan-1-one (2.5%); the imidazole is essential (P. Battioni *et al., J. Amer. Chem. Soc.*, 1988, 110, 8462). Asymmetric oxidation of indene using the optically active manganese porphyrin shown below occurs quantitatively to give the [1R,2S(-)] epoxide at 41% e.e. (R.J.Halterman and S.T.Jan, *J. Org. Chem.*, 1991, 56, 5253):-

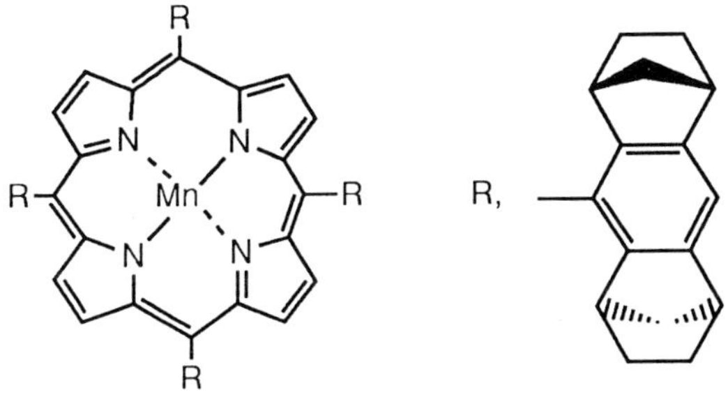

Using the even more elaborate, chirally-vaulted bis[binaphthyl] Fe III porphyrin (**24**) and iodosobenzene as the oxidant both indene and indane afford optically active oxidation products (J.Groves and P.Viski, *J. Org. Chem*, 1990, 55, 3628). Indene gives its [1S,2R(-)]-epoxide in 73% yield at 20% e.e., whilst indane is oxidised to mainly the R-(+)-1-ol in 72% yield with an e.e. of 54%.

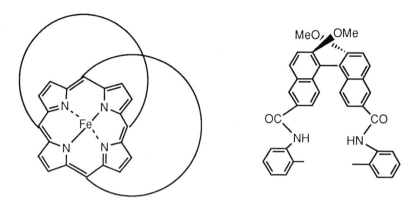

24, where each cyclic chain is as shown above

Second Supplements to the 2nd Edition of Rodd's Chemistry of Carbon Compounds, Vol. III F(Partial), G and H, by M. Sainsbury

147

Chapter 27

AROMATIC COMPOUNDS WITH CONDENSED NUCLEI: NAPHTHALENE AND RELATED COMPOUNDS

N.H. WILSON

1.   *Naphthalene*

Naphthalene continues to provide interesting chemistry as the first member of the acene series of compounds, (annelated benzene). Accordingly naphthalene nearly always features in any discussion of aromatic theory, (see for example M.J.S. Dewar and R.D. Dennington J. Amer. Chem. Soc., 1989, **111**, 3804, for detailed molecular orbital (MO) treatment of naphthalene and other polycyclic aromatic hydrocarbons, and for more general but very comprehensive accounts, P.J. Garratt, "Aromaticity", John Wiley and Sons Inc., New York 1986, and M.V.Gorelick, Russian Chem. Rev. 1990, **59**, 116, (English Translation)). The previous treatise (Suppl. to the 2nd edition IIIG pages 176, 177) discussed the partial 'localisation' of double bonds in the naphthalene system and the use of NMR couplings and X-ray data in estimating the extent of the double bond character.

Structure **1** still seems to be best model for representing naphthalene, since the 'real' structure is <u>mainly</u> derived from the three Kekulé forms (**1,2** & **3**) depicted, (M. Sironi *et al.*, J. Chem. Soc. Chem Comm., (1989), 675.). It will be apparent from the following review of the thermal reactions of $C_{10}H_8$ isomers that the naphthalene structure sits in a deep potential energy minimum and dominates the energy profile of these molecular species. Clar's view of the acene series, which correlates well with observed properties, states that linear annelation leads to a decrease in stability and an increase in reactivity. In the case of naphthalene the aromatic sextet of electrons, from which a great deal of the stabilisation is derived, is spread over two rings, since in the canonical forms shown, only one ring has 4n+2 π electrons. Linear addition of further rings leads to even more labile systems, culminating in hexacene **4** which reacts with air. Heptacene (seven rings) has only been observed in an impure state and octacene is unknown. (see E. Clar, "The Aromatic Sextet", John Wiley & Sons, London 1972).

4   **5**

Angular ring addition, allows more aromatic sextets in the structure, as in phenanthrene **5** (two), and a consequent increase in stabilty. The central ring "double bond" does in fact show very high double bond character reflecting the two adjacent complete aromatic sextets. This hypothesis has been endorsed by gas phase proton transfer measurements in the mass spectrometer, (M. Moet-Ner, J.F. Liebman, and S.A. Kafafi, J. Amer. Chem. Soc., 1988, **110**, 5937). This study shows that naphthalene is more acidic than benzene, the proton transfer only perturbing a 4, rather than a 6, π electron system . The "secondary" ring in naphthalene is therefore shown to be less stable than the "primary" by 6 kcal/mol. Deuterium quadrupolar couplings have also measured arene magnetic anisotropies as an index of aromaticity, (P.C.M. van Ziji *et al.*, J. Amer Chem. Soc., 1986, **108**, 1415). Pertinent to this subject is a review of the mechanisms of pyrolytic formation of polyaromatic hydrocarbons, (S.E. Stein, Acc. of Chem. Res., 1991, **24**, 350).

2. Valence Isomers of Naphthalene

This area was covered insome detail in the previous supplement, (*loc. cit.*), and the isomerisation of hemi-Dewar to the more dominant Kekulé form discussed. The reverse reaction, induced by steric overcrowding, has now been reported, (S. Miki, T. Katayama and Z. Yoshida, Chem. Lett., 1992, 41; S. Miki *et al., ibid.*, 61; *idem.*, Tetrahedron Letters 1992, **33**, 953, 1619), however, even in these examples the isomer equilibrium is more in favour of the Kekulé structure **6** than the tricyclic Dewar form **7**. Several examples of this type of "thermally forbidden" interconversion of valence isomers are given in these references.

6   **7**

Also of interest is the reaction of naphthvalene (benzobenzvalene) **8** with sulphur dioxide to give sultone **9** and sultine **10**. These isomeric compounds interconvert on irradiation and SO_2 is extruded to regenerate naphthvalene along with naphthalene. Pyrolysis of the **9/10** mixture gives 1H-indene-1-carboxaldehyde **11**, and naphthalene **1** (U. Burger, S.P. Schmidlin, and J. Mareda, Phosphorus Sulfur and Silicon and the Related Elements, 1993, **74**, 417; Chem. Abstr., 1994, **120**, 164132s).

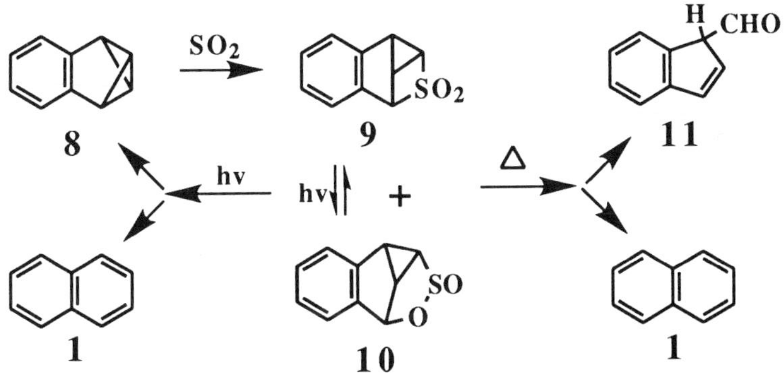

Related to these reactions are the azulene to naphthalene conversion (the so called AN reaction) and the naphthalene automerisation (the AUN reaction) where the ring carbons in naphthalene are moved relative to each other. In the latter case the apparent carbon atom movements are observed by ^{13}C labelling. L.T. Scott has reviewed these processes, (Acc. Chem Res., 1982, **15**, 52), and ^{13}C labels have also been used to elucidate the mechanisms of the AN process. A likely pathway is depicted here but is by no means totally certain as azulene to azulene automerisations can occur prior to the devolvement to naphthalene, and carbon scrambling cannot be fully explained by the mechanisms proposed, (M.J.S. Dewar and K.M. Merz, Jr., J. Amer. Chem. Soc., 1986, **108**, 5142).

Scott (*loc. cit.*) has pointed out that the reverse AN reaction does not occur to any detectable degree, since no azulene was detectable (to <1ppm) on flash vacuum pyrolysis (FVP) of naphthalene, and therefore at even higher temperatures azulene is not an significant intermediate in the AUN reaction. The most likely carbon atom scrambling mechanism in naphthalene pyrolysis seems to involve the benzofulvene species shown, (M.J.S. Dewar and K.M. Merz, Jr., J. Amer. Chem. Soc., 1986, **108**, 5146). This paper reviews all the theoretical, and known, $C_{10}H_8$ systems and calculates the lowest energy intermediates to be benzofulvene **12** and possibly isobenzofulvene **13**. The former allows scrambling of the 1- and 2-positions, the latter the 2- and 3-positions. This mechanism finds support by comparison with benzene automerisation (K.M. Merz Jr., and L.T. Scott, J. Chem. Soc. Chem. Comm., 1993, 412, and other work cited in this paper).

Naphthvalene **8** does not seem to be a contending intermediate in the above thermal reactions, but contrary to the previous report which stated that it rearranges to benzofulvene **12** (Suppl. to the 2nd edition IIIG pages 177, 178) **8** does in fact give naphthalene on FVP or thermolysis in basic solution media. This suggests that the formation of **12** is a catalysed process (D.P. Kjell and R.S. Sheridan, Tetrahedron Letters, 1985, **26**, 5731). Benzofulvene **12** can still be formed in the gas phase due to surface catalysis, and this fascinating paper serves to illustrate some of the experimental difficulties and pitfalls of this field of research.

Kjell and Sheridan have also obtained naphthalene, naphthvalene and benzofulvene by nitrogen extrusion from diazabenzosemibullvalene **14** (J. Amer. Chem. Soc., 1986, **108**, 4111).

Before leaving high energy processes associated with naphthalene mention must be made of the use of naphthalene as a source of C_{10} units in the formation of fullerenes C_{60} **15** and C_{70} **16** in an electric arc discharge (R. Taylor, *et al.*, Nature, 1993, **366**, 728).

Finally a further interesting valence isomer bicyclo[6.2.0]deca-pentaene **17** has been synthesised, and found to be planar and weakly aromatic, having a resonance stabilisation energy of approximately 4 kcal/mol (J. Cremer, T. Schmidt and C.W. Bock, J. Org Chem., 1985, **50**, 2684; W. D. Roth *et al.*, Ber., 1986, **119**, 837). Presumably this is due to the 4n+2 (10) p electrons around the periphery, as in [10]annulene systems, which are known to be aromatic, (see P.J. Garratt, "Aromaticity", page 108 *et seq.*,John Wiley and Sons Inc., New York 1986)

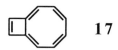 **17**

For azulene valence isomers see the azulene section of this chapter, and there is a very short-lived naphthalene isomer discussed on page 44,45.

3. General Properties of the Naphthalene Ring System

(a) Steric effects

Reports of steric effects particularly across the 1,8- (*peri*) positions of naphthalene continue. Restricted rotation can give rise to conformational enantiomers. An example, where spontaneous optical resolution can occur, is provided by the chirally stable binaphthyl system **18** (G. Gottarelli and G.P. Spada, J. Org. Chem., 1991, **56**, 2096). Seeding the racemate solution with one enantiomer, crystallised material of 90% enantiomeric purity. A general review of this area giving many examples of the use of "BINAP" compounds of this type as chiral resolution agents is available (G. Bringmann, R. Walter and R. Weirich, Angew. Chem. Internat. Edn., 1990, **29**, 977), This type of chiral auxilliary seems to be efficient, and of wide applicability. Asymmetric induction is observed in many systems (eg., hydrocarboxylation ,H. Alper and N. Hamel, J. Amer. Chem. Soc., 1990, **112**, 2803; aldol condensation, T. Mukaiyama *et al.*, Chem. Letters Jpn., 1990, 1015). A NMR and X-ray crystallographic study of conformational enantiomers, racemisation, and deformation of ring system bonds, including molecular calculations, has been made for 1,8-diacyl-naphthalenes **19**. The accumulated evidence shows that the carbonyl groups are directed inwards as illustrated (D. Casarini *et al.*, J. Org. Chem., 1994, **59**, 4637).

In accord with this, NMR long range coupling experiments (S.R. Salman, Org Mag, Res., 1983, **21,** 672) and nuclear Overhauser effects (NOE) (L.I. Kruse and J.K. Cha, Tetrahedron Letters 1983, **24,** 2367) show the conformation (highlighted) in molecules like **20** to be preferred, where the carbonyl has the s-trans orientation vis á vis the *ortho*-position of greatest double bond character. Studies of the dipole moments of *peri*-, and other disubstituted naphthalenes, have been reported in assessing the conformational preference of substituents. For example, in counterpoint to the carbonyl systems, a +I group such as methoxy prefers a cisoid conformation with respect to the *o*-position of highest double bond character (S.B. Bulgarevich *et al*., J. Molec. Structure, 1994, **326,** 17;   V.Baliah and V. Balasubramaniyan, J. Indian Chem. Soc., 1993, **70,** 755). This can have steric ramifications in substitution reactions (see section on sulphonation of methoxy naphthalenes page 13). Finally the structures of 1,8-bis(silyl)naphthalenes **21** incorporating different sizes of silyl groups have been analysed by X-ray diffraction (R. Schröck *et al*., Organometallics 1994, **13,** 3399). When R = H the aromatic system is essentially planar with the SiH$_3$ groups splayed outwards. Larger groups, R = *p*-anisyl, cause bending of the aromatic rings in order to relieve strain.

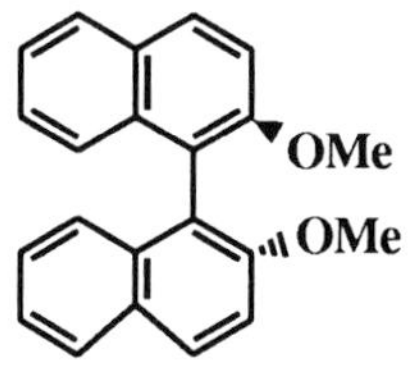

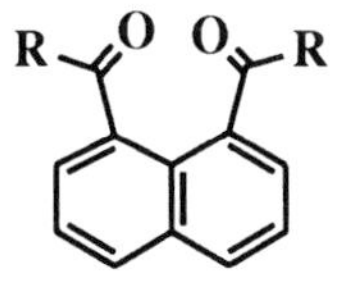

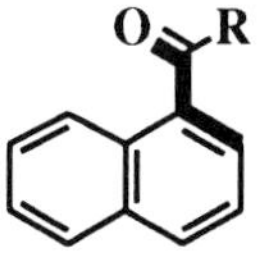

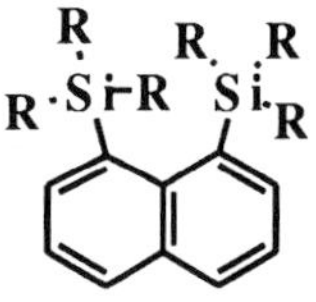

Steric compression across the *peri*-positions of naphthalene and related molecules produce the special basic properties of "proton sponges". The parent substance N,N,N',N'-tetramethyl-1,8-diaminonaphthalene **22** was briefly mentioned in the previous edition (Suppl. to the 2nd edition IIIG page 179) and since then proton sponge technology has become a large subject. The very powerful basic strength of these compounds depends on the relief of strain, the contribution of H-bonding and the charge delocalisation in the monoprotonated species . The proton is thus chelated between the nitrogen atoms and the base will not add a second proton, as this would disrupt a very favourable arrangement. These effects were illuminated by examination of analogues including the examples **23-26**, (pK$_a$ values shown in brackets) depicted below which appear to have the highest basicity (H.A. Staabe and T. Staupe, Angew. Chem. Internat. Edn., 1988, **27,** 865).

Me Me
Me−N   N-Me

22   (pK$_a$ = 1 2.1)

H$^+$ →

Me,   H ,Me
Me−N + N−Me

Me,   ,Me
N   N
(pK$_a$ = 1 2.1)
23

N, N,
Me   Me Me Me
(pK$_a$ = 1 2.8)
24

Me−N   N-Me
(pK$_a$ = 1 4.1)
25

N   N
(pK$_a$ = 1 2.8)
26

Apart from their ability to add a single proton, these bases have virtually no other nucleophilic properties. This is undoubtedly due to the crowded reaction centre. The more powerful proton acceptors based on the fluorene nucleus **24, 25** suffer from the ability of the fluorene system to take part in base catalysed reactions, because of the reactive methylene group. Thus the literature abounds with examples of the use of naphthalene proton sponges, where a non-nucleophilic base is essential, (*eg.*, isocyanate formation from diphenylphosphoryl azide and carboxylic acid, J.W. Gilman and Y.A. Otonari, Synth. Comm., 1993, **23**, 335; palladium catalysed alkenylation {Heck reaction} of olefins, F. Ozawa, Y. Kobatake and T. Hayashi, Tetrahedron Letters, 1994, **34**, 2505). A phosphorus analogue of proton sponge has been made, protonation of which gives a fluxional molecule proposed to have a P-P bond (R.D. Jackson *et al.*, J. Organomet. Chem., 1993, **458**, C3-C4). Normal coordination compounds with platinum and palladium were obtained with this system. Reactions involving proton transfer across the *peri*-positions in the rate determining step are facilitated due to favourable transition states similar to that in proton sponges (A.J. Kirby, and J.M. Percy, J. Chem, Soc. Perkin Transactions 2, 1989, 907; F. Hibbert and K.J. Spiers, *ibid.*, 1989, 377).

(b)   Electrophilic substitution reactions

Preparative reactions in naphthalene have been well covered in previous editions. However naphthalene compounds continue to be essential in synthesis as well as substrates in many mechanistic studies. For an excellent treatise on this subject see Roger Taylor, "Electrophilic Aromatic Substitution", John Wiley and Sons Ltd., Chichester 1990.

(i)   Nitration
The nitration process under various conditions has been extensively studied and it is now apparent that with more readily oxidised, reactive arenes electron transfer processes can become important. This is more likely when +I (electron donating) groups are present (J. Feng, X. Zheng and M.C. Zerner, J. Org. Chem., 1986, **51**, 4531). A significant contribution from the electron transfer process has been shown by chemically induced dynamic nuclear polarisation (CIDNP) monitoring of ^{15}N label signals in the nitration of naphthalene (J.F. Johmston, J.H. Ridd and J.P.B. Sandall, J. Chem. Soc. Chem. Comm., 1989, 244).

Classical Nitration Mechanism

$$ArH \; + \; NO_2^+ \longrightarrow Ar\overset{+}{\cdot}\overset{H}{\underset{NO_2}{}} \longrightarrow ArNO_2 \; + \; H^+$$

Electron Transfer Mechanism

$$ArH \; + \; NO_2^+ \longrightarrow ArH^{+\cdot} \; + \; N\overset{\cdot}{O_2}$$

A detailed paper on the nitration of naphthalene and methyl-substituted naphthalenes comparing nitration procedures has been published (L. Eberson and R. Finn, Acta Chem. Scand., 1986, B **40**, 71). The substitution patterns are very much as expected, but the regioselectivity alters dramatically depending on the reaction conditions. The reaction of the arene radical cation salt with nitrogen dioxide gives the highest selectivity, and thus to obtain unusual isomers acetyl nitrate offers the least selective method, but not in every case. For example, 1,4-dimethylnaphthalene **27** is anomalous in several ways. The arene radical cation / NO_2 (method 1) gives the 2-isomer **28** in high yield, acetyl nitrate (method 2) gives the 5-isomer **29** exclusively, while nitrogen dioxide in dichloromethane (method 3) gives **30** by nitration of a methyl group via free radical species (see also below), again in very high yield. The nitration of the methyl group(s) in such a system has been observed by several researchers.

Me

29

NO$_2$Me

Me

Me
NO$_2$ 1

Me
2

Me
Me

28

27

Me
3

Me

30

CH$_2$NO$_2$

The same authors have produced a general review of electron transfer mechanisms in aromatic nitration (Acc. Chem. Res., 1987, **20**, 53). Eberson and his colleagues have also used the photolysis of the charge transfer complex of arene and tetranitromethane to generate aryl radical cation. It was possible, at low temperatures, to isolate the unstable intermediate adducts, in this reaction. The electron transfer mechanism contribution was assessed by ESR and the utility / selectivity of the reaction, with and without, a protic acid present, was explored for several arenes including naphthalene, and its methyl- and methoxy-derivatives (L. Eberson et al., J. Chem, Soc. Perkin Transactions 2, 1994, 1719).

Finally nitration of naphthalene via a totally uncharged free radical mechanism (NO$_2$ in CCl$_4$) has given rise to isomer distributions similar to that obtained in the gas phase of the environment due to pollution. With naphthalene itself, 1-nitronaphthalene is usually the dominant product, but significant amounts of the 2-isomer are formed along with 1,3-dinitro- and 2,3-dinitro-naphthalenes. Initially it may be surprising that dinitro compounds appear at these low level conversions, but the free radical process has no electrostatic repulsion costraints and allows further attack on the already nitrated ring, (G.L. Squadrito et al., J. Org. Chem., 1989, **54**, 548). This may have important consequences relating to formation of mutagens in the environment (see also *idem.*, J. Org. Chem., 1990, **55**, 2616).

$-HNO_2$ → 1-nitronaphthalene

2-nitronaphthalene

$-2HNO_2$ → 1,3-dinitronaphthalene

2,3-dinitronaphthalene

(ii) Acylation

The Friedel-Crafts acetylation mechanism in naphthalene has also been reassessed (D. Dowdy, P.H. Gore and D. N. Waters, J. Chem. Soc., Perkin Transactions 2, 1991, 1149). Analysis of the traditional reaction (AcCl / AlCL$_3$) in dichloroethane showed the usual interpretation of the mechanism, whereby the 2-isomer formation can be enhanced, to be wrong in this system. The reaction reversiblity leading to the more thermodynamically stable 2-isomer on prolonged reaction times still applies to polyphosphoric acid (PPA) mediated acylations (Suppl. to the 2nd edition IIIG page 180, 181) but in the traditional reaction the rate of formation of the 2-isomer is first order in naphthalene and acylating complex, as expected, whereas the rate for the 1-isomer is second order in acylating complex. This surprising result is explained by AcCl being involved in a reverse reaction step with the intermediate complexes. Excess AcCl augments the reverse reaction and thus favours the formation of more 2-isomer. Conditions for obtaining either isomer can therefore be defined. However, it was pointed out in this outstanding investigation that the results apply only to dichloroethane solutions, and that other solvent systems and conditions for preparation of either isomer are known. A further paper propounding regioselectivity to be due to reaction reveriblity with very reactive substrates allows preparation of 2-propanoyl-6-methoxynaphthalene **33** (R=Et), an important intermediate in the synthesis of the antiinflammatory drug Naproxen, (C. Giordano, M. Villa and R. Annunciata, Synth. Comm., 1990, **20**, 383). Other acyl groups used in this investigation were butanoyl and phenacyl (**33** R=*n*-Pr and R=CH$_2$Ph respectively). 1-Acyl-2-methoxy-naphthalenes **32** were also obtained from **31** as the kinetic products.

A useful alkylation of naphthalene has been devised (A. Jaxa-Chamie, V.P. Shah and L.I. Kruse, J. Chem. Soc., Perkin Transactions 1, 1989, 1705) via acylation, followed by *in situ* reduction with silane. Polyalkylation is thus avoided in this "one pot " process.

Another electrophilic substitution employing the easily generated O,O'-dication of a nitro-alkene such as 2-nitropropene yields the α-arylated ketone **34**. This reaction shows great promise for the preparation of these compounds (K. Okabe *et al.*, J. Org. Chem., 1989, **54**, 733).

(iii) Sulphonation

The control of sulphonic acid isomer ratios by rapid heating with microwaves has been researched as an efficient method (D. Stuerga, K. Gonon and M. Lallemant, Tetrahedron, 1993, **49**, 6229). There is also an elegant series of papers on sulphonation of hydroxy- and dihydroxy-naphthalenes (naphthols), and their alkoxy and mesyloxy derivatives, by Dutch chemists (H.R.W. Ansink *et al.*, Recueil des Trav. Chim., 1992, **111**, 499; J. Chem. Soc , Perkin Transactions 2, 1993, 721; Recueil des Trav. Chim., 1993, **112**,

210 *et seq.*). This first citation compares sulphonation with other sustitution reaction of these naphthalenes. With equimolar amounts of sulphonating agent, naphthol ring sulphonation occurs on the naphthol itself, whereas with excess reagent, ring sulphonation occurs to a major extent on the aryl hydrogen sulphate which is formed in the equilibrium reaction shown.

$$ArOH + SO_3 \rightleftharpoons ArOSO_3H$$

This group is deactivating *para* directing in contrast to the hydroxy group which is activating and *ortho / para* directing. Hence the product distribution depends on the amount of reagent used, temperature etc. In the methoxy derivatives the high steric requirements of the sulphonation reaction and the preferred conformations of the methoxy group (*cf.,* page 7) can lead to good control of substitution position. In the example depicted 2,3-dimethoxy-naphthalene **35** gives **36** and **37**. There is no substitution at the 1-position due to the methoxy groups, but in the ethylenedioxy derivative where the alkyl moieties are "tied back" a large proportion of homonuclear substitution takes place.

35   **36**   **37**

(iv) Miscellaneous Reactions

Naphthalene has been involved in a study of electrophilic substitution reactions in an effort to determine whether the electron transfer mechanism is a contributory pathway (see nitration section). For halogenation, acetylation, mercuriation and thallation the traditional mechanism was vindicated (C. Galli and S. Digiammarino, J. Chem. Soc , Perkin Transactions 2, 1994, 1261).

Iodination (G.C. Tustin and M. Rule, J. of Catalysis, 1994, **147**, 186), and *iso*-propylation (P. Moreau *et al.*, Studies in Surface Science and Catalysis, 1993, **78**, 575; J. Chem. Soc. Chem. Comm., 1991, 39)) have been made regioselective by careful choice of zeolite catalyst. High yields of 2-iodo- or 2,6-diiodo- and 2,6 di-*iso*-propyl-naphthalenes respectively were achieved. Selective chlorinations are also claimed (A.P. Singh and S.B. Kumar, Catalysis Letters, 1994, **27**, 171).

(c)   Nucleophilic substitution reactions

This reaction class has been studied for electron transfer mechanisms in a similar way to electrophilic sustitution, but most evidence points to the traditional $S_{RN}2$ pathway. For a large list of references see A.P. Chorlton, Ann. Reports of the Royal Soc. Chem., 1993, **90**, Sec.B, 162, 163.

Reaction activation by nitro groups is a well known phenomenon. For a theoretical discussion of mechanism and leading references see S. Sekiguchi, T. Aizawa and N. Tomoto, J. Org. Chem., 1984, **49**, 93). The hydrolysis of a halo-dinitronaphthalene **38** to **39** has been researched, and kinetics determined, (R. Germani *et al.*, Langmuir, 1993, **9**, 55). Phase-transfer catalysis assisted hydrolysis in this system was discussed and the optimum micelle character and size for rapid reaction determined, (C.A. Bunton *et al.*, Langmuir, 1993, **9**, 117).

A displacement of the nitro group itself by thiophenoxide anion has been demonstrated (M. Nori, C. Dell'Erba and F. Sancassan J. Chem. Soc , Perkin Transactions 1, 1983, 1141) in a naphthalene **40** with two activating nitro groups . This reaction giving **42** and **43** is similar to one reported in the previous edition (Suppl. to the 2nd edition IIIG pages 182, 183) where a nitro group is lost and some or all of the incoming nucleophile attacks a methyl group.

A nitro group can also activate an adjacent unsubstituted position to nucleophilic attack to give a potentially very useful reaction as in **44** to **45** (W. Danikiewiig and M. Makosza, Tetrahedron Letters, 1985, **26**, 3599). This reaction does not appear to occur in benzenoid systems.

The methoxy group, suitably activated, can also be a good leaving group. Syntheses of binaphthyl systems (eg **48**) which are extremely useful as chiral resolution agents (*cf.* **17** page 6) have been devised by several groups. 1-Naphthylmagnesium halide **46** reacts with a hindered ester of 1-methoxy-naphthalene-2-carboxylic acid **47** in a displacement of the methoxyl (T. Hattori *et al.*, Bull. Chem. Soc Japan, 1993, **66**, 613). If the ester group R in **47** is enantiomeric, the binaphthyl produced can be up to 98% optically pure. This binaphyl preparation is a modification of a synthesis by A.I. Meyers and K.A. Lutomski (Synthesis 1983, 105), where the ester was protected as the oxazoline derivative (see also below).

The use of *ortho*-lithiation of 1-methoxynaphthalene **49**, allowed introduction of a diphenylphosphine oxide as an activating group making displacement of the methoxyl in **50** by nucleophiles facile (T. Hattatori *et al.*, Synthesis, 1994, 199). In a similar way to the synthesis of **48**, structures **51-53** can yield useful chiral phosphine ligands. There is a related process where naphthyl phosphate esters **54** and **56** on treatment with lithium amides rearrange to the *o*-hydroxyphosphonates **55** and **57** respectively (B. Dhawan and D. Redmore, J. Org. Chem., 1991, **56**, 833). This paper discusses similar useful reactions.

The reduced aromaticity of naphthalene, with respect to benzene, allows organometallic nucleophiles to give 1,4-addition by participation of a "fixed" double bond in the ring. This 1,4-addition appears to be unique to Grignard reagents, which act on the imine alcohol open-chain tautomer of the oxazolidine **58** via a tight complex involving the halomagnesium alkoxide salt. Therefore two equivalents of the Grignard reagent are required. This has produced a useful synthesis of chiral dihydronaphthalenes **60** via the aldehydes **59**. Lithium, cerium and even copper reagents give 1,2-addition in this system (L.N. Pridgen, M.K. Mokhallalati and M.J. Wu, J. Org. Chem., 1992, **57**, 1237).

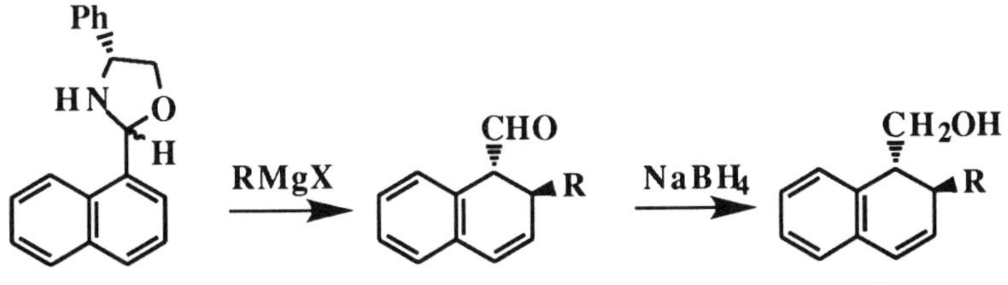

Ph

HN O
H

58   RMgX   CHO

R

59   NaBH₄   CH₂OH

R

60

> 90% ee

Displacement of both iodo groups by cyclopentadienyl in 1,8-diiodonaphthalene **61** has led to synthesis of the novel ferrocene structure **62** (B.M. Foxman et al., J. Org. Chem., 1993, **58**, 4078)

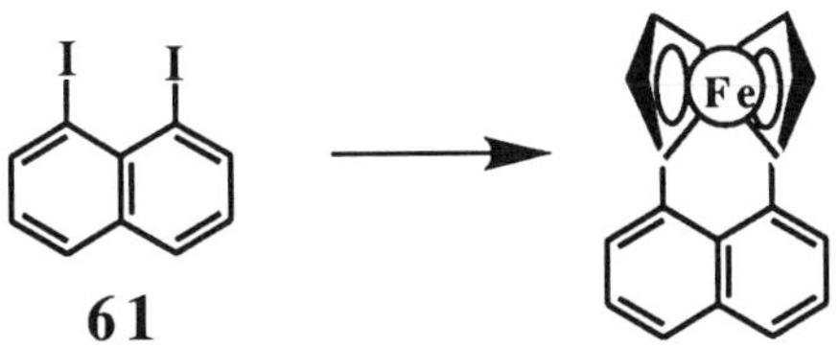

I I

61   Fe   **62**

The Heck coupling reaction of **61** with acenaphthylene **63** gives the beautifully symmetrical acenaphth[1,2-a]acenaphthylene **64**, which is an excellent dienophile (G. Dyker, Tetrahedron Letters, 1991, **32**, 7241).

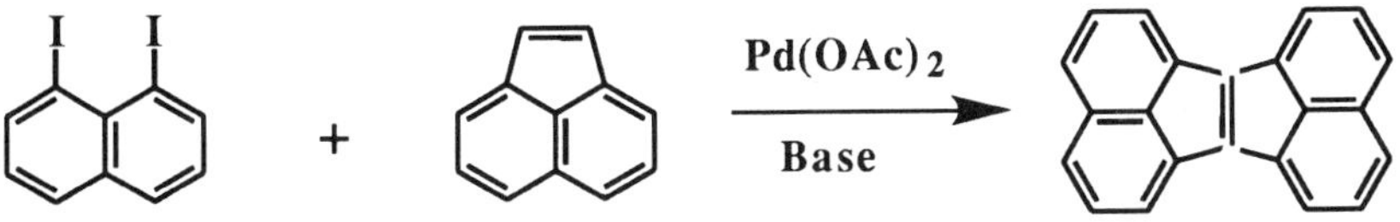

I I

61   +   **63**   Pd(OAc)₂

Base   **64**

For a review of this type of reaction and many other palladium mediated reactions see R.F. Heck, "Palladium Reagents in Organic Syntheses", Academic Press Ltd., London, 1985.

Finally reaction with cobalt carbonyl in the presence of dimethyl sulphate and concomitant displacement of tosyloxy group permits the conversion of hydroxyl to carboxyl **65-66** (G. Cometti *et al.*, J. Organo-metallic Chem., 1993, **451**, C13-C14).

65   **66**

(d)   Other substitution reactions including metallation and free radical

(i)   Lithiation

It is appropriate in the light of the previous section, to discuss the lithiation of naphthalenes in general. The mechanism of formation of these species is generally regarded as a mutual electron transfer. The reaction of organometallic reagents and the regiospecific formation of naphthyl derivatives as in the above examples are typical of this area of naphthalene chemistry. Naphthalene itself in reaction with metallic lithium forms a naphthalenide radical anion. This reaction is not usually regarded as important other than as a catalyst system to bring the lithium more readily into play in other reactions. A useful example is the ready formation of primary or secondary alkyllithiums from the inexpensive dialkyl sulphates. In the absence of a catalytic quantity of naphthalene the reaction is very inefficient (<30% yield). Organolithium reagents can often be made and reacted *in situ* after the manner of Barbier the founder of organometallic chemistry (eg., D.J. Ramon and M. Yus, Tetrahedron, 1993, **49**, 10110 and references therein).

$$ArH + Li \longrightarrow ArH^{-\bullet} Li^{+} + R_2SO_4$$

catalytic   excess

$$RLi + RLiSO_4 + ArH$$

The metal-halogen exchange reaction is a major route to organolithium reagents. In the case of 1,5-diiodonaphthalene it has been shown that iodo groups can be replaced, one at a time, using *tert*-butyllithium (W. Wang *et al.*, J. Org. Chem., 1991, 56, 2914) and the resulting lithium compounds reacted to give aldehydes, alcohols etc. However in many aromatic systems, especially naphthalene lithiation by *tert*-butyllithium can be substituent directed as in the example **49** discussed earlier. This is a very convenient method of functional group introduction. 1-Naphthol **67** directs lithiation to the 2- and 8-positions (**68,69**) whereas 2-naphthol **70** gives exclusively the 3-isomer **71** (G.A. Suñer, P.M. Deyá and J.M. Saá, J. Amer. Chem. Soc., 1990, **112**, 1467). 1-Methoxy-naphthalene can be made to give either *ortho-* or *peri*-lithiation by choice of solvent, TEMEDA and reaction time (D.A. Shirley and C.F. Cheng, J. Organomet. Chem., 1969, **20**, 251; see also J.B. Christensen et al., Synth. Metals, 1993, **56**, 2128).

An elegant use of *ortho*-lithiation in the synthesis of benzochromone **76** via compounds **72** to **75** has been described (R.G. Harvey *et al.*, J. Org. Chem., 1990, **55**, 6161). The methoxymethyl protecting group has a powerful role in directing the initial lithiation reaction.

For a list of other functional groups which have been used to direct this type of reaction see G.A. Suñer, P.M. Deyá and J.M. Saá, *loc. cit.* and B. Dhawan and D. Redmore, J. Org. Chem., 1991, **56**, 833.

(ii) Arylation

Several studies have been made in this field. The reactivity of the naphthalene positions to phenyl radical in the gas phase at 400°C has shown that the 2-phenyl isomer **78** is forned preferentially, in contrast to the reaction in the liquid phase which favours **77** (R.H. Chen, S.A. Kafafi and S.E. Stein, 1989, J. Amer. Chem. Soc., **111**, 1418, : see this citation for leading references).

	minor	major
1	**77**	**78**

Perhaps the most unusual reaction is the photochemically induced phenylation of 2-naphthoxide anion in liquid ammonia to yield **79** and **80** (M.T. Baumgartener, A.B. Pierini and R.A. Ross, Tetrahedron Letters, 1992, **33**, 2323). At higher dilutions product **80** is virtually exclusive

	79	**80**

4.   Didehydronaphthalenes (naphthynes)

A convenient and important advance in obtaining heats of formation of short-lived species such as didehydroarenes (arynes) has been made (P.G. Wenthold, J.A. Paulino and R.R. Squires, J. Amer. Chem. Soc., 1991, **113**, 7415). The use of ion-molecule collisions, in the mass spectrometer, allows estimation of this fundamental parameter. Chloroaryl anion is allowed to collide with an argon target at different energies to produce the aryne by loss of chloride ion. Naphthalene does not seem to have been investigated by this technique, but results for 1,2-, 1,3- and 1,4- dehydrobenzenes gave heats of formation of 106, 116 and 128 kcal/mol respectively. Bicyclic isomeric forms of the *meta-* and *para-* arynes were estimated to be of much higher energy.

(a)   1,2- and 2,3- Didehydronaphthalenes

Elegant syntheses of the biologically significant methylphenanthrenes **83** has been performed via 1,2-naphthynes generated from **81** (K-y. Jung and M. Koreeda, J. Org. Chem., 1989, **54**, 5667). The removal of the oxygen bridge in **82** was achieved by cleavage with iodotrimethylsilane. These 1,4-endoxy compounds can also be reduced by titanium tetrachloride / lithium tetrahydridoaluminate in the presence of triethylamine, (H.N.C. Wong, Acc. Chem. Res., 1989, **22**, 145) a very useful procedure, since these furan adducts are common in the Diels-Alder methods of ring synthesis, (see later). For a review of other aryne mediated cyclisations see E.R. Biehl and S.P. Khanapure, Acc. Chem. Res., 1989, **22**, 275.

Another interesting paper (S.L. Buchwald and S.M.King, J. Amer. Chem. Soc., 1991, **113**, 258) expounding the synthesis of highly substituted naphthoquinones 91, via naphthalenes **84, 85** and **86**, involves the formation of a zirconocene complex of 2,3-naphthyne **87** (called naphthalyne by these authors). The reaction of **87** with a nitrile is regiospecific, and the zirconium moiety can be replaced by hydrogen or iodine to give **89** and **90** respectively. The R protecting group can be methoxy-ethoxymethyl (MEM) or benzyl (Bn), and the more highly substituted ring can be oxidised to the 1,4-quinone by Jones Reagent.

(b)   *1,8-Didehydronaphthalenes*

The generation of this intermediate **93** from **92** was discussed in the previous review (Suppl. to the 2nd edition IIIG pages 186, 187) and its diradical nature noted. Attempts to trap **93** by reaction with C_{60} fullerene in benzene led to reaction with the solvent to produce 6b,10a-dihydrofluoranthene **94** (J. Averdung and J. Mattay, Tetrahedron Letters, 1994, **35**, 6661. This in turn entered into a Diels-Alder reaction with the C_{60} fullerene to produce the novel 4+2 adduct **95** in 27% yield. The dienophilic character of fullerene was thus confirmed and the reaction exposes interesting ways of functionalising this molecule.

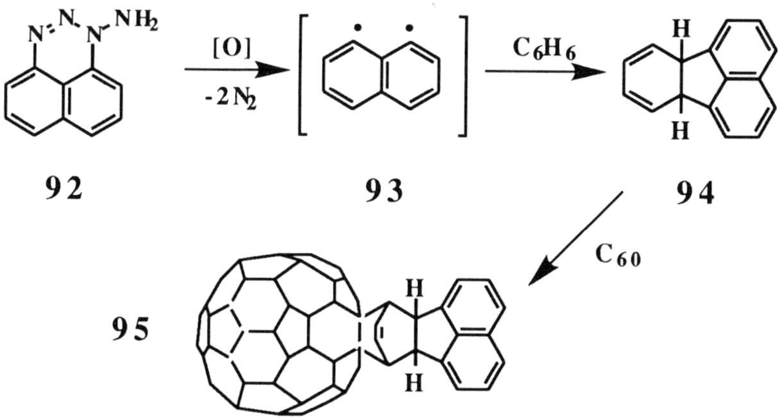

92 **93** **94**

95 C_{60}

(c) *1,4-Didehydronaphthalenes*

This field, which warranted a brief desciption in the previous treatise (Suppl. to the 2nd edition IIIG page 187), has blossomed into a major research area over the last decade, since the discovery of natural products with the structural feature, an enediyne unit, required to form 1,4-dehydroarene and other similar diradical species. This reactive intermediate type **97** is derived, as shown, from the enediyne system **96** and this is now referred to as Bergman cycloaromatisation. The naturally occurring molecules with this feature are the calicheamicins and the esperamicins, which are of antibiotic origin, and some have oligosaccharide addenda which promote specific association with parts of the DNA in living organisms. The generation of the 1,4-diradical then disrupts the DNA at this specific site. These substances are extremely toxic. For a review of this important field see K.C. Nicolaou and W.-M. Dai, Angew. Chem. Internat. Edn., 1991, **30**, 1387.

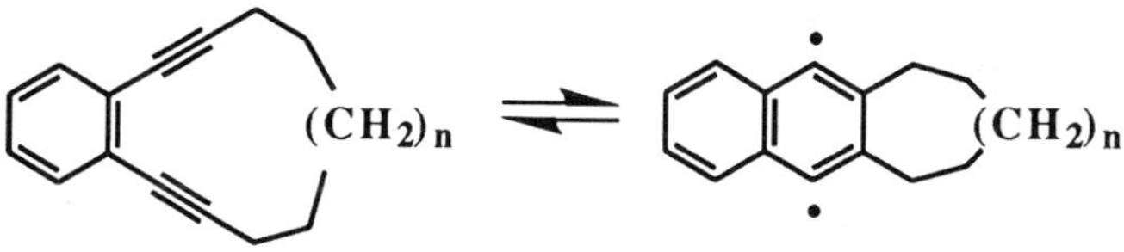

96 **97**

The generation of naphthalene systems by this method has been explored to determine the ease of molecular reorganisation required. The quinone **100** is much faster than **98** at forming the diradical intermediates (**101** as opposed to **99**) (K.C. Nicolaou *et al.*, J. Amer. Chem. Soc., 1992, **114**, 9279). Synthesis of simplified versions of the calicheamicins has been achieved and the DNA cleaving properties mimicked. Several naphthalene structures were generated in the course of this work (K.C. Nicolaou *et al.*, J.Amer. Chem. Soc., 1992, **114**, 7360).

A more specific example of naphthalene formation is illustrated in the efficient ring closure of **102**, via **103**,to give **104** in 72% yield (J.W. Grissom and T.L. Calkins, Tetrahedron Letters, 1992, **33**, 2315). Tetralin has also been formed from cyclodeca-1,5-diyn-3-ene by the Bergman cyclisation (M.F. Semmelhack, T. Neu and F. Foubelo, Tetrahedron Letters, 1992, **33**, 3277).

(d)   Other Didehydronaphthalenes

Other cycloaromatisations, have been discovered by analogy with the Bergman cyclisation. Bergman's group have generated *p*-benzyne **106** from **105** which has an additional adjacent enyne system, and this leads to 2,6-didehydronaphthalene **107**. In the presence of deuterated solvent, deuterium appears at the 2,6-positions in the final product proving the mechanism (K.N. Bharucha *et al.*, J.Amer. Chem. Soc., 1992, **114**, 3120). However, the yield in this reaction is low (10%).

105   **106**   **107**

Finally a very efficient generation of 1,5-didehydronaphthalene **109** forming the naphthalene **110** in 85% yield, after hydrogen abstraction from the solvent, has been documented (A.G. Myers and N.S. Finney J. Amer. Chem. Soc., 1992, **114**, 10986; A.G. Myers and P.S. Dragovich, *ibid.*, 1993, **115**, 7021). The precursor 1,6-dihydro[10]annulene **108** appears to be "aromatic" on the basis of its ^{1}H NMR, but is stable only at low temperatures. It rapidly rearranges as shown. The naturally occurring toxic antibiotic neocarzinostatin has a very similar structure to **108**, and undergoes the same type of rearrangement. It is indeed interesting that the Bergman cycloaromatisation was discovered prior to the isolation of these natural substances, and this reactivity has engendered a ready understanding of their toxicity.

108   **109**   **110**

5. Formation of the Naphthalene Nucleus

Synthesis of the naphthalene system is obviously an enormous subject and has been addressed in previous volunes. Highlights of particular importance or interest are covered in this section.

(a) Cyclisation of benzene derivatives

(i) Aryne mediated ring formation

o-Benzyne mediated cyclisation was discussed previously (Suppl. to the 2nd edition IIIG pages 188, 189) and the use of this versatile, high reactivity intermediate shows no sign of diminishing. The cyclisation process can take place by a nucleophilic attack on an aryne such as **112** as in the example below (**111-113**). Note that the bromo function does not need to be *ortho* to the ring forming chain for the reaction to succeed. This could be of importance if the required aromatic substitution pattern is inaccessible.

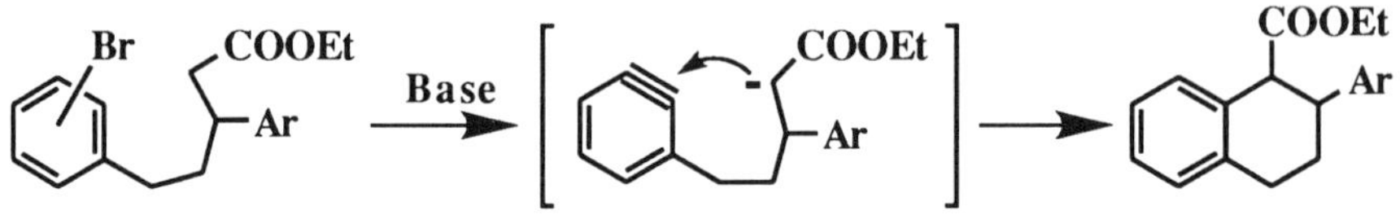

Arynes can also participate in a 4+2 cycloaddition with dienes (see section 4). Furans are often employed as the diene component and new methods for removal of the oxygen bridge have been cited already. It is also possible to retain the bridge oxygen in the form of the naphthol (see next section). For further information on the aryne cyclisation methods and related reactions see E.R. Biehl and S.P. Khanapure, Acc. Chem. Res., 1989,, **22**, 275, R.H. Levin in "Reactive Intermediates" Vol. 3, pp. 1-4, eds., M. Jones Jnr., and R.A. Moss, John Wiley & Sons, N. York 1985.

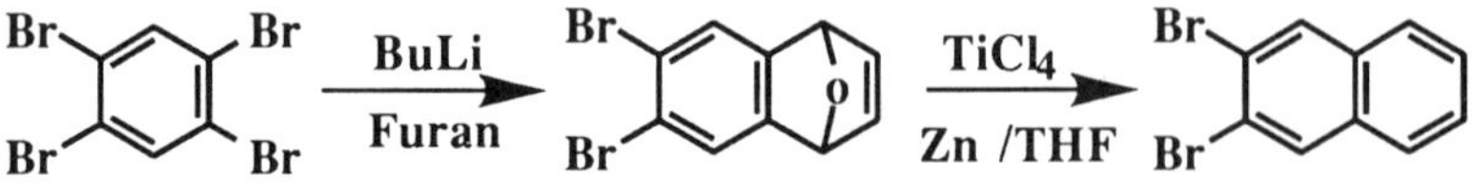

The above reaction **114-116** is taken from H. Hart *et al.*, Tetrahedron, 1986, **42**, 1641. Note that in this example the sequence can be repeated using the remaining bromo groups in **116**, to give an anthracene system.

(ii) Traditional cyclisation methods

Bradsher has reviewed the area of six-membered carbocycle construction from aldehydes and ketones, illustrating that this approach is far from outmoded (C.K. Bradsher, Chem. Rev., 1987, **87**, 1278). Acid catalysed ring closure of molecules of types **117-119** was discussed, and there is a specific section about naphthalene synthesis.

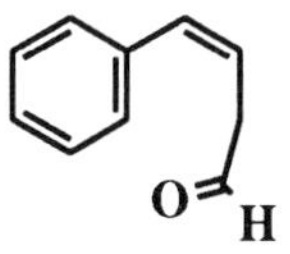

117 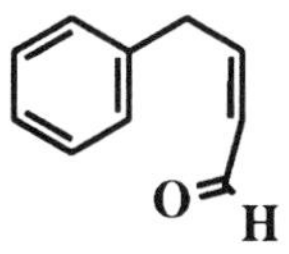   **118** 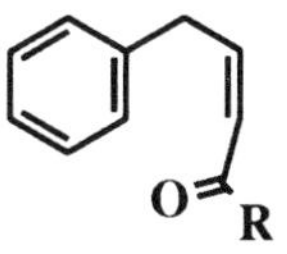   **119**

(b) Diels-Alder reactions with o-xylylenes (o-quiniodimethanes)

Several areas of this large field have been reviewed in the period of this treatise. A major work is that dealing with the useful diene isobenzofuran **121** (R. Rodrigo, Tetrahedron, 1988, **44**, 2093). An example of its formation from 2-acylbenzyl alcohols **120** is illustrated.

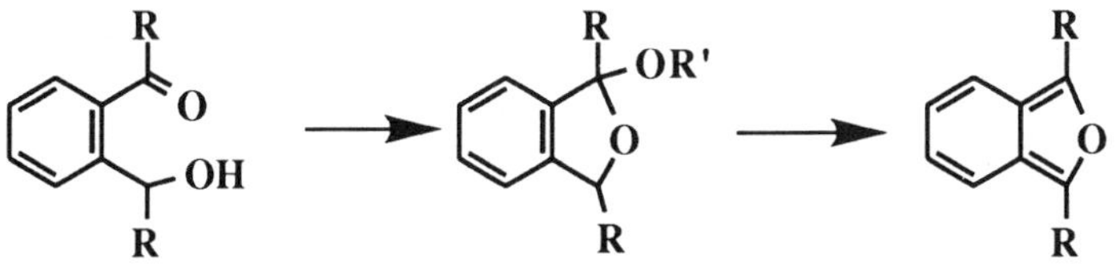

120   **120**   **121**

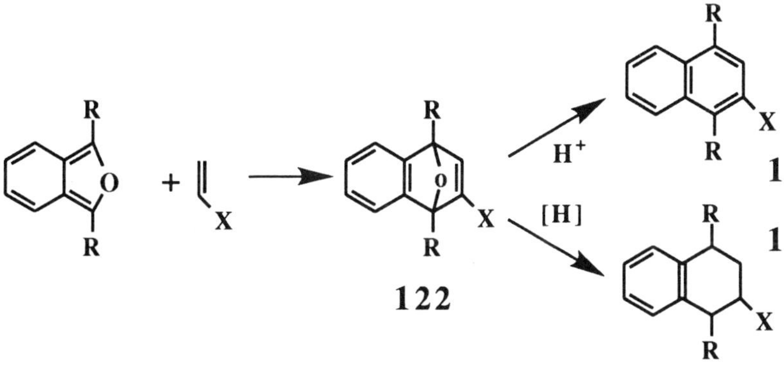

122

123

124

This is a very large and comprehensive report covering all aspects, including synthesis, use in formation of Diels-Alder adducts **122,** and conversion of these adducts to naphthalenes **123** and tetralins **124.**

Naphthalene 'hydrates' and naphthols are also possible products from this approach. The easiest method for obtaining the latter is by use of an acetylenic dienophile with the isobenzofuran. This field, and other Diels-Alder methodologies, are also the subject of an excellent book (F. Fringuelli and A. Taticchi, "Dienes in the Diels-Alder Reaction" John Wiley & Sons Inc., New York 1990). This work alludes to other ways of obtaining *o*-xylylenes (see also Suppl. to the 2nd edition IIIG pages 189-192). Two literature examples, dated after these comprehensive reviews, serve as illustrations.

The usual way of generating *o*-xylylene **126** itself is by thermolysis of benzocyclobutene **125.** Diene **126** is then trapped by the dieneophile to form tetralin **127** (K. Kobayashi *et al.*, Perkin Transactions 1, 1992, 3111)

A transition metal mediated or catalysed cyclisation of dialkynes **128** gives a metal stabilised *o*-xylylene type species **129** which can then react with a dienophile (J. Inanaga, Y. Sugimoto and T. Hanamoto, Tetrahedron Letters, 1992, **33**, 7035; S. Torii, H. Okumoto and A. Nishimura, *ibid.*, 1991, **32**, 4167). This reaction type, reminiscent of the Bergman cyclisation, was also discussed in Suppl. to the 2nd edition IIIA , pages 47-49.

A common variation on this theme is the metal complex mediated trimerisation of alkynes to form aromatic derivatives (C. Bianchini *et al.*, J. Amer. Chem. Soc., 1991, **113**, 5127; E. Lindner and H. Kühbauch, J. Organomet. Chem., 1991, **403**, C9), and this method can easily be adapted to give a useful synthesis of a large number of naphthalene systems eg., **131** from **130** (P. Bhatarah and E.H. Smith, J. Chem. Soc. Chem. Comm., 1991, 277).

130   **131**

(c)   Other Diels-Alder reactions

The timely monograph by Fringuelli and Taticchi (*loc. cit.*) covers reactions of dienes with benzoquinones **132** and cyclohexenones **134** to give adducts **133** and **135** respectively. The adducts can be easily aromatised to naphthoquinones and naphthalenes.

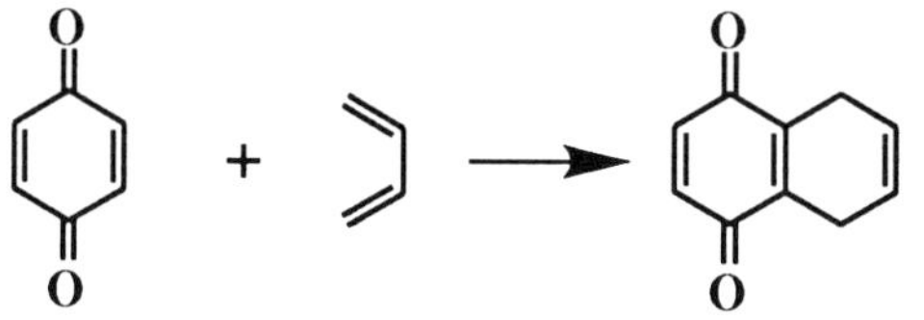

132   **133**

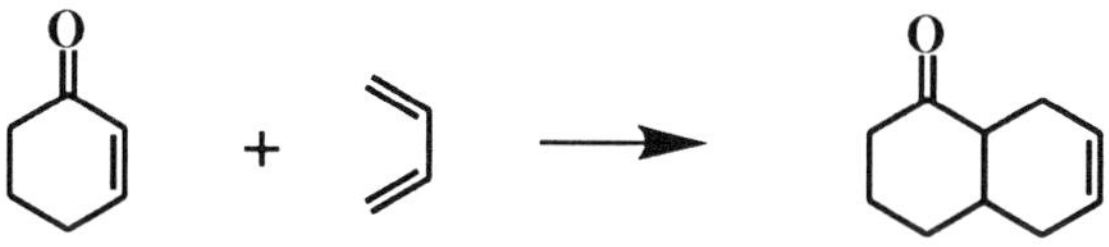

134   **135**

A useful variant of these reactions, is the oxidation of a phenol **136** to quinone hemiketal **137**, whereupon an intramolecular cycloaddition forms the naphthalene skeleton **138**. Other analogous reactions where the dienophile is an *o*-benzyne are known (see W.M. Best and D. Wege, Aust. J. Chem., 1986, **39**, 635, 647).

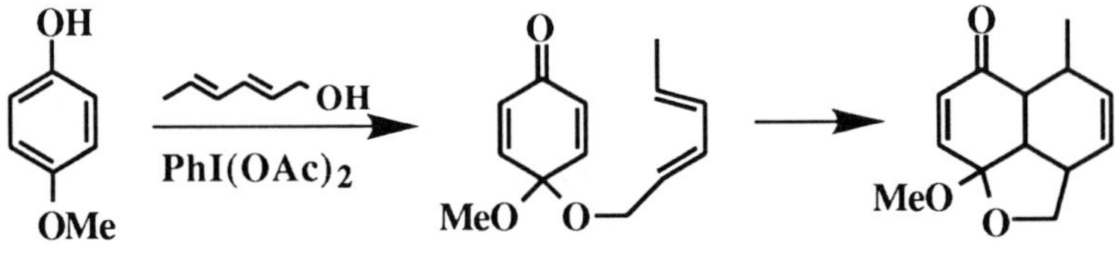

136 **137** **138**

Dienes of the vinylcyclohexene type **139** (T.R. Hoye and M.J. Rother, J. Org. Chem., 1979, **44**, 458) and certain styrenes **141** (Y. Kita et al., Tetarhedron Letters, 1984, **25**, 1813) are also useful precursors to the hydronaphthalenes **140** and **143** respectively. In the latter case the substrate must be highly activated such that an 'aromatic' ring double bond participates in the reaction giving the initial adduct **142** which can be O-desilylated and acetylated in "follow-up" steps.

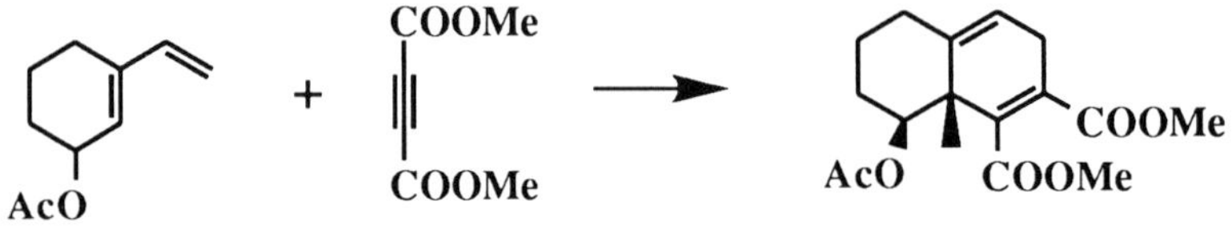

139 **140**

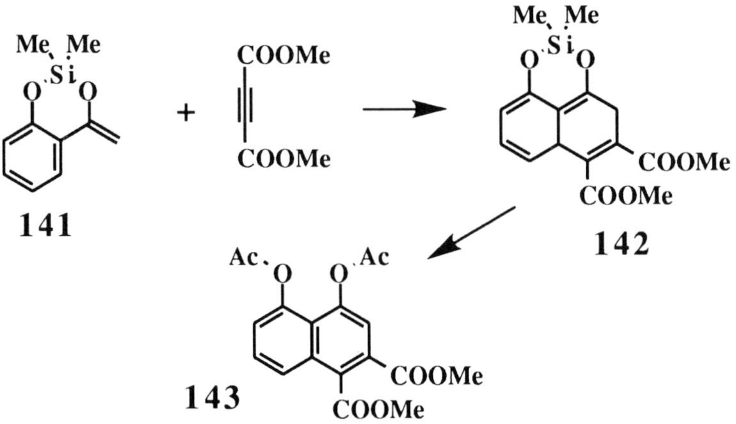

141 **142**

143

(d)   *Other cyclisation reactions*

Several new electrophilic aromatic cyclisations have been explored, leading to dihydro- or tetrahydro- naphthalenes **145**. Aryliodonium salts **144** react rapidly at 60°C in the presence of boron trifluoride etherate (M. Ochiai *et al.*, J. Chem. Soc. Chem. Comm., 1986, 1382). Analogous reactions with positively charged sulphur or selenium as activating groups are known (E.D. Edstrom and T. Livinghouse, J. Chem. Soc. Chem. Comm., 1986, 279; *idem.*, J. Amer. Chem. Soc., 1986, **108**, 1334; *idem*, Tetrahedron Letters, 1986, **27**, 3483. The full potential of these reactions remains to be investigated.

Cyclisations proposed to take place via a ketene intermediate are of interest but are quite idiosyncratic. The example depicted **145-147** involves a Wolff rearrangement where the aryl group migrates, followed by ring closure on an adjacent alkyne moiety (A. Padwa *et al.*, Tetrahedron Letters, 1991, **32**, 5923).

A similar remarkable reaction is the ring closure of ketadiene **150** to a relatively unactivated benzene ring to form naphthalenediol **151** (S.T. Perri and H.W. Moore, Tetrahedron Letters, 1987, **28**, 4507; K. Chow and H.W. Moore, *ibid.*, 1987, **28**, 5013; Y.-S. Chung *et al.*, J. Org. Chem., 1987, **52**, 1284; L.S. Lieberskind, S. Iyer and C.F. Jewell, *ibid.*, 1986, **51**, 3065; S.T. Perri *et al.*, *ibid.*, 1986, **51**, 3067). The ketadiene **150** can be generated the cyclobutene **149** which is accessible by a Grignard reaction on cyclobutendione **148**. The naphthoquinone can be obtained by oxidation of the diol **151**.

Another naphthalene synthesis by small ring cleavage has merit in the availability of the starting material, the cyclopropane **153**, which can be obtained by dihalocarbene addition to the allylic alcohol **152** (S. Seko, Y. Tanabe and G. Suzukamo, Tetrahedron Letters, 1990, **31**, 6883). The stability of the cationic intermediates **154** determines which bond of the cyclopropane system cleaves. When R_1=alkyl **155** is obtained exclusively, but when R_1= aryl, only **156** is formed. This reaction was used in the synthesis of two naturally occurring lignans, Justicidin E and Taiwanin C.

Finally there are a great many reactions usually involving alkyne coordination with a transition metal and insertion of carbon monoxide and/or carbene. An introductory example is the formation of the dihydronaphthalene lactone **158** from **157** (G. Wu, I. Shimoyama and E.-i. Negishi, J. Org. Chem., 1991, **56**, 6506; for similar reactions see W. Oppolzer, J.-Z. Xu and C. Stone, Helv. Chim. Acta., 1991, **74**, 465; T. Mandai *et al.*, Tetrahedron Letters, 1991, **32**, 7687).

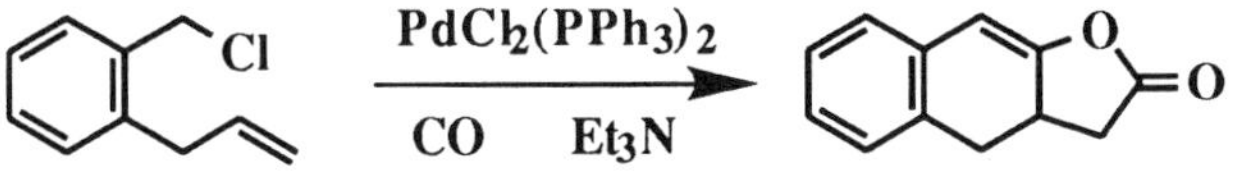

157   **158**

Other examples leading to naphthoquinones **160** from carbene complex **159** (K.H. Dötz and V. Leue, J. Organomet. Chem., 1991, **407**, 337, and naphthols **162** from **161** (C.A. Merlic and D. Xu, J, Amer. Chem. Soc., 1991, **113**, 7418) are illustrated below. For a review of this process referred to as Dötz cyclisation see D.B. Grotjahn and K.H. Dötz, Synlett, 1991, 381.

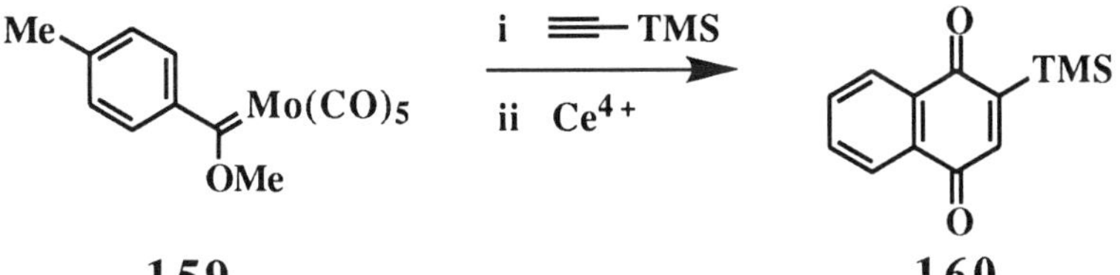

159   **160**

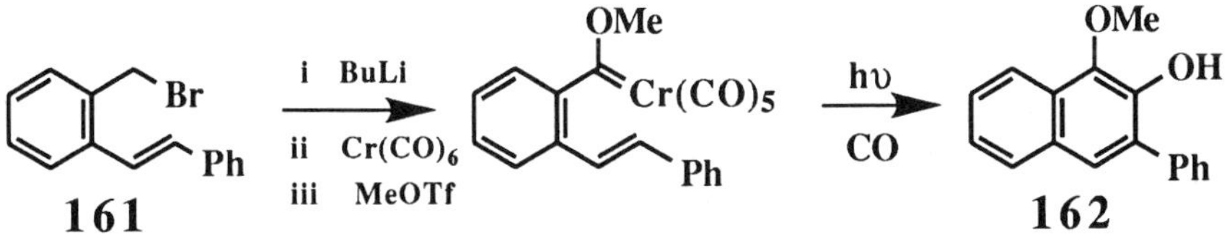

161   **162**

6   Reduction and Oxidation of Naphthalenes

(a)   Reduction

Very little of great significance has appeared in this field. There have been some improvements in catalytic hydrogenations making tetralins easily available, but perhaps the most interesting reaction is the reduction with potassium graphite intercalate (C_8K) in tetrahydrofuran at 0°C (I.S. Weitz and M. Rabinovitz, J. Chem. Soc. Perkin Transactions 1, 1993, 117). The products are like those derived from the Birch reduction (2nd edtion Vol IIIG, page 132) but the C_8K technique may be ultimately more convenient. In the example shown (**163-165**) the relative amounts of the two products can be controlled to some extent by the reaction conditions. The cited paper gives leading references.

| 163 | 164 | 165 |

(b)   Oxidation

(i)   Oxidation to quinones

Simple substututed naphthalenes on oxidation with a variety of reagents form the 1,4-naphthoquinone **166** (R=H or electron donating group) (indexed in Chemical Abstracts as naphthalenediones). In the reaction shown the yield is quatitative with ceric ammonium sulphate (M.V. Bhatt and M. Periasamy, J. Chem. Soc. Perkin Transactions 2, 1993, 1811). This paper gives a full mechanistic study.

166

The oxidation can be carried out electrolytically with or without assistance from an oxidant like dichromate (S. Ito *et al.*, J. Appl. Electrochem., 1993, **23**, 677). 2-Methylnaphthalene gives 2-methyl-1,4-naphthoquinone in this dichromate assisted method (S. Chocron and M. Michman, J. Molec. Catalysis, 1993, **83**, 251).

The quinone bis- and monoketals, which are useful in synthesis of anthracyclinones, are easily prepared by anodic oxidation of methoxynaphthalenes (**167-169**), and in some cases a monosubstituted naphthalene can be further substituted and/or oxidised to a quinone (*cf.*, next page). This field has been reviewed (J.S. Swenton, Acc. Chem. Res., 1983, **16**, 74).

The formation of 1H-cyclopropa[b]naphthalene-3,6-dione **171** from the compound **170** provides the first example of this type of tricyclic quinone (B. Halton, A.J. Kay and Z. Zhi-mei, J. Chem. Soc. Perkin Transactions 1, 1993, 2239).

Oxidation of suitable substrates to produce the labile 2,3-quinone have been studied. The diol **172**, where the 1- and 4- positions are blocked by aryl substituents, is transformed to the 2,3-quinone **173** by lead tetraacetatae (D.W. Jones and A. Pomfret, J. Chem. Soc. Chem. Comm., 1983, 703).

The 2,3-quinones, of type **173**, are very powerful oxidants, more powerful at hydride abstraction than *o*-chloranil, and have been characterised by trapping with norbornadiene (D.W. Jones and F.M. Nongrum, J. Chem. Soc. Perkin Transactions 1, 1990, 3357).

Substitution of the naphthalene nucleus by oxygenated functions has been touched on in the anodic oxidation to form **168**. A similar reaction, promoted by argentic ion, with dicyclohexylperoxy dicarbonate as oxidant gives 1-acetoxynaphthalene (F. Maspero and U. Romano, Gazz. Chim. Italiana, 1993, **123**, 177). Easily oxidised substrates such as 2,7-dimethoxy-naphthalene **174** can be converted to several useful derivatives (**175-178**) by an analogous manganese(III) oxidative system (H. Nishino, K. Tsunoda and K. Kurosawa, Bull. Chem. Soc. Jpn., 1989, **62**, 545).

(ii) *Oxidation leading to ring cleavage or substituent oxidation*

This subject has been well covered in previous volumes (*loc. cit.*) and only a few recent developments are included here. The oxidation of 1,2-dimethylnaphthalene **179** by ceric ion gives the surprisingly selective conversion of the 1-methyl group to the aldehyde **180**. The other isomer **181** is only a minor byproduct (L.K. Sydnes *et al.*, Tetrahedron, 1985, **41**, 5205). Since the most encumbered position is oxidised, the reaction is electronically driven, the positively charged intermediates being more stable when in the 1-position due to more effective charge delocalisation.

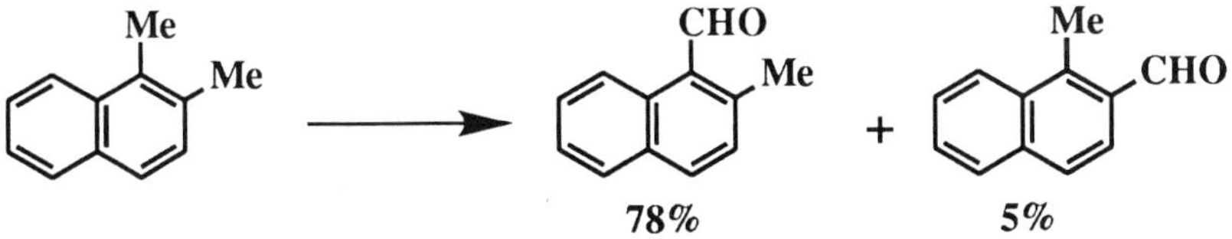

The degradative oxidation of 1,4-dimethylnaphthalene seems to reflect the higher double bond character of the 1,2- and 3,4 -bonds. The similar oxidation of the 6-acetyl derivative **182** to 1,2,5-triacetoxybenzene **183** is as shown below (R. Riemschneider and T. Wons, Monatsh. Chem. 1983, **114**, 1267).

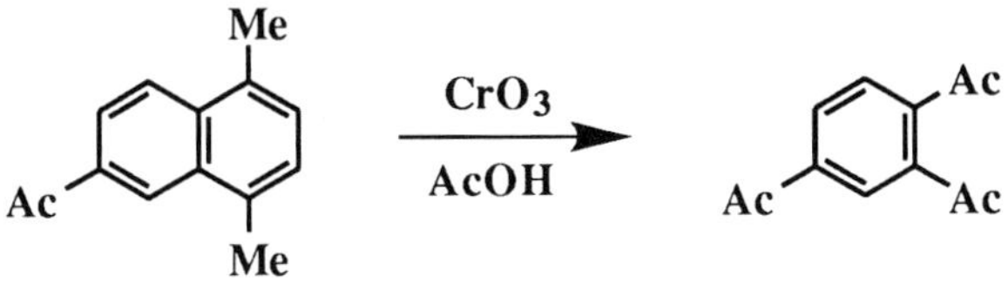

The cyanoanthracene sensitised photooxygenation of 2-methyl-naphthalene in the presence of tetraethylammonium acetate gives phthalic acid and 4-methylphthalic acid (T. Yamashita *et al.*, Bull. Soc. Chem. Jpn., 1993, **66**, 857). The mechanisms of these reactions are unclear, but conditions can be adjusted to produce also 2-methyl-1,4-naphthoquinone or naphthalene-2-carboxaldehyde.

(iii) Naphthalene Oxides, Naphthalenols(naphthols) and Related Compounds
    Oxides of naphthalene were covered in depth previously (Suppl. to the 2nd edition, pages 195-205), and since then there has been a comprehensive review of arene oxides (G.S. Shirwaiker and M.V. Bhatt, Adv. in Heterocyclic Chem., 1984, **37**, 65). A major impetus in this field is the biological metabolism of naphthalene and other polycyclic arenes, since this is the mechanism whereby they are activated as cancer causing agents. Mammalian and bacterial oxidation is currently under intensive study and naphthalene is often used as the model substance for these investigations. Naphthalene derivatives, while toxic, are not usually very active carcinogens, and can be used reasonably safely to explore chemistry and biochemistry, knowledge which can then be applied to much more carcinogenic systems such as benzo[a]pyrene, benzanthracene etc..

Naphthalene-1,2-oxide **184** appears to be a first metabolite formed by the cytochrome P450 liver enzyme systems. The enzymic oxidation is normally highly stereospecific. Several products (185-187)can then be formed form this; usually an increase in water solubility is induced which allows the naphthalene derived species to be cleared from the body. Naphthalene has been used as a probe of the various enzymic activities illustrated in the diagram (see for example C.H. Chichester *et al.*, Molecular Pharmacology, 1994, **45**, 664; M.D. Tingle *et al.*, Biochem. Pharmacol., 1993, **46**, 1529).

In highly carcinogenic aromatic polycycles the analogous species to **187**, the diol-epoxides, seem to be the main cancer causing agents. The mechanism involves intercalation of the DNA and covalent bond formation with a DNA base via epoxide ring opening. Assessment of the various mammalian enzyme actions is therefore important as is the metabolic actions of microorganisms on arenes, in view of the ubiquitous nature of these substances in the environment.

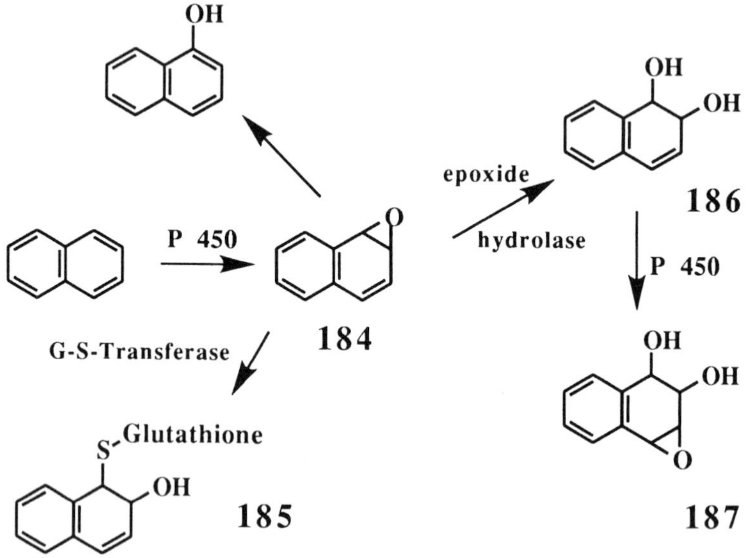

For details of this complex field see R.G. Harvey, "Polycyclic Aromatic Hydrocarbons, Chemistry and Carcinogenicity", Cambridge University Press, Cambridge 1991, and P. Garrigues and M. Lamotte, " Polycyclic Aromatic Compounds, Synthesis, Properties, Analytical Measurements, Occurrence and Biological Effects" Gordon and Breach Science Publishers, UK, USA 1993.

A comparison of the rates of acid-catalysed aromatization of naphthalene 'hydrates' and naphthalene oxides, as depicted in the scheme below, and other arene cogeners, has shown that the 'hydrates' aromatize faster. It is suggested that this surprising effect is due to homoaromatic stabilisation in the oxides (S.N. Rao *et al.*, J. Amer. Chem. Soc., 1993, **115**, 5458).

As discussed previously (Suppl. to the 2nd edition IIIG page 196) the Chemical Abstracts name for naphthalene-1,2-oxide is 1a,7b-dihydronaphth-[1,2-b]oxirene. This is derived from the fully aromatic system naphth[1,2-b]-oxirene **194**. This ring system is unknown as it is isoelectronic with the highly reactive ketocarbenes **190** and **191**. These species are intermediates in the Wolff rearrangement (Suppl. to the 2nd edition IIIG page 196) which has been further studied as in the diagram. The ^{13}C labels in the precursor diazoxides **188** and **189** are shown as an asterisk, and there is no scrambling of this during the thermal or photochemical rearrangement to the ketene(s) **192** (and **193**), (A. Blocker and K.-P. Zeller, Chem. Ber., 1994, **127**, 551). Even the transient formation of **194** is therefore unlikely.

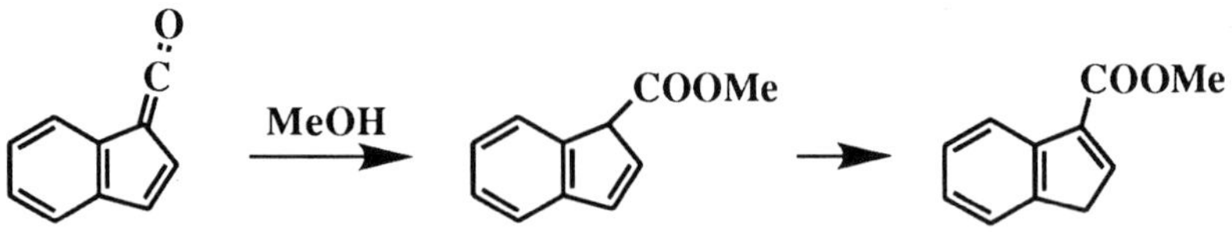

192   **195**   **196**

The ketene **192** reacts with methanol to give the ester **195** which isomerises to **196** (J. Andraos, A.J. Kresge and V.V. Popik, J. Amer. Chem. Soc., 1994, **116**, 961).

The 2,3-ketocarbene, similarly labelled, formed by photolysis, also shows no scattering of the label and hence the intermediacy of naphth[2,3-b]-oxirene is also discounted (A. Blocker and K.-P. Zeller, *loc. cit.*).

The so-called 'NIH Shift', where a hydrogen isotope label in a phenol shifts to an adjacent position has been portrayed previously, and occurs through an arene oxide and/or keto-enol mechanism (G.S. Shirwaiker and M.V. Bhatt,.*loc. cit*).

There is another possible tautomeric equilibrium between arene oxide and oxepine forms where the oxygen atom can move round the ring. This is referred to as the 'oxygen walk'. These processes allow for racemisation of the chiral oxides and there is considerable interest in this area (G.S. Shirwaiker and M.V. Bhatt, Adv. in Heterocyclic Chem., 1984, **37**, 106-125;). The migration of oxygen has been observed in several systems including triphenylene-1,2-oxide **197**. The **197** and **199** tautomers are favoured over **198** and **200**, as the delocalisation of the phenanthrene part is maintained (D.R. Boyd et al., J. Chem. Soc. Perkin Transactions 1, 1987, 369).

The energy barrier to oxygen walk in naphthalene-1,2-oxide is higher than in the triphenylene example, (D.R. Boyd *et al.*, J. Chem. Soc. Chem. Comm., 1987, 1633). In contrast to the naphthalene-1,2-oxide case a 'nitrogen walk' in the aziridine analogue has been observed (K. Satake *et al.*, J. Chem. Soc. Chem. Comm., 1987, 197). Boyd and his collaborators have reviewed these results and descibed standard methods for the chemical synthesis of arene oxides from cis-diols (D.R. Boyd *et al.*, J. Chem. Soc. Chem. Comm., 1994, 1693)

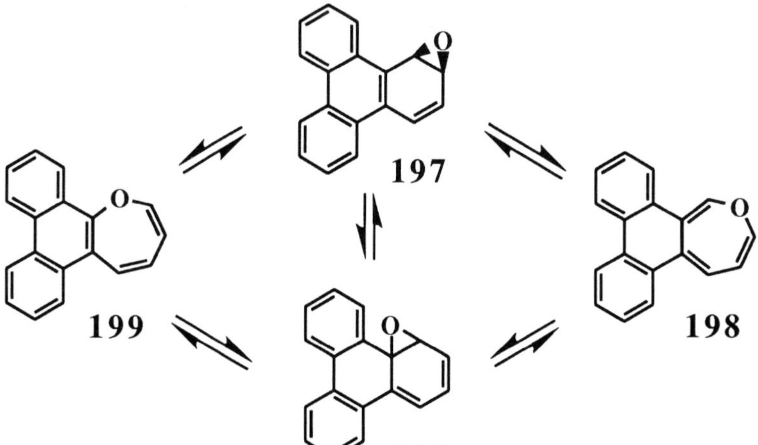
O
O
O
O

197 198 199 200

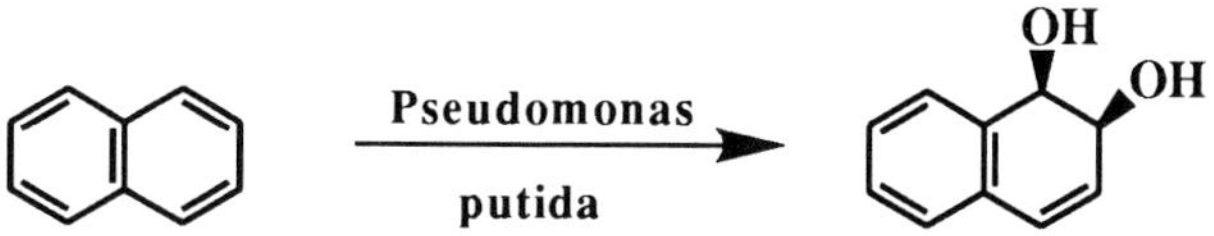
Pseudomonas
putida
OH
OH

201 202

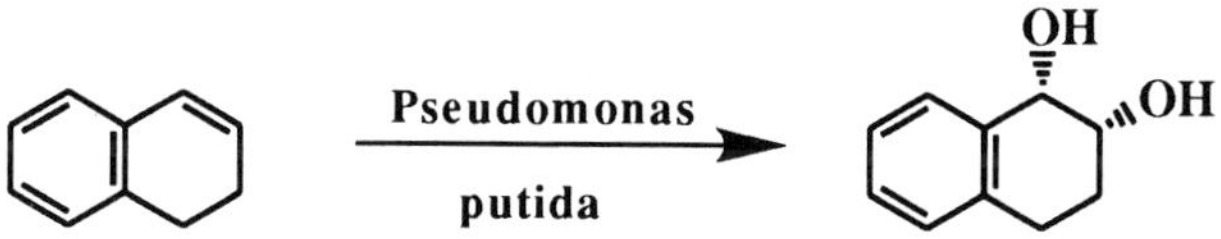
Pseudomonas
putida
OH
OH

203 204

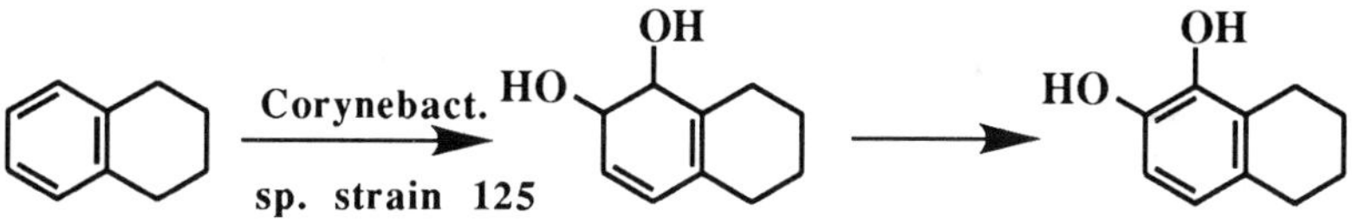
Corynebact.
sp. strain 125
HO
OH
HO
OH

205 206 207

This last citation also gives a lead into the vast literature concerning the microbial (*Pseudomonas putida* UV4) oxidation of naphthalenes 201, 203 and other ring systems in order to obtain the chiral cis-diols 202, 204. (See also D.R. Boyd et al., J. Chem. Soc. Chem. Comm., 1989, 339; M.E. Deluca and T. Hudlicky, Tetrahedron Letters, 1990, 31, 13). A microbe (Corynebacterium sp strain-C125), specially selected for its ability to grow on tetralin 205, can perform preferential hydroxylation of the aromatic ring to give the unusual tetralin 207 via 206 (J. Sikkema and J.A.M. Debont, Appl. and Envir. Microbiology, 1993, 59, 567).

Singlet oxygen attack on naphthalene to form endoperoxides 209 and 210 by 4+2 cycloaddition, has been shown to occur asymmetrically when chiral substituent groups are present. The chiral compounds 208 were synthesised from 1-bromo-4-methylnaphthalene. The asymmetric centre is denoted by an asterisk and a series of these substances, of varying "R", were examined. When R=Br or TMS, very high diastereoselectivities (d.r. > 90%) were achieved, whereas much lower discrimination was observed for R=tert-butyl or OR' (d.r.= 60% approx.). The reaction selectivity is therefore mainly determined by electronic rather than steric factors (W. Adam and M. Prein, Tetrahedron Letters, 1994, 35, 4331).

| 208 | 209 | 210 |

7   *Small Ring Annelated Naphthalenes*

These ring systems can often be converted to benzocycloheptanes and benzocyclooctanes and are included in these sections where appropriate.

(a)   *Naphthocyclobutanes*

A relative newcomer to this scene is 1H-cyclobuta[d,e]naphthalene 216, a colourless distillable liquid. The bromo compound 212 is a white solid m.p. 103°C. The synthesis of this strained ring system is as shown, making it much more accessible than the earlier synthesis. The best yield (43%) is obtained by the photolysis of 211, but the accessiblity of the diazo compound or the bromonaphthaldehyde tosylhyrazone salt is a limitation. This makes the other method from **213** via **214** and **215** attractive in spite of final stage yields of 20%. The lithium derivative **215** or the Grignard reagent from **212** also give access to the carboxylic acid, alkyl derivatives etc., (L.S. Yang, T.A. Engler and H. Schechter, J. Chem. Soc. Chem. Comm., 1983, 866; R.J. Bailey, P.J. Card and H. Schechter, J. Amer. Chem. Soc., 1983, **105**, 6096).

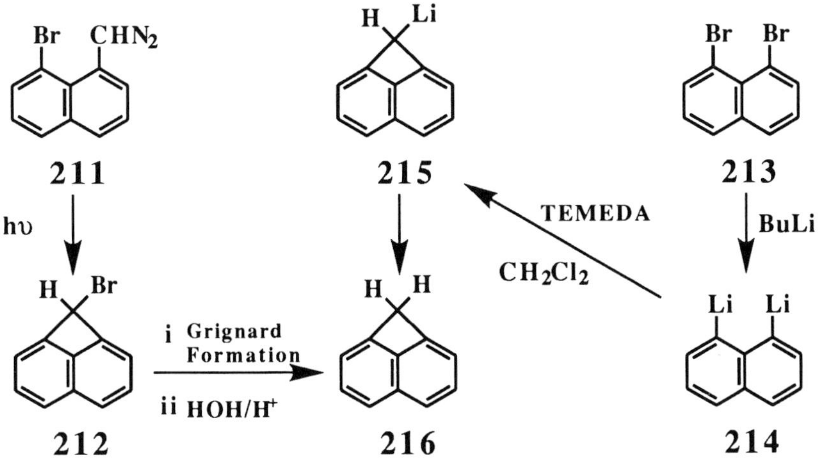

The chemistry of this strained system is dominated by ring opening reactions as shown below **212-217, 218-219,** but the bromo compound undergoes some normal nucleophilic substitutions without ring cleavage (P.J. Card, F.E. Friedli and H. Schechter, J. Amer. Chem. Soc., 1983, **105**, 6104).

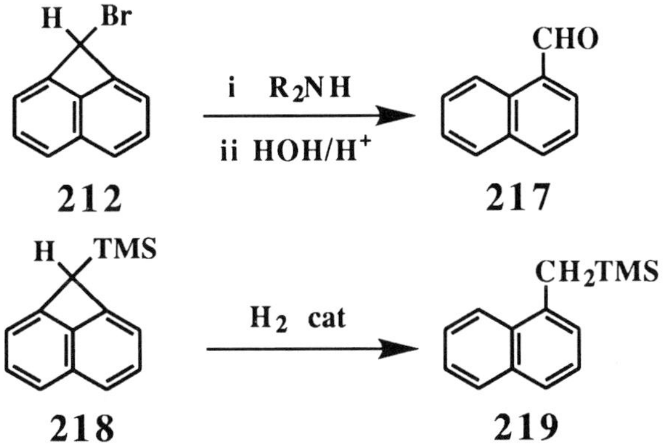

A number of cyclobuta[b]naphthalenes eg. **222** have been reported. A simple entry to this system is the photochemical addition of an alkene to a naphthoquinone **220**. The re-establishment of the quinoid system by elimination of a suitable leaving group from **221** is shown (T. Naito, Y. Makita and C. Kaneko, Chemistry Letters, 1984, 921).

220   **221**   **222**

Another approach utilises conventional electrophilic ring closure reactions to benzocyclobutene derivatives, **223-225**. The cyclobutanaphthalene was then used to generate the *o*-xylylene (*cf.*, previous section 5(*b*)) as an intermediate in the construction of anthracyclinones (T. Watabe and M. Oda, Chemistry Letters, 1984, 1791).

223   **224**   **225**

For an example of photchemical 2+2 cycloaddition of alkenes to the 1,2-bond of naphthalene to form cyclobuta[a]naphthalenes (see N.A. Aljalal, Gazz. Chim. Italiana, 1994, **124**, 205).

The formation of cyclobutanaphthalenes has shown the transient existence of the isonaphthalene species **227**. The intermediate, generated from **226**, was trapped by reaction with styrene in a 2+2 cycloaddition to produce **228** and **229** in a ratio of 14:1 (M. Christl, M. Braun and G. Müller, Angew. Chem. Internat. Edn., 1992, **31**, 473). The difference in reactivity of the two cyclic allene bonds is remarkable and a theoretical study of analogous systems to the isonaphthalene structure has been attempted (R. Janoschek, Angew. Chem. Internat. Edn., 1992, **31**, 476). A similar reaction sequence, where the initial dihalocarbene addition is performed on a 1,2-dihydronaphthalene, yields mainly dimers of benzocyclohepta-1,2-diene on treatment with organolithium via this less strained cyclic allene (H. Jelinek-Fink et al., Chem. Ber., 1991, **124**, 2569).

(b)   *Naphthocyclopropanes*

The elegant but convoluted synthesis of 1H-cyclopropa[a]naphthalene was reported previously (Suppl. to the 2nd edition IIIG page 225,226). Another more straightforward synthesis of the 1,1-difluoro derivative **233** has been performed based on the difluorocarbene addition to 1,2-dihydronaphthalene **230** to give adduct **231**. This was then brominated and the product **232** treated with base to aromatize the ring (P. Müller and H.C. Nguyen-Thi, Helv. Chim. Acta, 1984, **67**, 467)

The fluoro groups in **233** are solvolysed by acidic methanol to a mixture of the 1- and 2- methoxycarbonyl-naphthalenes **234** and **235**.

The more accessible 1H-cyclopropa[b]naphthalene **236** was also reported previously, and there has now been an extensive research into its chemistry. Ring cleavage by reactive dienophiles such as 4-phenyl-1,2,4-triazoline-3,5-dione, tetracyanoethylene or benzyne is observed **237-239**. The last reaction was complex but a low yield of 2,3-benzofluorene **239** was obtained (I. Durucasu, N. Saraçoglu and M. Balci, Tetrahedron Letters, 1991, **32**, 7097). The insertion reaction with dichlorocarbene gave the cyclobuta[b]-naphthalene derivative **240**.

8 *Phenalene, Phenalenone and Related Compounds*

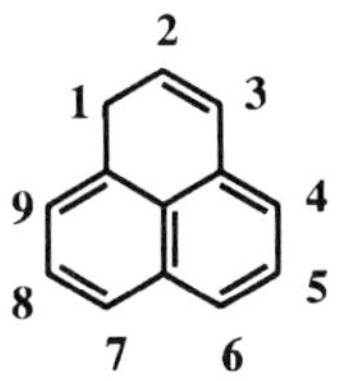

1

The structure of phenalene is as illustrated. However, only in substituted or labelled phenalenes is the phenomenon of tautomerism apparent in this system. F. Tort *et al.,* (Mag. Resonance. in Chem., 1992, **30**, 689) has reviewed this area and synthesized 1-(2-methylindol-3-yl)phenalene **3** from phenalenone **2.**

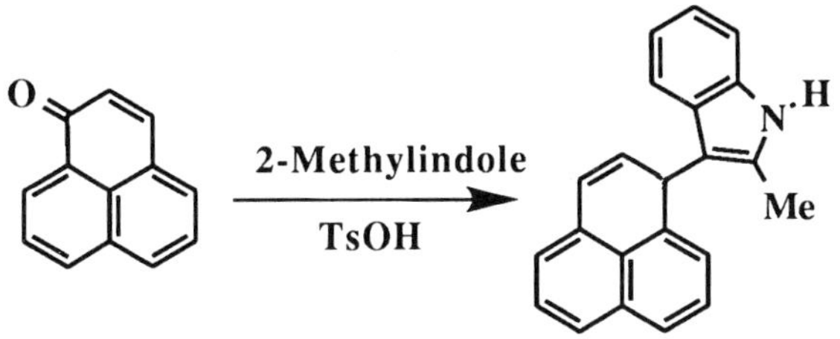

2 **3**

Five tautomeric forms were discernible for compound **3,** as shown to by ^{13}C NMR, and assigned structures as depicted below.

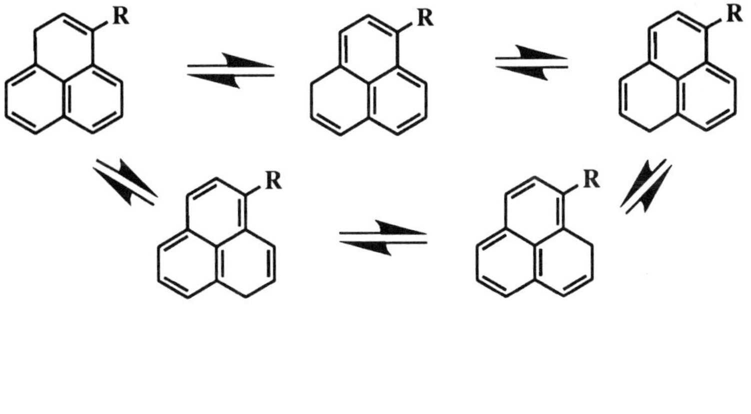

Another unusual facet of the phenalene ring system, and related to the tautomeric phenomena is the synthesis of phytoalexins (J.G. Luis *et al.*, Tetrahedron, 1994, **50**, 10963). These phenalenone derivatives seem to be a chemical defence response of certain plants to physical, chemical or microbiological attack. The Grignard reagent with phenalenone gives 1,4-addition but with participation of the 9,9a-double bond, not the 2,3-bond. Oxidation of the alcohol formed gave **4** in good yield, but on re-reduction with Dibal, the alcohol **5**, could not be isolated and rearranged to **6** and **7** (70:30).

Oxidation of this mixture gave **8** and **9** which were separated easily by chromatography. The desired compound **8** was oxidised, with *tert*-butyl hydroperoxide, to epoxide **10**, which on acidic workup, gave the phytoalexin **11**. Conventional transposition of the carbonyl group in **4** to give **11** was ineffective in this system.

This research is a fine example of the interesting chemistry of this area, and the citation gives leads to natural products which are phenalene derivatives.

Acid catalysed conventional ring closure reactions to give phenalenones gives rise to some interesting compounds (J.L. Carey and R.H. Thomson, J. Chem. Soc. Perkin Transactions 1, 1983, 1267). The phenalenone **13** arises from aldol condensation of the naphthaldehyde **12** with acetophenone followed by ring closure. The other compound **14** is formed by a complex series of reactions involving two equivalents of acetophenone one of which decarbonylates the aldehyde **14** to form benzoylacetaldehyde. This molecule reacts at the 8-position of the naphthaldehyde, the formyl group of which, continues to condense with acetophenone. Cyclisation to **14** then occurs.

Another similar conventional cyclisation has been reported for 1-cinnamoylpyrene in the formation of 5-phenyl-3H-benzo[c,d]pyren-3-one (S.I. Didenko and Y.E. Gerasimenko, Zh. Org. Khim., 1984, **20**, 213, Chem. Abstr., **100**, 191539y).

A novel entry to the reduced phenalene system **16** via a Diels-Alder reaction on **15** has been investigated with a view to demonstrating the regiospecificity and steric requirements of the reaction (E. Piers *et al.*, Can. J. Chem., 1993, **71**, 1463).

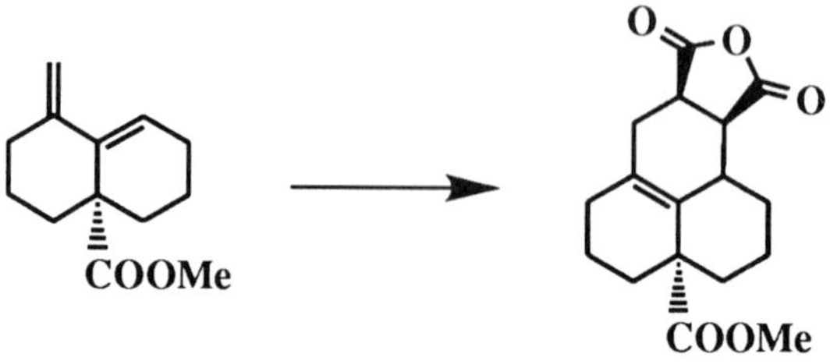

15   **16**

Cyclohepta[d,e]naphthalene-7,8-dione **19** has been made from acenaphthylene **17** (J. Tsunetsugu *et al.*, J, Chem, Soc. Perkin Transactions 1, 1984, 1465). Hydrolysis of the dichloroketene adduct of acenaphthylene **18** gives a 50% yield of **19**.

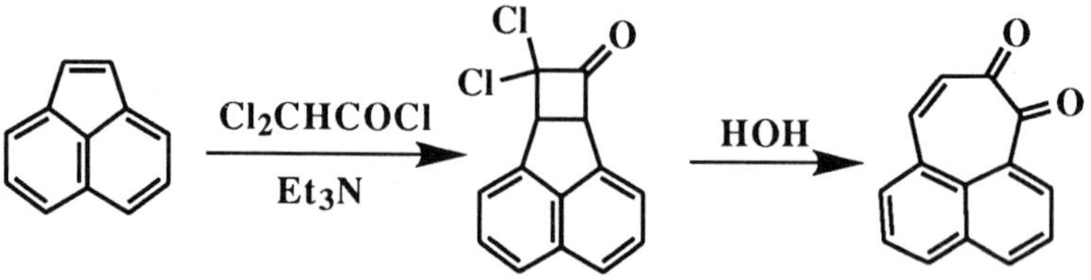

17   **18**   **19**

Structure **19** is thought to have considerable benzotropolone-type character. Alkaline hydrolysis of **19** gave 1-carboxy-1-hydroxyphenalene **20**.

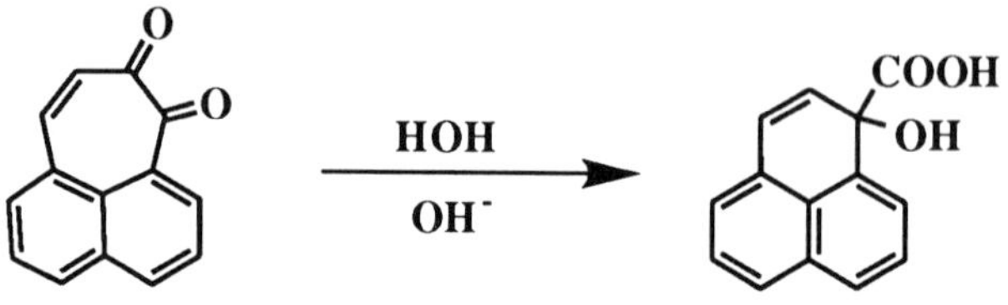

19   **20**

Finally full details of the synthesis of 1,3-dihydrophenalen-2-one **26** have been published (F.S. Jorgensen and T. Thomsen, Acta Chem. Scand., 1984, B **38**, 113). The route is shown below and the structure of **26** appears to have the carbonyl group out of the plane of the aromatic system. There is therefore very little electronic interaction between these parts of the molecule.

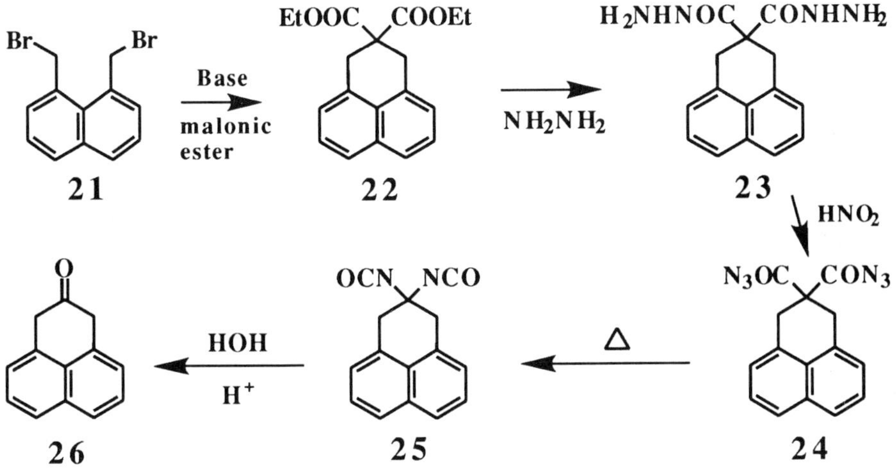

9 Azulene and Related Compounds

(a) Structure and Properties of Azulene Systems

The structure, properties and synthesis of azulenes has been reviewed since the last edition of "Rodd", (D. Lloyd, "Non-Benzenoid Conjugated Carbocyclic Compounds" Elsevier Science Publ. Amsterdam 1984). The name azulene derives from the beautiful blue or green colour of this type of molecule. A blue colour is rare in organic molecules and very rare in a hydrocarbon. For a detailed discussion of the structure of azulene **1** and MO calculations for prediction of the considerable dipole moment in this nonalternant system see S. Grimme, Chemical Physics Letters, 1993, **201**, 67. For a less quantitative description of the electronic structure and transitions giving rise to the blue colour see D.M. Lemal and G.D. Goldman, J. Chem. Educ., 1988, **65**, 923. This article gives a detailed preparation of the parent azulene suitable for teaching laboratories (see also Suppl. to the 2nd Edn IIIG 245, 246).

1

As a consequence of the intense dipole in azulene the system can stabilise an adjacent positive charge. Reaction of 1-formylazulene **2** with two molecules of azulene itself gives tris(1-azulenyl)methane **3**.

Hydride abstraction with DDQ, and counter-ion exchange, produces the extremely stable cation **4**, isolable as a blue solid, as the hexafluorophosphate salt (S. Ito, N. Morita and T. Asao, Tetrahedron Letters, 1991, **32**, 773; for an extension of these azulene stabilised cation systems see *idem, ibid.*, 1992, **33**, 3773).

Also the weak aromatic character of azulene causes instability in some derivatives such as the hydroxyazulenes. Hydrolysis of the acetate ester **5** by aqueous base gives only polymeric material but lithium tetrahydridoaluminate reduction in dry THF with mildly acidic workup at 0°C produces the green coloured desired hydroxy compound **6**.

This substance is stable at low temperatures but, on warming, was observed, by NMR, to be in equilibrium with other forms, which did not appear to be keto tautomers. It is suggested that these forms, which are temperature and solvent dependent, are charge separated species such as **7** and **8** (T. Asao, S. Ito and N. Morita, Tetrahedron Letters, 1989, **30,** 6693; see also Suppl to the 2nd edn IIIG page 254).

By a similar strategy the same investigators managed to isolate 3-hydroxy-guaiazulene **9** as a very unstable green oil which changed on standing in ether solution to the keto form **10** and the dimer **11**.

(b)   Quinones of Azulenes (Azuloquinones)

Stability problems also attend the quinones of azulene (azuloquinones), and a comprehensive treatment of all possible isomers with regard to stability and formation has been published (T. Kurihara, S. Ishikawa and T. Nozoe, Bull. Chem. Soc. Jpn., 1992, **65**, 1151). This paper gives leading references to azulenes with quinoid derivative structures including some natural products. In regard to the the the parent compounds however, only the 1,5-quinone **12**, and 1,7-quinone **13** appear to be isolable as stable yellow crystalline solids both of m.p. 100°C [dec] (L.T. Scott and C.M. Adams, J. Amer. Chem. Soc., 1984, **106**, 4857). In contrast the very reactive 1,4- and 1,6-azuloquinones **14** and **15** have been generated and trapped as their Diels-Alder adducts with cyclopentadiene (L.T. Scott, P. Grütter and R.E. Chamberlain, III, *ibid.*, 1984, **106**, 4852).

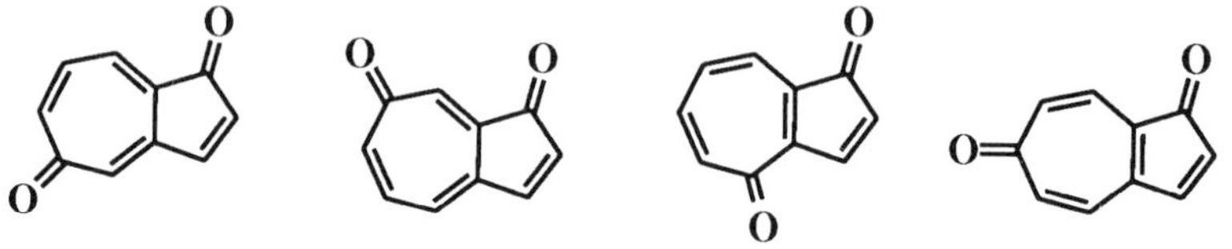

12   **13**   **14**   **15**

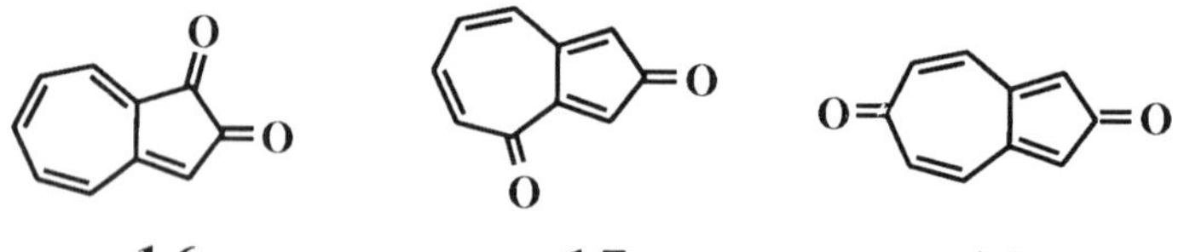

16   **17**   **18**

Azuloquinone structures with a tropone ring are more stable than those with a cyclopentadienone unit, and those with both these features are intermediate in stability. Therefore the structures **12-18** are arranged in order of stability (**12**, **13** being most stable and **17**, **18** least stable) as borne out by experiment.

The synthesis of the stable azuloquinones is shown below. The precursor **19** is readily available (see Suppl to the 2nd edn IIIG page 248 and for more recent background to this compound see also section *(f)* in this chapter on the synthesis of azulenes). Singlet oxygen addition to **19** followed by acetylation of the two endoperoxides **20** and **21**, gave the precursors **22** and **24**. The other diacetoxyazulene **23** was a minor product from **20**. The formation of quinones **12** and **13** from **22** and **24** respectively was identical and is only illustrated for **12**. The oxidation of the disilyloxyazulene **25** was best done by pyridinium chlorochromate but other oxidants also gave the desired quinone compound.

The two less stable azuloquinones **14** and **15** were generated and trapped as shown. Again the starting point was ketone **19**. Chromic oxidation gave the diketones **26** and **27** which were converted, by essentially identical steps, to the quinones **14** and **15** respectively, by the route illustrated for the former.

Treatment of **26** gave the diacetoxyazulene **28** which was brominated to **29**. Hydrolysis of the ester groups to give **30** allowed the transient quinone **14** to be generated in a very mild dehydrohalogenation step. The existence of **14** was proved by the isolation of the Diels-Alder adduct **31**. This circuitous route involving the very mild dehydrobromination as the final step was necessary, as other more conventional dehydrogenation methods applied to **26** or **27** failed.

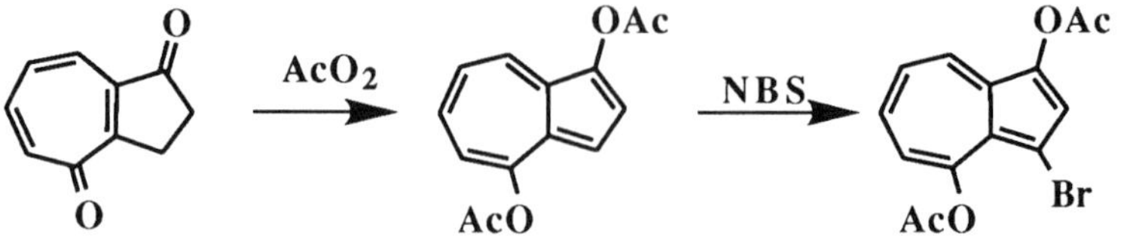

28   **29**

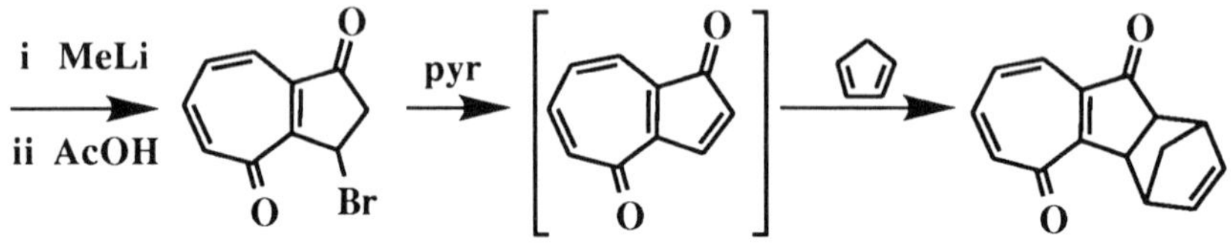

30   **14**   **31**

(c)   Diels-Alder Reactions of Azulenes

A large series of papers on this subject has appeared since the initial experiments reported previously (Suppl. to the 2nd edn., IIIG page 247) concerning the high temperature (200°C) reaction of azulene with dimethyl acetylenedicarboxylate (DMAD). The basic reaction products are portrayed in the diagram. Expansion of the five-membered ring takes place to give the heptalene **32**, after DMAD addition the loss of a different alkyne fragment yields another azulene **33**, and addition of two DMAD molecules give the complex ring structure **34** (Y. Chen and H.-J. Hansen, Helv. Chim. Acta, 1993, **76**, 168; A. Magussen, P. Uebelhart and H.-J. Hansen, *ibid.*, 1993, **76**, 2887; A.J. Rippert and H.-J. Hansen, *ibid.*, 1993, **76**, 2906). For related papers on the synthesis of heptalenes and their structure see H.-J. Hansen *et al.*, *ibid.*, 1993, **76**, 2876; *ibid.*, 1992, **75**, 2447, 2493. For a general review of the non-planar, rather unstable heptalenes see L.A. Paquette, Isr. J. Chem., 1980, **20**, 233.

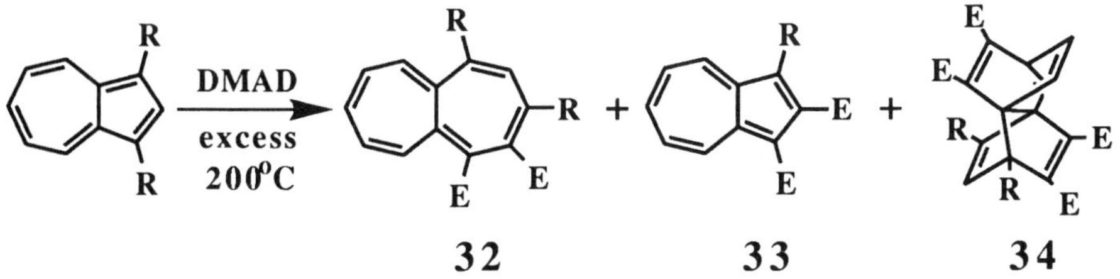

32 **33** **34**

(d) Valence Isomers of Azulene

Both azulvalene **42** and Dewar azulene **45** have been synthesized, as illustrated by fairly long and elaborate sequences **35-42** and **43-45** respectively.

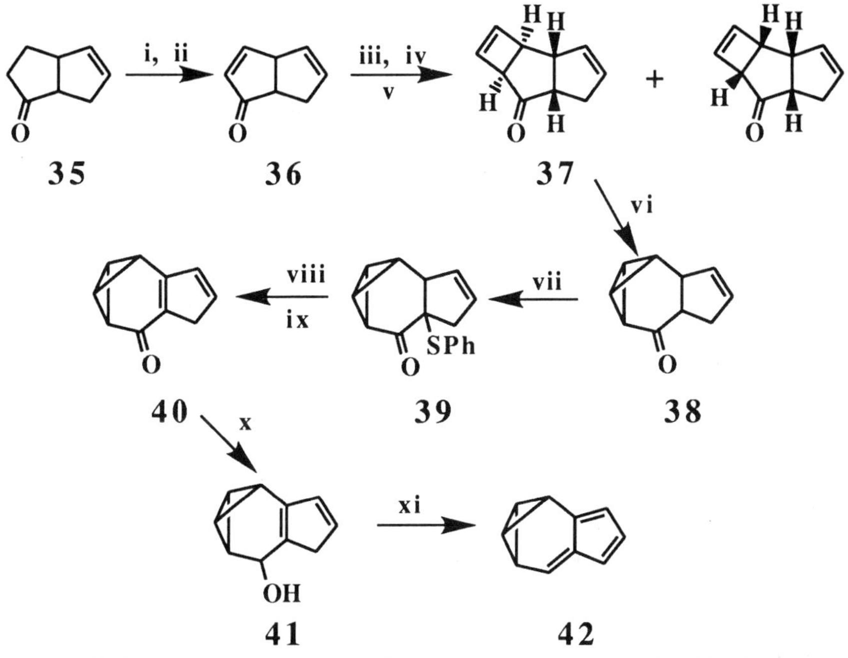

35 **36** **37**

40 **39** **38**

41 **42**

i = NaH / ArSO$_2$Me; ii = C$_6$H$_6$ /80°C; iii = CH$_2$OHCH$_2$OH / TsOH; iv = hυ / CHCl:CHCl; v = Na / NH$_3$ Liq.; vi = hυ; vii = LDA /PhSSPh / -35°C; viii = MCPBA; ix = 45°C / CCl$_4$; x = NaBH$_4$; xi = AcCl / DMAP

The precursor **35** is available via the dichloroketene adduct of cyclopentadiene. Both azulvalene **42** and Dewar azulene **45** are converted readily, thermally or photolytically, to azulene (Y. Sugihara, T. Sugimura and I. Murata, J. Amer. Chem. Soc., 1984, **106**, 7268). A follow up of this work portraying the synthesis of 6-dimethylamino Dewar azulene from **44** has been reported (Y. Sugihara and T. Sugimura, Chem. Letters, 1993, 785).

In addition to these valence isomer conversions reference to the conversion of azulene to naphthalene has been made in this chapter (Section 2). Since the review by L.T. Scott (Acc. Chem. Res., 1982, **15**, 52) two ^{13}C scrambling studies of azulenes have been reported. 1-Phenyl- and 2-phenyl-azulene with ^{13}C labels at the azulene 4-position thermally rearrange to azulenes with high proportions of label at the 3- and 3a-positions respectively (A. Wetzel and K.P. Zeller, Zeit. fur Naturforschung, B 1987, **42**, 903). 6-^{13}C-Azulene was rearranged under harsh conditions (1050°C) to yield naphthalene with most of the label at the 2-position. Small amounts of azulene with label at the 5- and 2-positions were also detected (H. Gugel, K.P.Zeller and C. Wentrup, Chem. Ber., 1983, **116**, 2775. 1-^{13}C-Azulene has also been rearranged by IR laser. The label in the resulting naphthalene is in the positions similar to that from the FVP study reported earlier (L.T. Scott, M.A. Kirms and B.L. Earl, J. Chem. Soc. Chem. Comm., 1983, 1373).

(e)   Substitution Reactions of Azulene

(i)   Electrophilic Substitution

Azulene undergoes most normal electrophilic aromatic substitution reactions, and these have been covered in previous editions (*loc. cit.*). All these reactions take place at the highly reactive 1-position in the electron rich five-membered ring. The reactivity of this position has been further defined and compared to other aromatic hydrocarbons (A.P. Laws and R. Taylor, J. Chem. Soc. Perkin Transactions 2, 1987, 591.

(ii)   Nucleophilic Substitution

A review paper, giving leading references, on this type of reaction is available (R. Bolton and A.J. Pilgrim, Aust. J. Chem., 1993, **46**, 1119). This excellent article discusses the reactivity of the seven-membered ring to nucleophilic attack, particularly at the 4-, 6- and 8-positions as well as describing the ability of azulene to stabilise nucleophilic displacement transition states in adjacent aromatic systems. Thus the fluorine in **46** is replaced easily by methoxy to give **47** under mild conditions. The effect of substituents at the azulene 1- and 1,3-positions was measured, and shown to be transmitted across the azulene system with unusually high efficiency.

Similar effects were observed in the methanolysis of 5,6-dichloroazulene. The 6-chloro group was replaced much more readily and the effects of remote substituent were efficiently transmitted (R. Bolton, D.G. Hamilton and J.P.B. Sandall, J. Chem. Soc. Chem. Comm., 1990, 917).

A very unusual introduction of a nitro group at the azulene 2-position has been observed using cupric nitrite in pyridine. Since a nucleophilic mechanism is proposed for this potentially very useful reaction, presumably the metal coordinates to the five-membered ring region of azulene and promotes reaction at this site which is not normally susceptible to nucleophilic attack (V.A. Nefedov, N.A. German and G.I. Nikishin, Zh. Org. Khim., 1983, **19**, 1123, Chem. Abstr., 1983, **99**, 87689x).

206

(iii) Other Substitution Reactions

Aryl radicals, formed by oxidation of arylhydrazines, attack azulene to give a mixture of 1-, 2-, 4- and 6-arylated azulenes reflecting the relatively small effect the dipole in azulene has on this type of reaction (V.A. Nefedov *et al.*, Zh. Org. Khim., 1987, **23**, 172, Chem. Abstr., 1987, **107**, 236145e).

Palladium acetate in benzene with sodium acetate gives oxidation of azulene to 1-acetoxy- and 1,3-diacetoxy-azulene (K. Saito *et al.*, Chem. Pharm. Bull., 1991, **39**, 1843).

The Heck vinylation reaction catalysed by palladium (*cf.*, section 3 *(c)*) is viable in azulene. Halogen atoms in both five-and seven-membered rings were replaced, in very good yield, by alkenyl (H. Horino, T. Asao and N. Inowe, Bull. Chem. Soc. Jpn., 1991, **64**, 183).

(f) Formation of Azulenes

Synthetic routes to azulene have been reported in previous editions and only novel, unusual, or specially versatile methods are covered here. Although azulenes were originally derived from terpenoid oils by acidic dehydration and oxidation, most modern syntheses create the azulene ring at an oxidation level at or close to the final aromatic state. If the synthesis requires a dehydrogenation step serious losses are often incurred at this stage.

An example is the synthesis of guaiazulene **53,** where the cyclisation step is achieved by a novel electrochemical technique of general applicability. This is shown above **48-53**. The final sulphur / decalin dehydrogenation step yield is only 35% (T. Shono *et al.*, J. Org. Chem., 1992, **57**, 7175).

The high yield synthesis from 2H-cyclohepta[b]furan-2-ones **55** has been further developed. This versatile intermediate is accessible by addition of an active methylene reagent such as malonate to a tropone **54** (X=Cl, OTs OMe). Further active methylene compound reaction gives **56**; ortho esters or vinyl ethers give **57**; enamines, generated *in situ*, similarly give **58** (T. Nozoe *et al.*, Heterocycles, 1989, **29**, 1225; T. Nozoe, H. Wakabayashi and K. Shindo, *ibid.*, 1991, **32**, 213; M. Yasunami *et al.*, Bull. Soc. Chem. Jpn., 1993, **66**, 892).

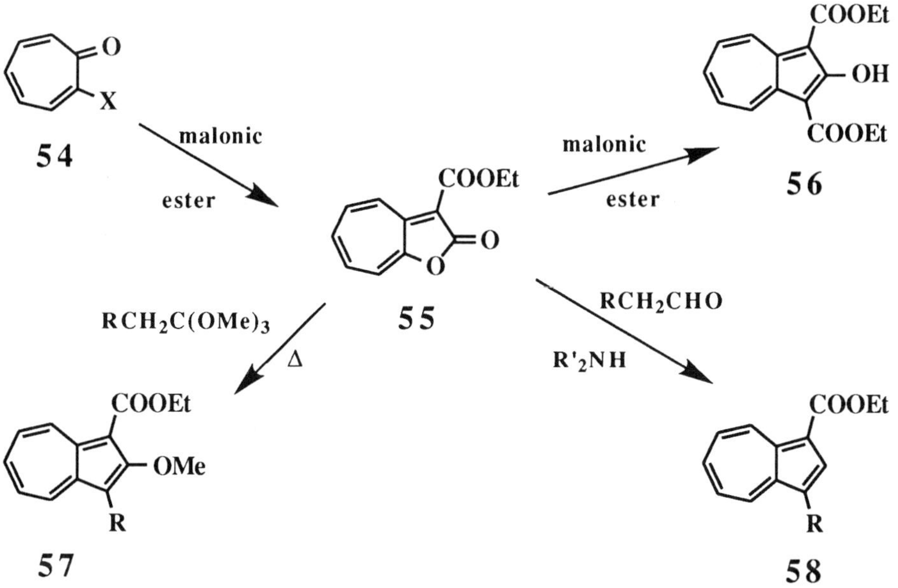

Another synthesis commencing with the seven-membered ring and cyclising the five-membered ring on to it, involves the reaction of allenylsilanes **60** with tropylium fluoroborate **59**. This reaction takes place in acetonitrile at room temperature, by way of the vinyl cations **61** and **62**, and is improved by the addition of poly(4-vinylpyridine) which scavenges the acid formed in the final steps **63-64**. The dehydrogenation in the last step is performed readily by an extra equivalent of tropylium. Substituted tropylium rings can be used in this reaction but isomer mixtures are then obtained (D.A. Becker and R.L. Danheiser, J. Amer. Chem. Soc., 1989, **111**, 389).

Intramolecular cycloaddition of diene to fulvene provides an interesting route to azulenes. The fulvene 65 (R=Alkyl; n=1) gives the mixture of products shown 66, 67, 68. The more electron rich fulvene 59 (R=NEt$_2$) enhances the 6+4 cycloaddition to give dihydroazulenes 66 selectively. The molecules with n=5 and 6 were dehydrogenated to azulenes with sulphur in triglyme (T.-C. Wu et al., J. Amer. Chem. Soc. 1983, 105, 6996)

Attack by a transition metal carbene complex 69 on tropylium fluoro-borate has also been shown to give azulenes 75 with amino and alkoxy substituents. The intermediate complex 70 is converted by hydride abstraction to tropylium species 71 which emits a proton to give 72. Further reaction with an isocyanide gives 73, which rearranges to 73. Heating extrudes the activating transition metal moiety with concomitant molecular reorganisation to yield the azulene in approximately 30% yield.

A low yield method (20%) which has some potential, because of the ready availability of the starting material, is the formation of azulene from cobaltocene **76** and toluene in acetonitrile / 2-methylpyridine. The mechanism of this reaction is obscure (E.N. Kayushina *et al.*, Zh. Org. Khim., 1982, **31**, 1931, Chem. Abstr., 1982, **98**, 34355h).

A standard azulene preparation (see Suppl. to the 2nd Edn IIIG 248,) via molecule **19** (see section *(b)* on the azuloquinones) has been improved greatly by using rhodium acetate as catalyst in the decomposition of the diazoketone **77** (A.M. McKervey, S.M. Tuladhar and F.M. Twohig, J. Chem. Soc. Chem. Comm., 1984, 129).

Finally, a review of free radical mediated ring expansion reactions to form hydroazulenes, and other exotic molecules such as benzocycloheptanes and benzocyclooctanes has appeared (see also next section 10) (P. Dowd and W. Zhang, Chem. Rev., 1993, **93**, 2091). In this work the hydroazulene part (pages 2112-2114) illustrates several schemes involving expansion and rearrangement of cyclopropyl rings in good yield. An example is given below **78-80** where the two isomers are formed in equal amounts, an overall conversion of 83%.

10   Benzocycloheptatrienes, Benzotropones, and Related Compounds

(a)   Formation

This area was well covered in the previous editions and highlights from the more recent literature are included here.

Standard electrophilic cyclisations to form the seven-membered ring ketone are exemplified by the synthesis of the cytotoxic plant product Faveline methyl ether **1** (A.K. Ghosh, C. Ray and U.R. Ghatak, Tetrahedron Letters, 1992, **33**, 655).

The related diterpenoid tropolone Salviolone **4** was formed in an aldol condensation (G. Haro *et al.*, Bull. Chem. Soc. Jpn., 1991, **64**, 3422). Dialdehyde **2** condenses with 1-methoxybutan-2-one to give the desired compound **3** along with the other isomer where the methoxy and the methyl groups are transposed. These compounds are of interest as possible treatments for leukacmia.

Bis(phenylthio)-7H-benzocyclohepten-1,4,7-trione **5** and its derivatives, synthesized by the familiar procedure illustrated below (*cf.*, Suppl. to the 2nd edn., page 260) are currently under investigation as organic electrical conductors (M. Kato *et al.*, Chem. Letters 1991, 1973; K. Furuichi *et al.*, J. Chem. Soc. Perkin Transactions 2, 1992, 2169; H. Tada *et al.*, *ibid.*, 1993, 1305).

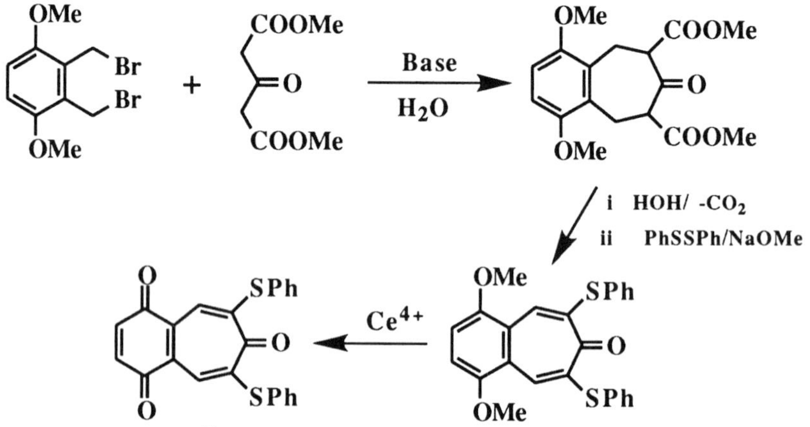

Free radical ring expansion reactions have already been discussed in the previous section (see review by P. Dowd and W. Zhang, Chem. Rev., 1993, **93**, 2091) and are viable here. Two examples are depicted below **6-9**:

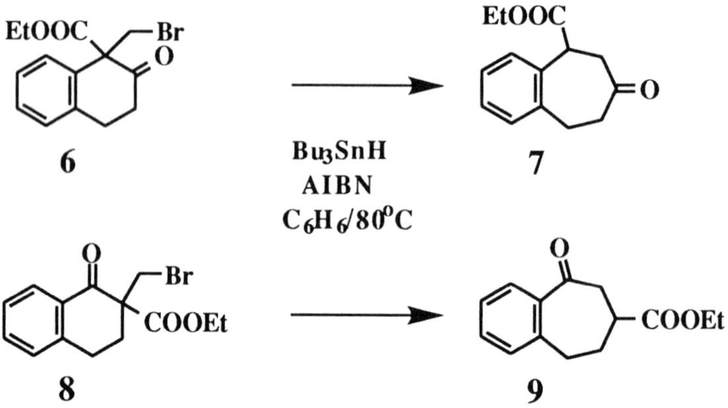

Several interesting cycloaddition reactions leading to the benzocycloheptane skeleton have been explored. Esterification of phthalic anhydride with pentyn-5-ol gives **10**, a simple entry to the tropolone **13**. Rhodium catalysed decomposition of the diazoketone **11** gave the oxygen bridged system **12** which was rearranged to the final product by Lewis acid treatment (C. Plüg and W. Friedrichsen, Tetrahedron Letters, 1992, **33**, 7509).

Intramolecular Diels-Alder cycloadditions to furan have also been shown to be viable for the construction of the benzocycloheptane skeleton. In the intermolecular Diels-Alder reaction, where two molecules are united into one, a large negative volume of activation is expected, and therefore high pressure favours the addition reaction. In the reaction shown, **14-15,16**, high pressure is required to force the reaction to occur, in spite of the intramolecular nature of the process. A negative volume of activation of approximately half that for the intermolecular reaction was observed. Both the isomers slowly revert to the starting material at normal pressure. Surprisingly, the endo isomer **15** is the thermodynamic product in contrast to the usual situation in these reactions (S.J. Burrell et al., Tetrahedron Letters, 1985, **26**, 2229).

The creation of the seven-membered ring with an oxygen bridge, derived from a furan, has also been accomplished by intramolecular cycloaddition of allyl cation **17-18** (B. Föhlisch and R. Herter, Chem. Ber., 1984, **117**, 2580).

Another possible cycloaddition approach to benzotropones and related compounds is through the highly reactive intermediate 4,5-didehydrotropone **20**, which can be generated from **19** in an analogous manner to *o*-benzyne. 4,5-Didehydrotropone **20** has the reaction characteristics of a typical aryne.

Diene addition should produce a range of 4,5-benzotropone systems **21**, and dienophile addition to the isofuran system **22** derived from the didehydrotropone is a further possibility (T. Nakazawa *et al.*, Tetrahedron Letters, 1994, **35**, 8421).

A remarkable reaction is a carbene addition to the 2,3-bond of naphthalene. This is the first time that this has been observed and the reaction produced the tropylidine **24** in 10% yield. The main product (>30%) of the reaction, however, is the addition to the 1,2-bond **23** (M. Pomerantz and M. Levanon, Tetrahedron Letters, 1991, **32**, 995).

The hitherto unknown 3,4-benzotropone **26** has been made by photochemical rearrangement of the ketone **25**, but is stable only in a glass matrix at -196°C. Warming of the matrix to -78°C causes a change in the typical benzotropone UV spectrum, due to dimerisation to **28** and **29**. The benzotropone can also be generated thermally and trapped as the Diels-Alder adduct **27** with maleic anhydride (M. Ohkita, T. Tsuji and S. Nishida, J. Chem. Soc. Chem. Comm., 1989, 924). For the related benzotropolone rearrangement see Suppl. to the 2nd edn., IIIG page 299.

An earlier related study of the benzocycloheptatrienes contained in a frozen argon matrix, and correlation of the results with semiempirical calculations, showed that the benzotropylidene **30** photochemically rearranges via the norcaradiene **31** to a mixture of methylnaphthalenes. A similar rearrangement was observed for the radical cation of this system (B.J. Kelsall, L. Andrews and C. Trindle, J. Phys. Chem., 1983, **87**, 4898).

The formation of the cyclic allene **32** has been convincingly demonstrated, by an analogous method to that in the similar naphthalene system, see section 7(*a*)), and leads to the exotic chemistry illustrated. In the absence of a suitable trapping agent the allene dimerises to **34** and the "valene" structure **33**. Dimer **34** can undergo various thermal reorganisations to **36** and **37** via the diradical **35** (H. Jelinek-Fink *et al.*, Chem. Ber., 1991, **124**, 2569) *(b)* *Reactions of benzocycloheptatrienes and benzotropones*

The apparently simple condensation of malononitrile with 2,3-benzotropone to give the heptafulvene **38** has hidden subtleties. Deuterium labelling adjacent to the carbonyl function showed that a Michael addition mechanism may be operating as well as the straightforward Knoevenagel reaction.

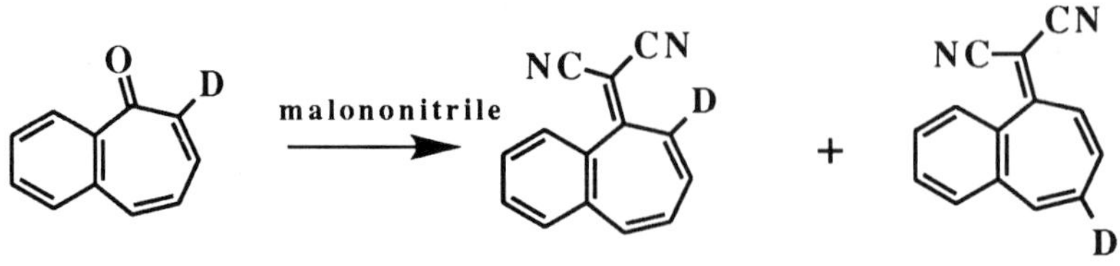

38

This mechanism finds some support in the nucleophilic amination of 2,3-benzotropone, which occurs with ring contraction to yield the 4-amino-1-naphthaldehyde **39** (T. Machiguchi, Chem. Letters, 1994, 1677).

39

In a paper reviewing the isomerisation reactions of benzocyclohepta-trienes such as **40**, the Diels-Alder reactions with dienophiles were investigated. The main product was the 2+2 adduct **41**, the other products **42,43,44** being formed in low yield. The isomerisation mechanisms giving rise to these minor products were considered (W. Adam *et al.*, Chem. Ber., 1991, **124**, 383). This reference gives leads to reactions with other dienophiles such as singlet oxygen and 4-phenyl-1,2,4-triazoline-3,5-dione.

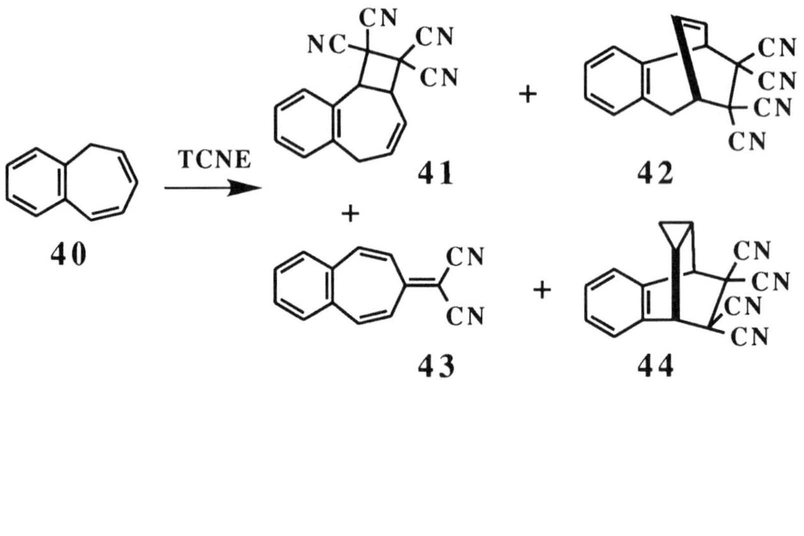

40   **41**   **42**   **43**   **44**

In a similar study dienophiles were shown to give mixtures of 2+4 adducts with the ethylene acetal of 4,5-benzotropone (I. Heise and J. Leitich, Chem. Ber., 1985, **118**, 332).

On the other hand in the reaction of benzotropones and benzocyclo-heptenes with cyclopentadiene, the latter acts as the 4 π component and often gives *exo*-addition depending on the balance of steric and electronic factors. An example is depicted below **45-46** (S. Sarkar, G. Saha and S. Ghosh, J. Org. Chem., 1992, **57**, 5771).

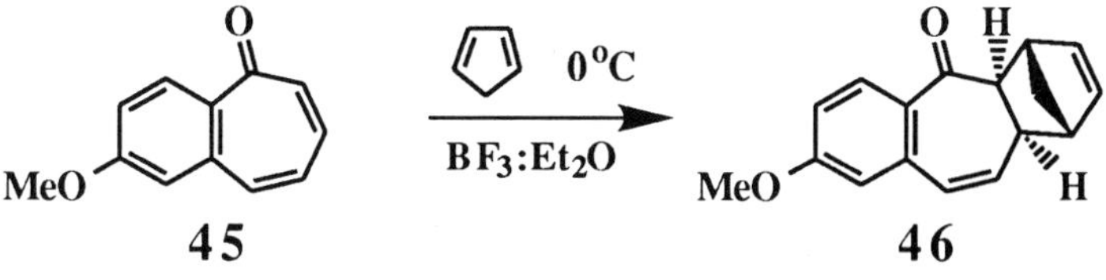

Lastly there is the interesting photoisomerisation phenomenon associated with the benzotropylidenes which is thought to take place via a transient bisnorcaradiene structure **49**.

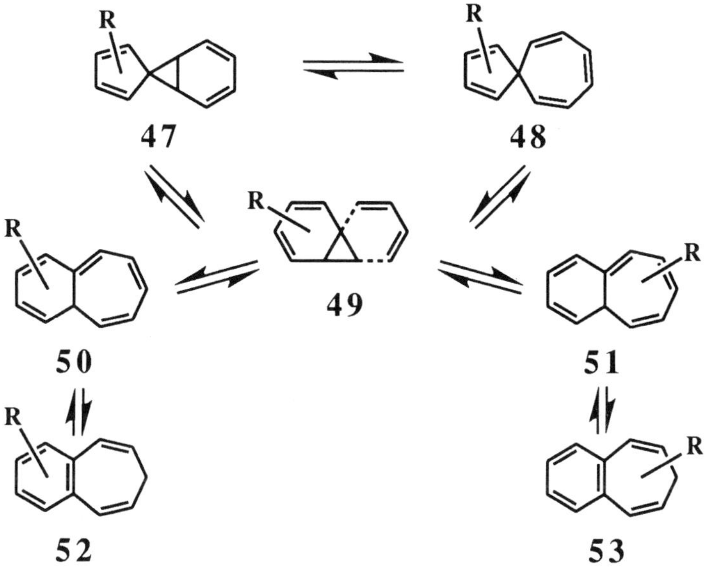

Thus it is possible for substituents in the six-membered rings **50, 52** to appear in the seven-membered rings **51, 53** after irradiation. The mechanism shown is given further credence by the isomerisation of the two spiro ring systems, **47** and **48**, which were produced independently (H. Dürr, K.-H. Pauly and K. Fischer, Chem. Ber., 1983, **116**, 2855).

11   *Benzocyclooctatetraenes and Related Compounds*

Ring expansion reactions continue to be explored, and many new versions of reactions already mentioned in previous editions have been published. The review by P. Dowd and W. Zhang (Chem. Rev., 1993, **93**, 2091) has examples of benzocyclooctane synthesis by analogous reactions to those forming benzocycloheptanes [see previous section 10*(a)*]. Two other examples from this source are given below **1-2** and **3-5**. The yields in these reactions are generally lower than for the smaller rings, probably due to the larger entropy factor, but where this is reduced as in the latter example the yield can be as high as 80%.

Ethylenic and acetylenic additions to naphthalene have featured in a review of aromatic photoadditions (J.J. McCullough, Chem. Rev., 1987, **87**, 811). Ring expansions of cyclobutanaphthalenes and naphthobarrellenes to benzocyclooctenes are discussed (see also Suppl. to the 2nd edn., IIIG page 264 *et seq.*). A particularly interesting aspect of this type of reaction was exposed by

W. Grimme, J. Lex and T. Schmidt (Angew. Chem. Internat. Edn., 1987, **26**, 1268). The benzobicyclo[4.2.0]octa-2,4,7-triene-2'-carbonitrile **7** was optically resolved, and then a single enantiomer was thermally rearranged to **9**, presumably via **8**. The rate of formation of **9** and the decrease in optical rotation of **7** are the same, within experimental error, showing that there is no back reaction of **8** to **7**. Bond reorganisation of **8** to **9** is so fast that **8** cannot be trapped successfully by a dienophile.

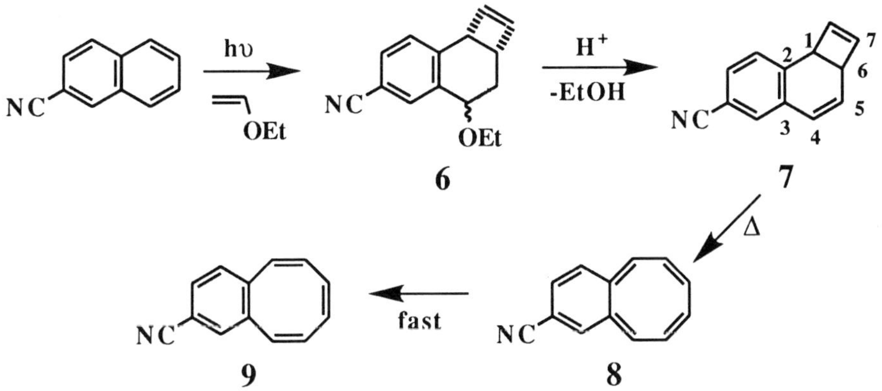

The chemistry of 8-methyl-2,3-benzobicyclo[4.2.0]octa-2,4,7-triene **10** has been researched, and quantitatively thermally rearranges to the benzocyclo-octatetraene **11** in an identical way to the reaction illustrated above. In contrast photolysis produces very little benzocyclooctatetraene. The main products are the semibullvalene **12** and the benzotetracyclooctene **13** (C.O. Bender *et al.*, Can. J. Chem., 1994, **72**, 1556, 1999)

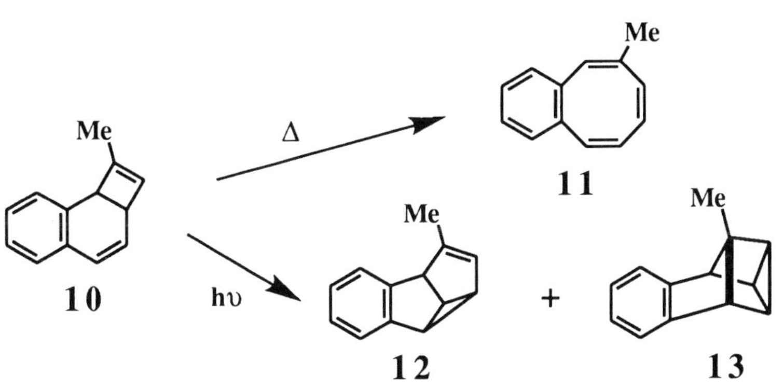

Related to these processes is the theoretically fascinating, but expensive, formation of benzene from the dimers, *anti-o,o'*-dibenzene **14** and *cis,syn-o,o'*-dibenzene **15**. The direct *retro*-dimerisation of **15** is orbital symmetry forbidden and the reaction is thought to proceed via the dihydrobenzocyclooctatetraene **16**, which is known to perform a disrotatory ring closure to **15** even below -10°C. The direct split of **14** into benzene would involve the creation of one molecule of benzene in the triplet state, and this was not detected (N.-c.C. Yang, B.J. Hrnjez and M.G. Horner, J. Amer. Chem. Soc., 1987, **109**, 3158).

Eight-membered rings have been made by nickel catalysed cyclisation of tetraenes. The precursor **17** was made as indicated, in 65% yield, and the cyclisation to **18** in 84% yield, making the overall method quite efficient (P.A. Wender and N.C. Ihle, J. Amer. Chem. Soc., 1986, **108**, 4678).

The nickel catalysed process is stereospecific giving only *trans*-fused ring junctions. On the other hand the related, triplet sensitised, photochemical process explored by the same group gave a mixture of isomers of **18** by the route illustrated. The intermediacy of, the isolable, **19** is interesting as this type of molecule is attractive as a trapping agent for exotic radicals etc., (P.A. Wender and C.R.D. Correia, J. Amer. Chem. Soc., 1987, **109**, 2523).

isomer mixture

17   **19**   **18**

Another highly unsaturated acyclic system, of theoretical importance, that can lead to benzocyclooctene is the endiyne **20**, which undergoes a Bergman cyclisation to various products including up to 20% of the benzocyclooctene **21** (T.P. Lockhart, C.B. Mallon and R.G. Bergman, J. Amer. Chem. Soc., 1980, **102**, 5976; T.P. Lockhart, P.B. Comita and R.G. Bergman, *ibid.*, 1981, **103**, 4082; T.P. Lockhart, and R.G. Bergman, *ibid.*, 1981, **103**, 4091).

20   **21**

Two ring expansion of benzocycloheptene systems are known which lead to substituted benzocyclooctatetraenes. Of great practical utility is the carbene insertion reaction derived from the diazo compounds **23** and **24** which can be obtained from the bromotropylium bromide **22** as illustrated. Decomposition of either of these diazo compounds by cupric acetylacetonate in hot benzene gives the common carbene intermediate which collapses to the product benzocyclooctatetraene **25** in high yield (M. Böhshar, H. Heydt and M. Regitz, Chem. Ber., 1984, **117**, 3093).

Dichlorocarbene (*cis*-fused) adduct of 6,7-dihydro-2,3-benzotropone **26** on prolonged photolysis gave a mixture of the *trans*-fused isomer **27** and the dichlorobenzocyclooctadienones **28** and **29**. Chromatography of the benzo-cyclooctadienone with the *trans* double bond **28** caused rapid isomerisation to the diol **30** (J.L. Wood, P.J. Carroll and A.B. Smith, J. Chem. Soc. Chem. Comm., 1992, 1433).

Finally a derivative of benzoquinone with trimethylsiloxybutadiene produced a Diels-Alder adduct **30**, which rearranged to a bright red benzocyclooctatetraene derivative **31** during chromatography on silica gel. A fuller investigation of this unusual rearrangement is awaited. The normal acid catalysed rearranged product of **30** is the dihydrobenzofuran **32** (F. Farina, M.C. Paredes and J.A. Valderrama, Tetrahedron, 1992, **48**, 4629).

Chapter 28

ANTHRACENE, PHENANTHRENE AND DERIVATIVES

R. BOLTON

1. Phenanthrene

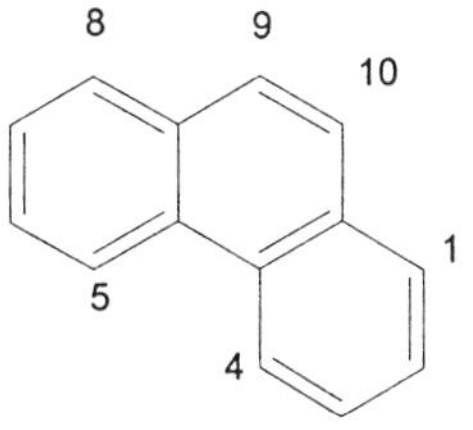

1 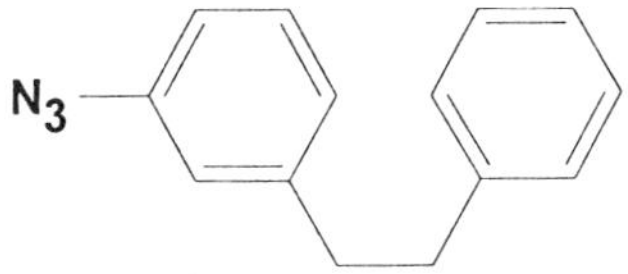   2

Substituted phenanthrenes have been obtained by the coupling of acetylenes R.C≡CH (R = Ph, SiMe$_3$, CO$_2$Et or CMe$_2$OH) with 2-biphenylyl diazonium ions in pyridine (R. Leandini *et al.*, *Synthesis*, 1988, 333). 3-Azidobibenzyl (2) gives a mixture of 2- and 4-amino-9,10-dihydrophenanthrenes with CF$_3$SO$_3$H (R.A. Abramovitch *et al.*, *Tetrahedron Lett.*, 1986, 27, 3705). 4,5-Disubstituted-9,10-dihydrophenanthrenes (X = F, Cl, CF$_3$) are prepared from the appropriate 3-X-anthranilic acid by Vorländer coupling of the diazonium ion, reduction of the resulting 6,6'-X$_2$-diphenic acid, and the PhLi-mediated Fittig reaction of the derivative bis-bromomethyl compound (R. Cosmo and S. Sternhall, *Aust.J.Chem.*, 1987, 40, 35 and 2137; *cf.* D.M. Hall, M.S. Lesslie and E.E. Turner, *J.Chem.Soc.*, 1950, 711; *ibid., idem,* 1951, 3072).   1-Chlorophenanthrene (63%) is formed by the photochemical cyclisation of *Z*-2-chlorostilbene, and the *E/Z*-mixture of 3-chlorostilbenes provides both

4-chloro- (29%) and 2-chloro-phenanthrenes (30%; J. Moursounidis and D. Waage, *Aust.J Chem.* 1988, <u>41</u>, 235).

Hydrogenation of phenanthrene with $RhCl_3$-Aliquot 336 gives only 9,10-dihydro-1 (I. Amer *et al., J.Mol.Catal.,* 1987, <u>39</u>, 185), but with $^iPr_2BB^iPr_2$ smaller amounts of 1,2,3,4,5,6,7,8- and 1,2,3,4,4a,9,10,10a-octahydro-phenanthrenes are also formed (M. Yalpani and R. Koester, *Chem.Ber.,* 1990, <u>123</u>, 719).

Phenanthrene undergoes nitration by acetyl nitrate solutions arising from HNO_3 or from $Cu(NO_3)_2$ in Ac_2O. In 14 solvents the 2-nitro derivative is always formed in the smallest amount but while, in general, the 9-nitro isomer is the major component the 1-isomer is preferred when these reactions are carried out in CH_2Cl_2, $ClCH_2CH_2Cl$ or $PhNO_2$ (L.I. Velichko, O.I. Kachurin and E.Yu. Balabanov, *Ukr.Khim.Zh.(Russ.Ed.),* 1988, <u>54</u>, 171; *Chem. Abstr.,* 1989, <u>110</u>, 38364). When $Cu(NO_3)_2$ in Ac_2O is the source of $AcONO_2$ the product isomer distribution (60 min., 22°C; 58% yield, 96% conversion) is 1- (22±2%), 2- (1±0.5%), 3- (19±2%), 4- (7±3%) and 9- (51±6%). Nitric acid in Ac_2O (30 min; 54% conversion) gave a nitrophenanthrene distribution of 1- (34±2%), 2- (>0.2%), 3- (15±3%), 4- (10±3%) and 9- (40±5%) (O.I. Kachurin, L.I. Velichko *Zh.Obshch.Khim.* 1989, <u>59</u>, 2766). These latter values compare with Dewar's earlier figures (P.M.G. Bavin and M.J.S. Dewar, *J.Chem.Soc.,* 1956, 164; Dewar and T. Mole, *idem,* 1441; Dewar, Mole, D.S. Urch and E.W.T. Warford, *idem,* 3572; Dewar, Mole and Warford, *idem,* 3576) of 1- (27%), 2- (7%), 3- (23%), 4- (6%) and 9-nitrophenanthrene (37%) upon which the commonly reported partial rate factors for nitration of **1** are usually based. The discrepancies between the two sets of Russian values suggests that the two reaction conditions cannot be viewed as identical. With NO_2 phenanthrene readily gives 9-nitrophenanthrene (D.K. Bryant, B.C. Challis and J. Iley, *J.Chem.Soc., Chem.Commun.,* 1989, 1027) although with styrene a range of addition and further oxidation products was found, but M.P. Hartshorn *et al. (Aust.J.Chem.,* 1990, <u>43</u>, 125) report that NO_2 in PhH gives 1-, 3-, and 9-nitrophenanthrenes along with the dimeric addition product **3** and the *cis-* and *trans*-isomers of **4**. These products support a homolytic mechanism of nitration involving the 10-nitro-9,10-dihydro-9-phenanthryl radical which may then be trapped by

229

other species present, an interpretation which is supported by the sensitivity of the product yields upon the concentration of phenanthrene in the reacting solution. The multiplicity of mechanisms associated with nitration by acetyl and benzoyl nitrates was acknowledged in early reports (V. Gold, E.D. Hughes and C.K. Ingold, *J.Chem.Soc.*, 1950, 2467; Gold, Hughes, Ingold and G.H. Williams, *idem*, 2452), and this makes such nitration mixtures less appropriate for studies of the relative reactivities of arenes towards electrophiles.

3   4

3,4-Dihydrophenanthren-1(2*H*)-one with PCl$_5$ is reported (O. Caamano *et al.*, *An.Quim.*, *Ser.C* 1989, <u>85</u>, 102) to give both 1- and 2-chloro-phenanthrenes, along with 1,2-dichlorophenanthrene and its 3,4-dihydro-derivative.

Phenanthrene-9-ol undergoes bromination (Br$_2$-CS$_2$) to give the 3,10-dibromo derivative (**5**; X = Br, Y = OH). Reduction gives 3-bromo-phenanthren-9-ol which undergoes the Bucherer reaction to provide the bromo-amine from which 3,9-dibromophenanthrene may be made by conventional diazonium processes (E. Ota and H. Shintani, *Nippon Kagaku Kaishi* 1987, 762). The corresponding chlorination (SO$_2$Cl$_2$-AcOH) of phenanthrene-2-ol provides the 1-chloro analogue; Cl$_2$-PPh$_3$ in CCl$_4$ gives 2-chlorophenanthrene (**6**; Z = Cl; 59%) and bromine analogously leads to 2-bromophenanthrene (E. Ota and J.T. Shimozawa, *idem*, 1987, 757).

5   6   7

2-Acetylphenanthrene forms 2-ethylphenanthrene (Huang-Minlon; 90-93%). The 3-isomer reacts similarly, and undergoes bromination (PBr$_5$-C$_6$H$_6$) to give 2-bromo-3-ethylphenanthrene (94%). 2-Ethylphenanthrene correspondingly provides a mixture of 1-bromo- and 3-bromo-2-ethylphen-anthrenes (2:1; 96% overall) but 2-*sec*-butylphenanthrene shows steric hindrance at C-1 and gives entirely (99.7%) substitution at C-3 (N.I. Tabashidze *et al., Izv.Akad.Nauk Gruz.SSR, Ser.Khim.* 1987, 13, 120; *Chem.Abstr.*, 1987, 108, 167092).

Phenanthrene may be chloromethylated to give the 9-isomer (F. Fernandez *et al., Synthesis*, 1988, 802). The composition of the mixture of 2-, 3-, and 9-isomers from the acetylation of phenanthrene in (CH$_2$Cl)$_2$ depends upon the method of mixing the reagents. In PhNO$_2$, 9-chlorophenanthrene gives mainly the 3-acetyl derivative with up to 11% of the 2-isomer (F. Fernandez *et al., J.Prakt.Chem.*, 1989, 331, 15) (*cf.* 65% 3- and 15% 2-acetylphenanthrenes in the attack of the parent hydrocarbon). 9,10-Bis(bromomethyl)phenanthrene provides dibenzo[*e,g*]isoindolines (R.P. Kreher and T. Hildebrand, *Chem.Ber.*, 1988, 121, 81) and 9,10-bis(methylthio)phenanthrene through nucleophilic attack by amines or by thiourea; in the latter case, the cyclic dithiane 7 results from oxidative coupling if the displacement is carried out in the presence of air (M. Banciu *et al., Rev.Roum.Chim.*, 1987, 32, 961).

4,5-Bis(bromomethyl)phenanthrene is still most conveniently prepared (K. Muellen *et al., Chem.Ber.*, 1990, 123, 2349) from pyrene by ozonolysis, treatment with DIBAL, and reaction of the resulting diol with PBr$_3$ (T.J. Katz and W. Slusarek, *J.Amer.Chem.Soc.,* 1979, 101, 4259).

3-Methyl- and 3-*t*-butyl-phenanthrene may be synthesised from the corresponding 4-alkylcyclohexanones by reaction with PhCH$_2$CH$_2$MgBr, cyclisation and dehydrogenation (Scheme 1; R = Me, CMe$_3$) (R.C. Bansal *et al., Org.Prep Proced.Int.* 1988, 20, 305).

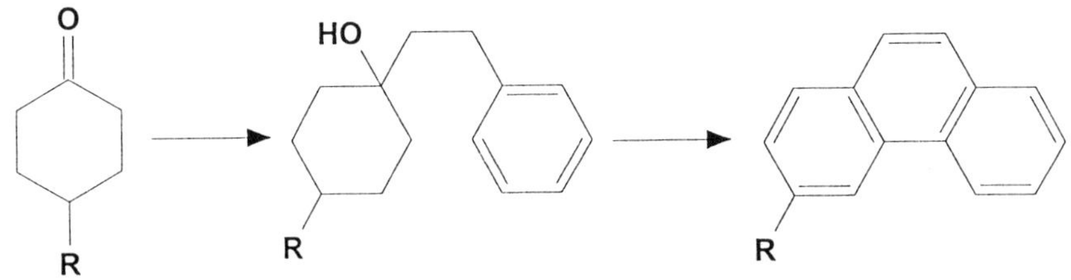

Scheme 1: 3-Alkylphenanthrene synthesis

The synthesis of 1,4-, 2,4- and 3,4-dimethylphenanthrenes has been achieved by the cycloaddition of furan across the appropriate dimethylnaphthynes, themselves derived from the dimethyl-2-bromo-1-tosyloxynaphthalenes (BuLi). The deoxygenation of the arene oxides is best achieved using Me$_3$SiI. Some unexpected orientations of attack are reported in the preparation of the required naphthalene derivatives (K. Yung and M. Koreeda, *J.Org.Chem.,* 1989, 54, 5667).

6-Methoxy-1,2,3,4,4a,10a-hexahydrophenanthr-9-one is made from 2-(3-methoxyphenyl)-2-cyclohexenone (**8**) through the Michael addition of diethyl malonate and cyclisation (Ac$_2$O-AcOH-ZnCl$_2$) of the resulting acetic acid derivative; 7-Br-**9** (X = Br) is the sole product of bromination of **9** (X = H) (M.Chini *et al., Gazz.Chim.Ital.,* 1988, 118, 369).

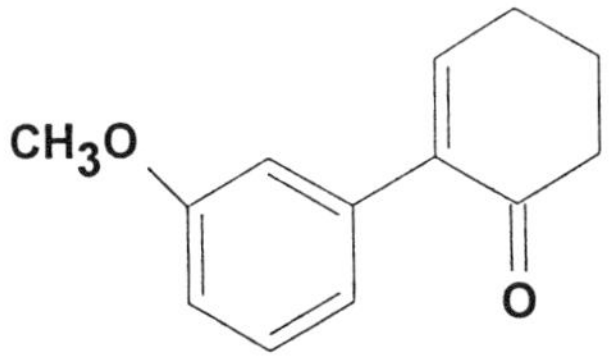

8 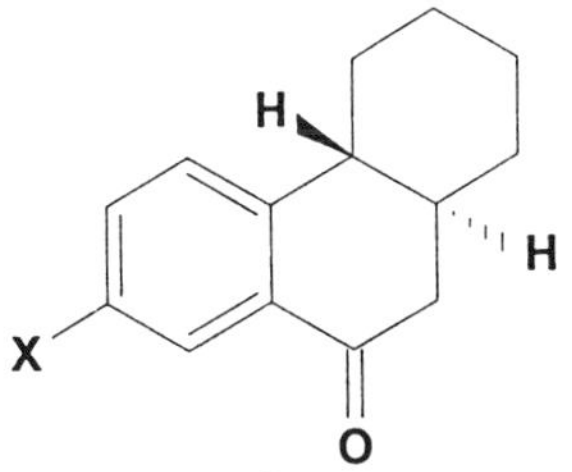   9

Methyl 2-bromoacrylate undergoes cyclocondensation with 5-methyl-6-tosyloxytetralin to give the 9,10-dihydrophenanthrene derivative (10) which is used in the synthesis of juncunol (11) (G.H. Posner, K.A. Canella and E.F. Silversmith, *Proc.-Indian Acad.Sci;Chem.Sci.*, 1988, 100, 81).

10   11

Both 1,8- and 4,5-phenanthrene dicarboxylic acids undergo oxidative decarboxylation (Li-NH$_3$, then Pb(OAc)$_4$) to give octahydrophenanthrene-1,8- or -4,5-diones, which may then be used to afford a number of hindered derivatives (P. Calune and T. Pepper, *J.Org.Chem.*, 1988, 53, 1786).

12

10,10'-Dihydroxy-9,9'-biphenanthryl (12) has been obtained by the oxidative dimerisation of 9-phenanthrol (Fe(III)acac - BF$_3$.Et$_2$O) and resolved into its enantiomers through the formation of diastereoisomeric cyclic phosphate esters (S. Xie, F. Wu and W. Huang, *Zhongshan Daxue Xuebao, Ziran Kexueban* 1986, 18; *Chem.Abstr.*, 1987, 107, 115344).

Anthracene and its compounds

13

High purity anthracene may be obtained by recrystallisation from DMSO (B. Marciniak, *Mol.Cryst.Liq.Cryst.*, 1988, <u>162B</u>, 301) as well as by codistillation with di- or tri-ethylene glycol.

Anthracene (**13**) characteristically undergoes attack at C-9 (C-10), the site(s) of highest electron density. Both addition and substitution may then occur. Y. Chung *et al* (*J.Org.Chem.*, 1989, <u>54</u>, 1018) have reported a detailed mechanistic study of the Diels-Alder process, which showed that the structure of the adduct affects the rate of the reversion, and some unusual features in the derived structure-reactivity profile. The ease of formation and destruction of such adducts plays an important part in some recent synthetic methods. Magnesium-anthracene systems have well marked uses in chemical synthesis (B. Bogdanovic, *Acc.Chem.Res.*, 1988, <u>21</u>, 261).

The hydrogenation of anthracene is well described. It occurs under a range of conditions in which isomeric tetra-, hexa- and octa-hydroanthracenes have been identified. Thus, under catalysis by Pr_2BBPr_2 anthracene gives 1,2-dihydro-, 1,2,3,4-tetrahydro- and (more slowly) 1,2,3,4,5,6,7,8-octahydro-anthracene; *cis*- and *trans*-1,2,3,4,4a,9,10,10a-octahydro- and dodecahydro-anthracenes then form in small amounts subsequently. 9-Phenyl- and 9,10-dimethyl-anthracene also undergo sequential saturation of the two outer rings of the anthracene structure; 9-bromoanthracene predominantly suffers reductive dehalogenation and then hydrogenation of the resulting anthracene, with the formation of 9-propylanthracene by propyldebromination as a prominant side-reaction. 1,4-Dimethylanthracene undergoes addition at rings C and then B with rearrangement of the alkyl groups within ring A (M. Yalpani and R. Koester, *Chem.Ber.*, 1990, 123, 719). In the presence of D_2O Na provides 9,9,10,10-d_4-9,10-dihydroanthracene in 94% yield with 87% D-incorporation (R.E. Pickering *et al., J.Labelled Compd.Radiopharm.* 1987, 24, 1503; 1988, 25, 83).

The chlorination of anthracene may be achieved using $CuCl_2$ in DMSO (A.S. Sahasrabhuddhe and B.J. Ghiya, *Indian J. Chem., Sect.B,* 1990, 29, 61) or SO_2Cl_2 in C_6H_5Cl, $CHCl_3$ or similar solvents. The kinetics of the reaction in PhCl indicates electrophilic attack by SO_2Cl_2; substituent effects are reflected by MNDO calculations of the energy of the intermediate complex and of the starting molecule (R.Bolton, D.B.Hibbert and S.Pirand, *J.Chem.Soc., Perkin Trans. 2*, 1986, 981). Alumina-supported $CuCl_2$ or $CuBr_2$ in CCl_4 at 50° also provide the appropriate halogenoanthracenes. All these methods are applicable to derivatives of anthracenes, and bring about substitution exclusively at the 9(10)-position(s). The substituent effect of the 9-halogen upon C-10 ensures stepwise halogenation.

Bromination is also brought about by $Ph_3P.Br_2$ bound to a poly(methyl methacrylate) backbone (M. Hassanein *et al.*, *Eur.Polym.* 1989, 25, 1083). and by $Br_2C(CN)NO_2$ which gives 9,10-dibromoanthracene (J.H. Boyer and T. Manimaran, *J.Chem.Soc., Perkin Trans. 1*, 1989, 1381).

Iodination of reactive polybenzenoid aromatics occurs when ICN is polarised by $AlCl_3$ in Et_2O-$MeNO_2$ (F. Radner, *Acta.Chem.Scand.*, 1989, 43, 481); 9-iodoanthracene is formed from anthracene in 48% yield. The trans-iodination method using 2,6-di-iodo-*p*-cresol and $GaCl_3$ (M. Tashiro, T. Makishima and S. Horie, *J.Chem.Res., Synop.*, 1987, 342) also seems to be applicable to anthracenes, although this has not been reported.

J.E. Baldwin *et al.* (*Tetrahedron* 1986, 42, 3943) have used the reversibility of Diels-Alder addition to anthracene to direct substitution. For example, 9,10-etheno-9,10-dihydroanthracene undergoes substitution at C-2 to give species which, on flash vacuum pyrolysis, provide 2-X-anthracenes. In this way 2,3-I_2-, 2,3-Br_2- and 3-NO_2-2-NH_2 anthracene may be obtained. 2,3-Dibromoanthracene is also made by the addition of isobenzofuran (from 2-ethoxy-2-H-isobenzofuran) to 4,5-dibromobenzyne (from 1,2,4,5-$C_6H_2Br_4$) followed by deoxygenation with low-valent Ti or with $Fe_2(CO)_9$ (M.R. Akula, *Org.Prep.Proced.Int.*, 1990, <u>22</u>, 102).

9-Fluoroanthracene is obtained from the reaction of $(Ph.SO_2)_2NF$ with 9-lithioanthracene, itself obtained from 9-bromoanthracene (E. Differding and H. Ofner, *Syn.Lett.*, 1991, 187), and is also obtained from the reaction between $CsSO_4F$ and 9-anthraldehyde (*Chem.Abstr.*, 113, 5837). In a series of syntheses of 9-fluorotriptycenes, W. Adcock and V.S. Iyer (*J.Org.Chem.*, 1988, <u>53</u>, 5259) prepared a number of derivatives of 9-fluoroanthracene, often using SF_4 in the sequence ArOH $\rightarrow$ ArF. 1,2,3,4,5,6,7,8-Octafluoroanthracene (**14**; X = F) is obtained by the reaction of $C_6F_5.CH_2.SO.CH_3$ with BuLi to give tetrafluorobenzoisothiophen, and subsequent addition of tetrafluorobenzyne and desulfurization (G.M. Brooke and S.D. Mawson (*J.Chem.Soc., Perkin. Trans.1*, 1990, 1919).

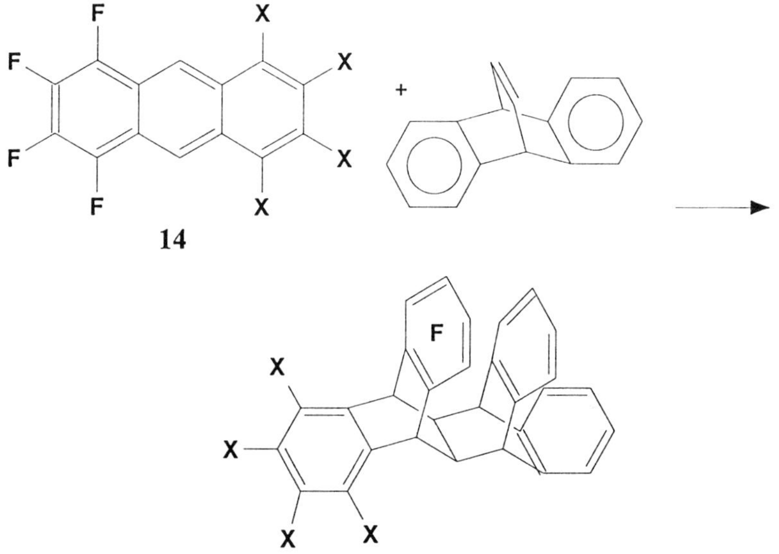

Scheme 1: Janusene structures from Diels-Alder reactions of anthracenes

The Diels-Alder adducts of polyfluoroanthracenes (**14**, X = H) with 9,10-etheno-9,10-dihydroanthracene (Scheme 1) gives janusenes such as **15** in which the proximity of benzene and polyfluorobenzene systems causes considerable donor-acceptor interactions (R Filler and G.L. Cantrell, *J.Fluorine Chem.*, 1987, 36, 407).

Similar such intramolecular interactions are reported by N. Kitagushi (*Bull.Chem.Soc.Jpn.* 1989, 62, 800) in triptycene-1,4-quinones (**16**), themselves derived from appropriately substituted anthracene derivatives. In general mono- and di-halogenoanthracenes are still made by the reduction of the corresponding 9,10-quinones; for example, 1,8-dibromoanthracene is obtained (M.W. Haenel *et al.*, *Chem.Ber.*, 1991, 124, 333) from potassium anthraquinone-1,8-disulfonate *via* the 1,8-dibromo-analogue (Br_2-H_2O, 260^0, 48h.; 70%) which may then be reduced (Al cyclohexoxide-$C_6H_{11}OH$).

The proliferation of triptycene structures by such Diels-Alder reactions of anthracene derivatives leads to "supertriptycene", $C_{104}H_{62}$ (K.Shahlai and H. Hart, *J.Amer.Chem.Soc.*, 1991, 112, 3687) and is discussed in Chapter 30, Section 13.

16   **17**

Large, linked phthalocyanines are obtained (H. Lam *et al.*., *Can.J. Chem.*, 1989, 67, 1087) by reacting iodophthalonitrile with 1,8-dichloro-anthracene so that the 1,8-anthracenylidene fragment acts as a spacer linking two phthalocyanine fragments. Nucleophilic attack of mono-halogeno-anthracenes may be achieved with Cu catalysis and OR⁻-ROH (1-Br; S. Kato and M. Ishizaki, JP 62,106,029; *Chem.Abstr.* 1988, 108, 93811). Correspondingly, **17** (X = Cl) reacts with KO*t*Bu in ethane-1,2-diol at 100°

to give the 2-hydroxyethyl ether (**17**; X = OCH$_2$CH$_2$OH. K.W. Blair, US 4,719,049; *Chem.Abstr.*, 1989, <u>110</u>, 134901).

Sulfur nucleophiles also react readily with 9-chloro- or 9,10-dichloro-anthracenes. In tetraglyme C$_{12}$H$_{25}$S$^-$ replaces Cl to give the RSAr system, but PhS$^-$ reacts with 9-Cl-**13** to give both PhSAr and also **13** by a reductive dehalogenation (S.D. Pastor, *Helv.Chim.Acta,* 1988, <u>71</u>, 859). 9-Iodo-anthracene is obtained (CuI; KI, HMPT) from the bromo analogue (H. Suzuki, *et al., Synthesis* 1986, 121) and may then react with Na$_2$Te (H. Suzuki *et al., Synthesis,* 1989, 468) and with alkynes (S. Okada *et al., JP* 01,233,247; *Chem.Abstr.* 1989, <u>112</u>, 178346). Reduction of halogeno-anthracenes by H$_2$-RhCl$_3$-Aliquat 336 removes only aliphatic and not aromatic halogen; 9-Br-**13** therefore gives 9-bromo-1,2,3,4-tetrahydro-anthracene (**18**) and 9-Cl-**13** behaves similarly. 2-Fluoroanthracene gives 6-fluoro-1,2,3,4-tetrahydroanthracene (I. Amer *et al., J.Mol.Catal.,* 1987, <u>39</u>, 185).

18

Succinoylation of anthracene derivatives remains an effective way to elaborate a new ring system; 4-(2'-anthracenyl)-2-methylbutyric acid leads to 2-methylbenz[*a*]anthracene by the Haworth process (L.M. Deck, G.H. Daub and A.A. Leon, *Org.Prep. Proced.Int.*, 1987, <u>19</u>, 277).

Chloromethylation and bromomethylation are achieved with (para)formaldehyde and the appropriate HX (*e.g.* M. Tanaka, F. Urano and R. Maruhashi, JP 62,108,864; *Chem.Abstr.,* 1987, <u>108</u>, P94407). These benzylic halides then allow conventional nucleophilic substitution to the corresponding 9,10-dialdehyde (Tanaka, Urano and Maruhashi, *loc.cit.*) and to the acetates (B.V. Suvorov, L.A. Krichevskii and A.K. Amirkhanova, *Izv.Akad.Nauk Kaz.SSR, Ser.Khim.,* 1990, 62). The elaboration of ArCHO to ArCH$_2$CHO may be brought about by condensation with CH$_3$NO$_2$ and reducing the ω-nitrostyrene so produced so that the intermediate aldoxime may be hydrolysed. Applied to the anthracene system and using $SnCl_2$ as the reducing agent, this process gives similar amounts of both E- and Z-oximes (**19**) (G.W. Kabalka, L.H.M. Guindi and R.S. Varma, *Tetrahedron*, 1990, <u>46</u>, 7443).

CH$_2$.CH=NOH

19

As with chlorination and other types of electrophilic substitution, chloromethylation proceeds mainly at C-9 and C-10 if these are available. The rate of attack may be affected by the electron demands of the substituent at C-9 but substitution is rarely diverted from C-10 even by substituents such as -NO$_2$ or -CHO. Substituents in the outer rings often influence the direction of attack at the *meso*-positions; thus, 2-ethoxyanthracene gives 2-methoxy-9-(bromomethyl)anthracene as the sole isolable product (C. Bilger, P. Demerseman and R. Royer, *J.Heterocycl.Chem.*, 1987, <u>24</u>, 565), suggesting a major contribution from structure **20**.

MeO

20

In contrast, 2,6-dimethoxyanthracene gives 1-formylation (Bilger, Demerseman and Royer, *Bull.chem.soc.Fr.*, 1986, 807) and 2-iodo-1,5-bis(methoxymethoxy)anthracene is obtained from 1,5-dihydroxyanthracene and the derived ether by metallation (Li), tributylstannylation (Bu$_3$SnCl) and treatment with I$_2$ (J. Gomez-Galeno and J.H. Zaidi, *Tetrahedron Lett.*, 1988, <u>29</u>, 6909) and this methodology is applied to the synthesis of some naturally occurring anthracene derivatives (M.A. Tius *et al.*, *J.Amer.Chem. Soc.*, 1991, <u>113</u>, 5775).

While chloromethylation of anthracene gives the 9-isomer, 1- and 2-(chloromethyl)anthracenes are obtained from the corresponding anthracene carboxylic acids by conventional chemistry, and undergo similar nucleophilic displacement processes (J. Goto, Y. Saisho and T. Nambara, *Anal.Sci.*, 1989, 5, 399).

Alkylation of anthracene (**13**) occurs when Me_3COH reacts in the presence of Lewis acids ($AlBr_3$, $AlCl_3$, $TiCl_4$, or $CF_3.CO_2H$), giving similar amounts of **21** and **22** along with some 2-^{t}Bu-**13**. In $MeNO_2$-CCl_4, $AlCl_3$ causes disproportionation, so that single dibutylanthracenes form 1:1-mixtures of the 2,6- and 2,7-isomers (Yu.V. Pozdnyakovich, *Zh.Org.Khim.*, 1986, 22, 590).

21

22

Nitration of anthracene to give 9-nitroanthracene may also occur by nitrosation, an interesting variant of the "nitration *via* prior nitrosation" mechanism (C.K. Ingold *et al.*, *J.Chem.Soc.*, 1950, 2628; L.R. Dix and R.B. Moodie, *J.Chem. Soc., Perkin Trans.2*, 1986, 1097; J.H. Ridd, *Chem.Soc.Rev.*, 1991, 20, 149; see also S. Sankaraman and J.K. Kochi, *J.Chem. Soc., Perkin Trans. 2*, 1991, 1). $NO^+PF_6^-$ forms a 1:1-complex with **13**. In CH_2Cl_2 this loses NO and forms $(ArH)^+ \cdot (PF_6)^-$, but in MeCN and in the presence of oxygen 9-NO_2-anthracene is formed along with anthraquinone (E.K. Kim and J.K. Kochi, *J.Org.Chem.* 1989, 54, 1692). Similarly, F. Radner (*Acta Chem.Scand.*, 1991, 45, 49) found that $NaNO_2$-CH_3OH in CH_2Cl_2-CF_3CO_2H converts **13** into 9,10-anthraquinone, observing electron-transfer between NO^+ and the substrate but deducing that the oxidation occurs through 1,4-addition across C-9-C-10. With N_2O_4

240

anthracene gives *cis* and *trans*-9,10-dihydro-9,10-dinitroanthracenes in an apparently free-radical addition; other products derive from these two (G. Squadrito *et al.*, *J.Org.Chem.*, 1990, 55, 4322). The oxidation of **13** and some derivatives to anthraquinones may be brought about using $Cu(NO_3)_2$-$Zn(NO_3)_2$ on silica gel, and might involve a similar mechanism (M Anastasia *et al.*, *Synthesis*, 1990, 1083).

9-Nitroanthracene, treated with KOMe-DMSO, gives the coloured Meisenheimer complex identical to that formed by warming 9-MeO-10-NO_2-9,10-H_2-anthracene (**23**) with 10% KOMe-MeOH (Yu.M. Atroshchenko, S.S. Gitis and A.Ya. Kominskii, *Zh.Org.Khim.*, 1988, 24, 6380).

23

Trifluoromethyl groups may be incorporated through a Grignard method in which CS_2 provides a dithiocarboxylic acid ($ArCS_2H$), and XeF_2 converts this into $ArCF_3$ (M. Zupan and Z. Bregar, *Tetrahedron Lett.*, 1990, 31, 3357)

Trifluoromethylation also is reported on the 0.3-2.0 mole scale through the thermal decomposition of $(CF_3COO)_2$ in solvents such as Cl_2CFCCF_2 at 50-100°; anthracene gives 9-CF_3-anthracene (H. Sadawa, M. Nakayama and K. Akusawa, JP 02 62832; *Chem.Abstr.*, 1990, 113, 77894). Hydroxylation of phenanthrene is achieved by CF_3CO_2OH in CCl_4 or CH_2Cl_2; a mechanistic study suggests that the slow step is the formation of the Wheland intermediate (E.S. Rudakov and V.L. Lobachev, *Kinet.Katal.* 1987, 28, 1335; *Chem.Abstr.*, 1988, 108, 109652).

Anthracene and especially its 9,10-substituted analogues have been widely studied as fluorescent or chemiluminescent materials and many of the literature references deal with such applications. Thus, 9,10-dibromoanthracene promotes the chemiluminescent decomposition of alkyl hydro-peroxides catalysed by Co phthalocyanines (H. Knopf and S. Ivanov, *Oxid. Commun.,* 1988, 11, 69), and the synthesis of a number of 1-, 2-, and 9-substituted anthracenes is reported in the preparation of materials in the fluorescent labelling of carbonyl steroid derivatives (J. Goto, Y. Saisho and T. Nambara, *Anal.Sci.,* 1989, 5, 399).

Similarly, the synthesis of electrically conducting materials has prompted the preparation of a great number of anthracene derivatives, many containing alkenyl or alkynyl groups at C-9 and/or C-10. These have been usually obtained through two processes. The first, exemplified by the synthesis of **24** (Scheme 2) (G.A. Kutikova *et al., Sb.Nauchn.Tr.-Vses.Nauchno-Issled.Inst. Lyuminoforov Osobo Chist.Veshchestv.* 1985, 28, 86; *Chem.Abstr.,* 1987, 107, 58625), treats the alkyne with a Grignard reagent to provide an alkynyl organo-metallic which then attacks the anthraquinone to form a 9,10-dihydro-9,10-anthracene diol; reduction is then achieved by $SnCl_2$ or other conventional reagents.

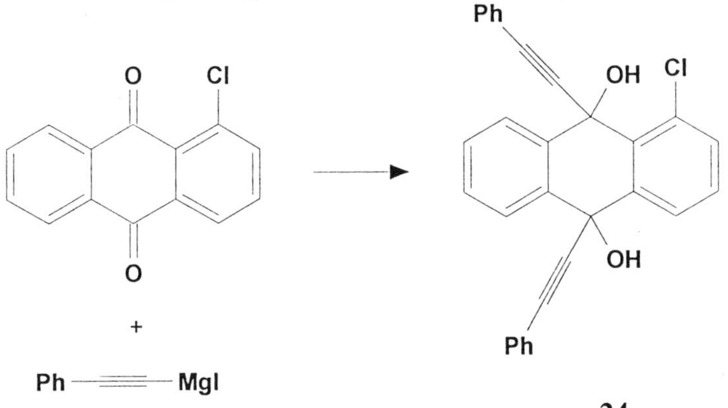

24

Scheme 2: Synthesis of 9,10-Dialkynylanthracene Derivatives

The second uses catalysed coupling of aryl halides; thus, 9,10-dibromo-anthracene and $RC\equiv CH$ react in Et_3N (CuI, $Pd(PPh_3)_2Cl_2$) to give 89% yield of the 9,10-dialkynylanthracene (A.V. Piskunov, A.A. Morea and M.S. Shvartsberg, *Izv.Akad.Nauk SSSR, Ser.Khim.,* 1990, 1441). Such a method is applicable in the synthesis of more complex systems by poly-

Heck methods (P.H. Weitzel and K. Muellen, *Makromol.Chem.*, 1990, 191, 2837). Studies of excitation hopping between two anthracene systems linked by a polymethylene chain (**25**) has used similar synthetic methods (T. Ikeda *et al.*, *J.Amer.Chem.Soc.*, 1990, 112, 4650; see also B. Becker *et al.*, *J.Amer. Chem.Soc.*, 1991, 113, 1121).

25

Birch reduction of 2-methoxyanthracene provides 1,4,5,8,9,10-hexahydro-2-methoxyanthracene (**26**) which is cleaved ($HClO_4$-THF) to give 4,4a,5,6,10,10a-hexahydro-2(3H)-anthracenone (**27**) (W.v.E. Doering and T Kitagawa, *J.Amer.Chem. Soc.*, 1991, 113, 4288). The methoxyanthracene is obtained by classical methods, starting with anisole and phthalic anhydride, followed by ring-closure and reduction of the resulting anthraquinone.

26   **27**

1-Aminoanthracene undergoes the Skraup reaction to give 1-azabenz[*a*]-anthracene ((*a*) H.Y. Wu and E.J. Reist, *J.Heterocycl.Chem.*, 1985, 22, 1597; (*b*) M.J. Tanga, R.M. Maio and E.J. Reist, *Polynucl.Aromat. Hydrocarbons Chem., Charact.Carcinog.,Int.Symp. 9th 1984* 901. Eds. M. Cooke, A.J. Dennis. Battelle Press. Columbus, Ohio); 9-azabenz[*a*]anthracene is obtained by the cyclisation (PPA) of the acetal **28** (Wu and Reist, *loc. cit.*).

28

Diazotisation of 9-aminoanthracene provides dyes through conventional coupling processes (V.I. Eroshkin and N.V. Pavlova, *Avtometriya*, 1988, 28; *Chem.Abstr.*, 1989, 110, 15841) as well as providing a route to 9-fluoroanthracene. Fluoro-arenes may also be made by the Halex exchange reaction; this is exemplified by the formation of 1-fluoro-8-chloroanthraquinone from the 1,8-dichloro compound (CsF-DMSO); selective reduction (Al-H_2SO_4) gives 1,8-dichloro-9-anthrone.

Ladder polymers have also been reported through the interaction of aminoanthracenes with "hexafluoroisopropylidene phthalic anhydride" (**29**) and *o*-phenylene diamine (M.A. Meador, *Polymer Prep. (Am.Chem.Soc., Div. Polym. Chem.)* 1987, 28, 375 and US 231026 (*Chem.Abstr.*, 1989, 111, 174864).

29

The double TCNQ analogue **30** may be obtained from 1,2,3,4,5,6,7,8 octahydroanthracene. Side-chain (benzylic) bromination provides 1,4,5,8-tetrabromo-octahydroanthracene; nucleophilic displacement by NaCH(CN)$_2$ and subsequent aromatisation (Br$_2$ with loss of HBr) then provides the 1,4,5,8-diquinotetramethane derivative (T. Mitsuhashi *et al.*, *J.Chem.Soc., Chem.Commun.*, 1987, 810). The further extended structure **31** is obtained from bianthrone and malononitrile in pyridine (S. Yamaguchi *et al.*, *Tetrahedron Lett.*, 1986, 27, 2411).

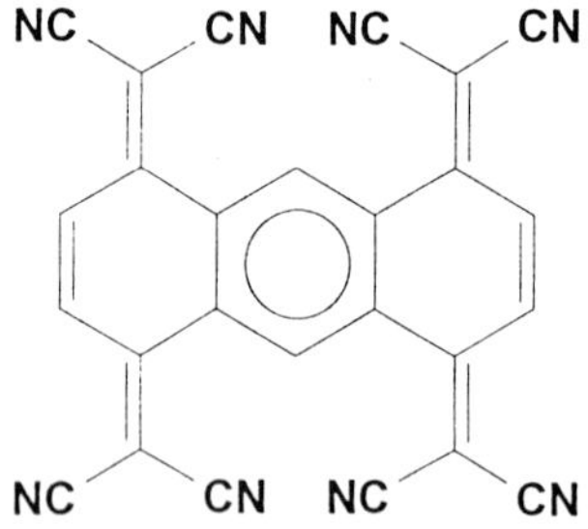

30

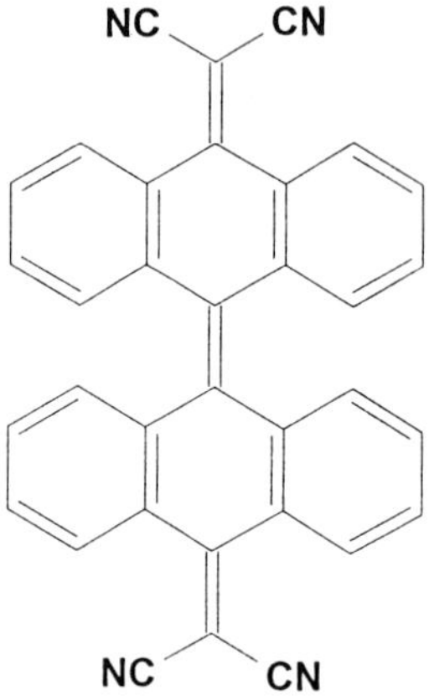

31

Conventional chemistry provides 2-fluoro-9,10-dihydrophenanthrene, a precursor of both 4-(2-fluoro-9,10-dihydrophenanthryl)-4-oxobutanoic acid and of 7-fluoro-9,10-dihydro-2-phenanthrylacetic acid (A. Eiren *et al.*, *Synth.Commun.* 1989, <u>19</u>, 1911)

Chapter 29

POLYANNULAR AROMATIC COMPOUNDS CONTAINING ONE OR MORE
   FIVE-MEMBERED RINGS

R. BOLTON

1.   The benzindene group

A range of isomeric benzindenes are reported in the recent literature, mostly
as their polyhydrogenated derivatives based on secocholestane and
secoergo-stane.

1*H*-Benzo[*e*]indene (**1**) and 1*H*-benzo[*f*]indene (**2**) are conveniently considered together. New methods of the synthesis of the **1** skeleton come in the
cyclisation of 2-(3'-butenyl)-1-naphthyl radicals which may be generated
either by the debromination of the 1-bromo compound by Bu_3SnH (**3**; X =
Br) (A.N. Abeywickrama, A.L.J. Beckwith and S. Gerba, *J.Org.Chem.,*
1987, 52, 4072) or the electrochemical reduction of the 1-PhO-analogue (**3**;
X = OC_6H_5) (T.A. Thornton *et al., J.Amer.Chem. Soc.,* 1989, 111, 2434).
In both cases 1-methyl-1*H*-benzo[*e*]indene is the product. In many
instances, however, conventional chemistry is still being used to construct
the ring system; for example, a range of biocompatible perfumes are
reported (N. Yamada and T. Kobayashi, JP 62,123,110; *Chem.Abstr.* 1987,
107, 140917). These are based upon the reactions of tetralin, isoprene, and
sulfuric acid which form 1,1- and 3,3-dimethyl-2,3,6,7,8,9-hexahydro-**1**.
 Correspondingly the cyclisation of 1,2-bis(trifluoromethyl)naphthalene
gives 1,1,3-trifluoro-2-CF_3-**1** (W.R. Dolbier *et al., J.Amer.Chem.Soc.,*
1991, 113, 1059).

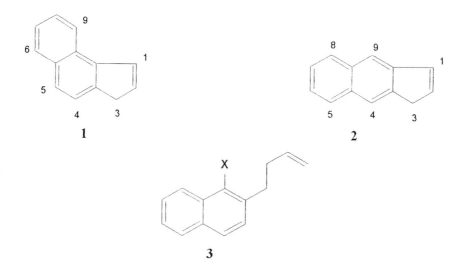

1

2

3

2. The acenaphthene group

(a) Acenaphthene and acenaphthylene - reactions

(4)

(5)

The bond between C1 and C2 in acenaphthylene (4) has considerable π-character, so that the chemistry of this system may be divided into (i) classical arene reactions such as substitution which occur in the naphthalene-like fragment, and (ii) classical alkene reactions such as addition. The chemistry of acenaphthene (5) shows a corresponding tendency towards alkane processes similar to that shown by toluene or ethylbenzene.

The recent reports of the synthesis and chemistry of acenaphthylene and acenaphthene support this generalisation, with relatively little novel chemistry to extend the synthetic methods described in earlier Chapters. Thus, the dehydrogenation of **5** to **4** is achieved (A.G. Russo and K.I. Kuchkova, *Izv. Akad.Nauk Mold.SSR, Ser.Biol. Khim.Nauk*, 1987, 76. *Chem. Abstr.*, 1988, 108, 37357) through bromodeprotonation by NBS and debromination (Zn/DMF) of the resulting 1,2-dibromide; 5-bromoacenaphthene reacts similarly, and Fe is an adequate substitute for Zn in the subsequent debromination (A.G.Russo *et al.*, *idem*, 1988, 73; *Chem. Abstr.*, 1989, 111, 57227). The dehydrogenation of derivatives of **5** may also be achieved with dichlorodicyanobenzoquinone (M.M. Teplyakov *et al.*, *Vysokomd.Soedin, Ser.A*, 1990, 32, 1683; *Chem.Abstr.*, 1991, 114, 7345).

Regioselective hydrogenation of **4** proceeds with ytterbium, either as the metal or as PhYbI, in MeOH; hydrogenation is specific to Ar.CH:CH.Ar systems (Z. Hou, H. Taniguchi and Y. Fujiwara, *Chem.Lett*, 1987, 305).

Acenaphthylene is industrially obtained by the dehydrogenation of acenaphthene (T. Hara and G. Takeuchi, JP 63,306,746; *Chem.Abstr.*, 1989, 111, 7090) or of tetrahydroacenaphthene followed by solvent extraction from 30% aq. NMP into isooctane (T. Hattori, Y. Tachibana, and K. Tate, JP 02,212,439; *Chem.Abstr.*, 1991, 114, 23566). It is formed in very small amount along with other aromatic hydrocarbons (benzene, styrene, naphthalene etc.) by the pyrolysis of ethylene; the corresponding pyrolysis of tetrachloroethylene (K. Ballschmeiter, P. Kirschmer and W. Zoller, *Chemosphere*, 1986, 15, 1369; *Chem.Abstr.*, 1987, 107, 6650) and the CO_2-laser promoted decomposition of Teflon (D.J. Doyle, J.M. Kokosa and D.G. Watson, *Polym.Prepr.(Am.Chem. Soc.,Div.Polym.Chem)* 1987, 28, 250) similarly yield polyhalogenoacenaphthylenes among many other aromatic products.

Flash vacuum pyrolysis of 1,8-bis(chloromethyl)naphthalene (**6**) gives predominantly **4** with a little **5** (E. V. Dehmlow and R. Kramer, *Z.Naturforsch., Anorg.Chem.,Org.Chem.* 1986, 41, 259), a process similar to the cyclodehydrogenation of 1,8-dimethylnaphthalene. The photodecomposition of 1-(diazomethyl)-8-methylnaphthalene provides acenaphthene *via* (8-methylnaphthyl)-carbene and 1,8-naphthaquinodimethane (M.C. Biewer *et al.*, *J.Amer.Chem.Soc.*, 1991, 113, 8069). 8-Methylcyclobuta[*a*]-naphthalene-1,2-dione on flash vacuum pyrolysis provides acenaphthylene in which

^{13}C at C-1 appears mainly at C-1 (65.9%) and C-4 (27.6%) in the resulting **4** (R.F.C. Brown, F.W. Eastwood and B.E. Kissler, *Aust.J.Chem.* 1989, 42, 1435).

The addition of bromine to **4** provides mainly the Z-1,2-dibromo adduct (**7**, X = Br), although the E/Z-ratio increases with lower solvent polarity; chlorine reacts analogously (V.F. Anikin and T.I. Levandovskaya, *Zh.Org. Khim.*, 1988, 24, 1064; *cf.* S.J. Cristol, F.R. Stermitz, and P.S. Ramey, *J.Amer.Chem.Soc.* 1956, 78, 4939). Dehydrohalogenation, expectedly, yields 1-bromoacenaphthylene (V.F. Anikin, T.I. Levandovskaya and Yu.E. Ganin, *Zh.Org.Khim.,* 1986, 22, 628).

CH$_2$Cl CH$_2$Cl

(6)

X   X

(7)

Fluorination of acenaphthylene may be brought about using CsFOSO$_3$ which, in the presence of HF in CH$_2$Cl$_2$, gives Z-1,2-difluoroacenaphthene (**7**; X = F). With MeOH as reaction solvent the corresponding fluoro-methoxy adducts are formed; here the preference in the adduct for the Z-configuration is not so marked (S. Stavber and M. Zupin, *J.Org.Chem.*, 1987, 52, 919).

Addition of Br$_2$ or of BrCl may also be brought about using a polymer-supported reagent. This behaves as a controlled source of halogen without the risk of over-substitution (J. Zabicky, M. Mhasalkar and I. Oden, *Macromolecules* 1990, 23, 3755) while *trans*-chlorination may be achieved using MnO$_2$ - Me$_3$SiCl (F. Bellesia *et al., J.Chem.Res., Synop.* 1989, 108).

Oxidation of >C=C< may be brought about using the expected reagents, or variants upon these; epoxidation occurs when **4** is treated with KO$_2$ and p-O$_2$N.C$_6$H$_4$.SO$_2$Cl in MeCN (H.K. Lee *et al., Chem.Lett.*, 1988, 561) while ozonolysis of **4** and its 1-Me derivative (but not the 1-Ph analogue) gave better yields of ozonide in protic solvents (T. Sugimoto, M. Nojima and S. Kusabayashi, *J.Org.Chem.*, 1990, 55, 3816). The light-catalysed oxidation of **5** is assisted by TiO$_2$. It gave **4**, 1-acenaphthone (**8**; X = O), and 1,8-naphthalide (D.N. Horng, H.K. Hsieh and J.J. Liang, *Hua Hsueh*, 1990, 48, 15; *Chem.Abstr.*, 1991, 114, 206720). and a number of syntheses of naphthalic anhydride (**9**; 1*H*,3*H*-naphtho[1,8-*c,d*]-pyran-1,3-dione) rely on conventional oxidation processes to convert either **4** or **5** to **9**.

8   **9**

Addition to acenaphthylene also occurs with NO (T. Okamoto *et al.*, *J.Org. Chem.*, 1987, 52, 5089) or EtONO (*ibid.*, *idem*, 1988, 53, 4897) in the presence of BH$_4^-$ and a cobalt catalyst such as Co(DMG)$_2$pyCl. The product is the oxime of 1-acenaphthone (**8**; X = NOH), involving sidechain nitrosation and proton-shift; the authors suggest a mechanism involving a metal-alkyl bond fission.

The reaction of sodio-acenaphthylene (**4**) with Cl$_2$BNHCHMe$_2$ appears to involve carbene or carbenoid intermediates to give **10** (A. Meller *et al.*, *Chem.Ber.*, 1987, 120, 1437). In contrast, tetraalkyldiboranes R$_2$BBR$_2$ simply add to give borylated acenaphthenes (R Koester, W. Schuessler and M. Yalpani, *Chem.Ber.*, 1989, 122, 677).

10   **11**

Diels-Alder reactions also occur. Hexachlorocyclopentadiene forms a 1:1-adduct with **4** from which the corresponding 1,1-(MeO)$_2$ derivative **11** may be prepared (H. Heitele, P. Finckh and M.E. Michel-Beyerle, *Angew.Chem.*, 1989, 101, 629)

Acenaphthylene undergoes photochemically excited addition of $CH_2=C(Cl)CN$ in EtBr. The product (12; *E*-isomer shown) may be hydrolysed to form the carboxylic acid (B. F. Plummer and M. Songster, *J.Org.Chem.*, 1990, 55, 1368).

12

Acenaphthylene reacts readily with $Cr(CO)_6$. The resulting tricarbonyl chromium species undergoes easy lithiation at C-3 to give an organometallic species which reacts with D_2O, MeI, Me_3SiCl, and RCHO to give the appropriate and expected products (R.U. Kirss, P.M. Trichel, and K.J. Haller, *Organometallics*, 1987, 6, 242). With carbanionic species such as $LiCMe_2CN$ addition takes place to give, after quenching, dihydroacenaphthylene derivatives; oxidation with iodine provides both the 3-substituted compound $R.C_{12}H_7$ and also dimeric species $R_2.C_{24}H_{12}$.

Hydroformylation of acenaphthylene occurs under very mild conditions, both in the presence of di-μ-chlorotetracarbonyldirhodiumtriphenylphosphine (80%; A. Raffaelli *et al.*, *Synthesis*, 1988, 893) or of [1,5-cyclo-octadiene] $Rh^+BPh_4^-$ (I. Amer and H. Aplen, *J.Amer.Chem.Soc.*, 1990, 112, 3674). The resulting racemic acenaphthylene-1-carboxaldehyde is converted to the acid *via* the oxime and nitrile (Raffaelli *et al.*, *loc.cit.*) suggesting that conventional mild oxidation does not give a pure product.

Acetoxyselenylation of 4 has been reported using PhSeBr in an acetate buffer (L. Engman, *J.Org.Chem.*, 1989, 54, 884)

4-Methoxyacenaphthylene (m.p. 46-8°C), 4-hydroxyacenaphthylene (m.p. 92-4°C) and their 1,2-dihydro-analogues (m.p.'s 89-90° and 148-50° respectively) have been prepared from 9 *via* the 3-sulfonic acid, phenol and methyl ether followed by reduction ($LiAlH_4$) and ring-closure (PhLi) to the methoxyacenaphthene. Dehydrogenation and cleavage of the ether requires carefully controlled conditions (R.F.C. Brown *et al.*, *Aust.J.Chem.* 1987, <u>40</u>, 107). Tyman's observation (J.H.P. Tyman *et al.*, *Synth.Commun.*, 1989, <u>19</u>, 178) that N-methylpyrrolidone greatly facilitated nucleophilic displacement of halogen in the 2- and 4-position of naphthalic anhydride offers an alternative route to the intermediates, and allows the synthesis of 3-bromoacenaphthene and a number of analogues by using a range of nucleophiles.

Anikin and his colleagues have published a number of papers on the preparation and reactions of the more accessible halogeno-acenaphthenes and -naphthylenes (*e.g.* V.F. Anikin *et al.*, *Zh.Org.Khim.*, 1988, <u>24</u>, 174, 181, 1064, 1517), while the preparation of 5,6-dichloroacenaphthylene (T.Otsubo *et al.*, *Synth.Met.*, 1987, <u>19</u>, 595; *Chem.Abstr.*, 1987, <u>107</u>, 198136) lead to the preparation of fluorantheno[3,4-*c,d*][1,2]dithiole and its Se and Te analogues H. Suzuki *et al.* report the preparation of 1,2-diiodoacenaphthylene (m.p. 147-9°; G. Felix *et al.*, *J.Org.Chem.*, 1982, <u>47</u>, 1423 gives 136-7°) by the reaction of the 1,2-dibromo analogue with CuI-KI in HMPT (6h; 150°) and its reaction with Na_2Te (*Synthesis* 1989, 468).

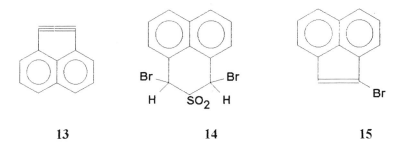

| 13 | 14 | 15 |

Acenaphthyne (**13**) is implicated in the Ramberg-Bäcklund reaction of the dibromo-sulfone **14**. Only 1-bromoacenaphthylene (**15**; 39%) is formed using NEt₃, but traces of the trimeric structure **16** is found with NaOMe, along with **15** (75%) and **4** (9%). With KO*t*Bu, however, **15** (36%) and **4** (27%) are accompanied by appreciable amounts (5%) of **16** (J. Nakayama *et al.*, *J. Org.Chem.*, 1983, <u>48</u>, 60).

4,5-Didehydroacenaphthene is prepared from 5-nitroacenaphthylene by the sequence 5-NH₂ → 5-NHAc → 4-NO₂-5-NHAc → 4-NH₂-5-NHAc → N-acetyltriazole analogue, followed by oxidation with Pb(OAc)₄. It may be

trapped by 7,9-diphenyl-8*H*-cyclopent[*a*]acenaphthylen-8-one (**17**) to give
the crowded hydrocarbon **18** (B.F. Plummer *et al., J.Org.Chem.,* 1991, 56,
3219).

16

17   **18**   **19**

The less accessible 4-bromoacenaphthene derivatives are also available
through substitution of 1,2,6,7,8,8a-hexahydroacenaphthylene (**19**). For
example, 4-Br-5-NH$_2$-**19** may be readily prepared (R. Gruber, E. Kirsch and
D. Cagniant, *Bull. Chem.Soc.Fr.,* 1987, 498). Dehydrogenation of the
substituted **19** should be achieved without loss of the aryl halogen substituent, since in the reverse reaction the hydrogenation of aryl halides by

RhCl$_3$-Aliquot 336 occurs only with loss of halogen from those rings which are hydrogenated (I. Amer *et al.*, *J.Mol.Catal.*, 1987, <u>39</u>, 185)

Acenaphthene-1,2-dione (acenaphthaquinone; **20**) condenses with ketones to form cyclopenta-2,4-dienone analogues (**21**). Diels-Alder reactions with diarylethynes followed by thermolysis of the intermediate adduct provides substituted fluoranthenes (**22**; S.R. Samanta and A.K. Mukherjee, *Indian J.Chem.*, 1987, <u>26B</u>, 26).

20   **21**   **22**

The mono-oxime of **20** affords *cis* and *trans* 1-amino-2-acenaphthenols (aceanthrene, *v.i.*, gives the analogous *trans*-amino-alcohol) which, by reaction with 6-chloro-9-β-ribofuranose, gives easily separable diastereomers of N-modified adenosine. These are equivalent to the RNA adducts formed from the corresponding epoxide metabolites of these hydrocarbons (A.W. Bartczak *et al.*, *Tetrahedron Lett.*, 1989, <u>30</u>, 3251).

23   **24**

The mono-oxime also reacts with phosphonium ylids. With Ph$_3$P=CH.CO.-CH$_3$ the azafluoranthene **23** results, but with *o*-C$_6$H$_4$(CH$_2$PPh$_3{}^+$)$_2$ and

LiOEt benzo[*k*]fluoranthene **24** is formed (G. Papageorgiou, D. Nicolaides and J. Stephanidou-Stephanatou, *Liebigs Ann.Chem.*, 1989, 397).

(b) Aceanthrylene and acephenanthrylene

Aceanthrylene (**25**) and acephenanthrylene (**26**) are respectively linear and angular derivatives of acenaphthylene. Much of their chemistry is predictible from that of acenaphthylene; 1,2-dihydro-**25** and 4,5-dihydro-**26** similarly resemble acenaphthene (**5**). Nitroarenes from both **25** and **26** (among many others) are reported by the action of HO· in the presence of oxides of nitrogen (B. Zielinska *et al.*, *Environ.Sci.Technol.*, 1988, 22, 1044).

25

26

27

MNDO calculations suggest that **4**, **25**, and **26** all behave as though the five-membered ring fragment is isolated from the acene moiety. This is especially evident in the dianions, where the smaller ring has considerable cyclopenta-dienyl anionic nature (C.L. Glidewell and D. Lloyd, *J.Chem.Res.*, *Synop.* 1989, 283). In the reduction of aceanthrylene (Na/NH$_3$) the major product is the 2,6-dihydro derivative (**27**). MNDO calculations here suggest that monoanion stability, and not dianion stability, is important. Further reduction then produces 1,9-ethano-9,10-dihydroanthracene by addition across the 1,10b bond. This rigid, boat-shaped, 9,10-dihydroanthracene shows long-range coupling in the 1H-NMR spectrum (*e.g.* $^5J_{9,10}$ = 4.7 Hz for dipseudoequatorial interactions; P.W. Rabideau *et al.*, *J.Org.Chem.*, 1988, 53, 589).

The carcinogenicity of these hydrocarbons has been discussed in terms of the derived epoxides and carbocations which bring about the formation of DNA adducts during metabolism (R. Sangaiah *et al.*, *Polynucl.Aromat.Hydrocarbons, Chem Charact. Carcinog.,Int.Symp. 9th*, 1984, 795; S. Nesnow *et al., Report 1988. EPA/600/D-88/ 026.* Order No. PB88-170071). The formation of 1,2-dihydro-1,2-dihydroxy-**25**, and the 7,8- and 9,10- analogues, as metabolites of liver microsomes suggests the usual hydroxylation processes (S. Nesnow *et al., Mutat.Res.,* 1989, 222, 223).

Aceanthrene (1,2-dihydro-**25**) is obtained from the quinone directly by Huang-Minlon reduction with N_2H_4 in 95% yield; dehydrogenation (DDQ-MeOH; 75%) then provides **25** (J.C.O. Boerrigter *et al., Recl.Trav.Chim. Pays-Bas,* 1989, 108, 79). Aceanthrylene, like acenapthylene, undergoes sensitised photopolymerisation (Rose-Bengal in MeOH). Head-to-head and head-to-tail linkage in both *syn* and *anti* configurations were detected (B.F. Plummer and S F. Singleton, *Tetrahedron Lett.,* 1987, 28, 480).

Derivatives of both **25** and **26** may be obtained by the ring-opening of α-arylcyclobutenones, themselves obtained from the addition of 1,3-cyclohexadiene to ketenes derived from indane or fluorene. Treatment of the [2+2]-adducts with $MeSO_3H$ causes ring-opening, cyclisation of the resulting carboxylic acid to adjacent aryl rings and, in some instances, dehydration. In such a way 4,5,7,8-tetrahydroacephenanthrylene is formed from indane precursors (Y.S. Chung *et al., J.Org.Chem.,* 1987, 52, 1284). The formylation of 4,5-dihydroacephenanthrylene at C-6 is reported ($SnCl_4$-Cl_2CHOCH_3 in CH_2Cl_2 at 0°C); dehydrogenation (DDQ-PhMe) then gives the fully aromatic alde-hyde (K.W. Bair *et al., J. Med.Chem.,* 1991, 34, 1983).

(c) Benzaceanthrylenes and benzacephenanthrylenes

256

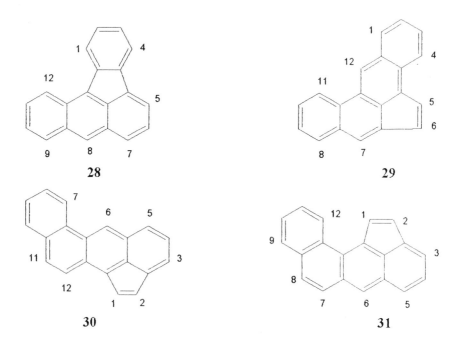

28

29

30

31

Benz[*a*]-, [*e*]-. [*j*]-, and [*l*]-aceanthrylenes (**28-31**, respectively) have each been obtained by new applications of standard methods. Y.S. Chung *et al.* (*J.Org. Chem.*, 1987, 52, 1284) used the ring-opening of α-arylcyclobutan-ones to prepare not only derivatives of **25** and **26**, but also **28** and benz[*e*]-acephenanthrene. Similarly, Harvey's general method of ring synthesis (M. Minabe, B. P. Cho and R. G. Harvey, *J.Amer.Chem.Soc.*, 1989, 111, 3809) has been successfully applied to the synthesis of **28**, when it was also shown that electrophilic bromination occurred at C-8, equivalent to the 9-position of anthracene.

Benz[*e*]aceanthrylene (**29**) has been prepared by a variant upon the use of partially hydrogenated aromatic skeletons in order to achieve selective sub-stitution or cyclisation. Thus, benz[*a*]anthra-7,12-quinone (**32**) is attacked by BrCH$_2$CO$_2$Et (Reformatski reaction) specifically at C-7. The resulting keto-alcohol is reduced to the 7,12-diol (NaBH$_4$) and rearomatised (SnCl$_2$-HCl). In principle, cyclisation to form the new ring may proceed in either direction, but selective reduction to **33** ensures the direction of cyclis-ation to yield **29** (H. Lee and R.G. Harvey, *J.Org.Chem.*, 1990, 55, 3787).

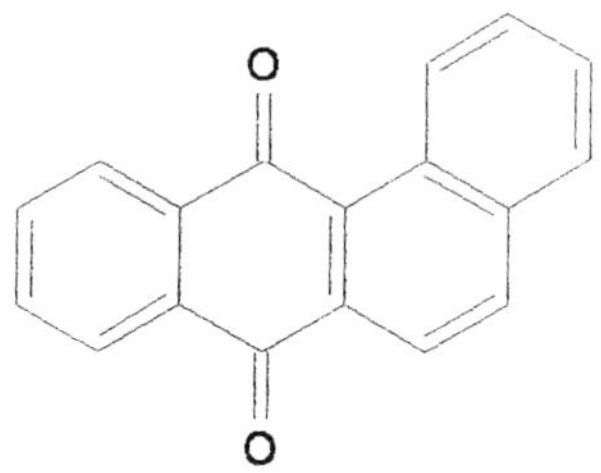

32

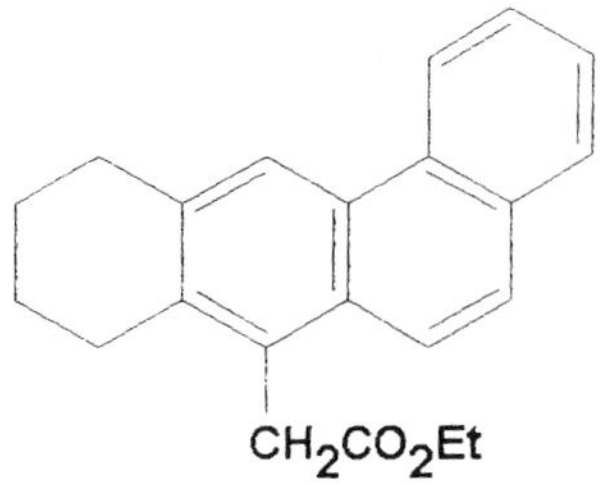

CH$_2$CO$_2$Et

33

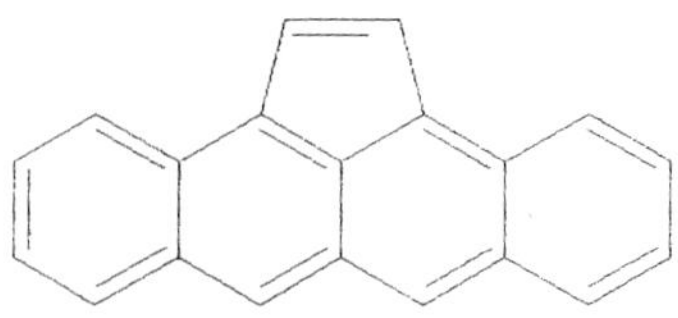

34

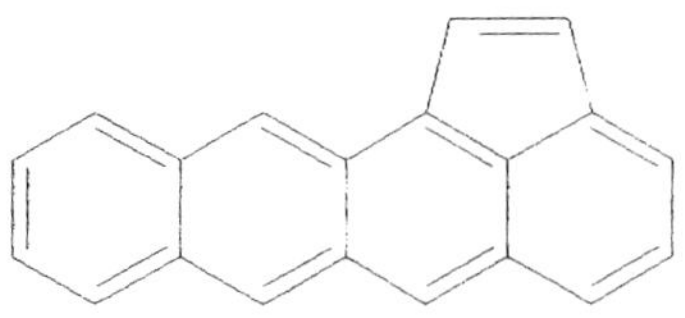

35

Benz[*d*]aceanthrylene (**34**) and benz[*k*]aceanthrylene (**35**) have also been prepared; the starting material in the synthesis of **34** is 1,2,3,4-tetrahydro-naphthacene which, using either ClCH$_2$COCl or ClCO.COCl, directs attack to form the new ring structure after internal Friedel-Crafts cyclisation. The synthesis of acenaphthylene-3,4-dicarboxylic anhydride (**36**) and a Haworth-type synthesis allows the formation of 1,2,7,12-tetrahydro-**35** (R. Sangaiah and A. Gold, *J.Org.Chem.*, 1987, <u>52</u>, 3205). In the same report, a synthesis of benz[*j*]acephenanthrylene (**37**) is achieved using MeCOCH=CH$_2$ as the source of the new ring system attached to the acephenanthrylene fragment.

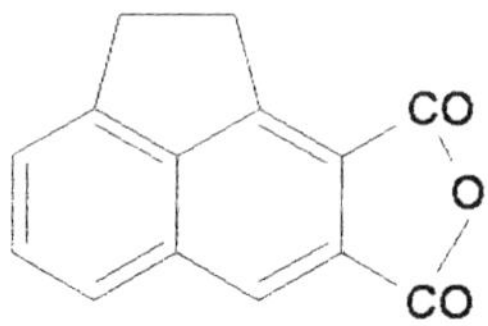

36

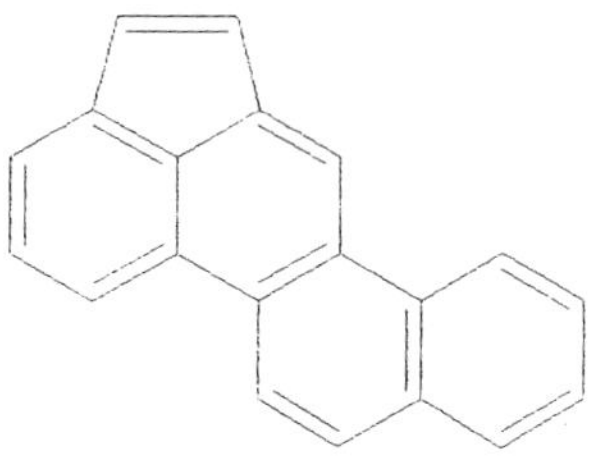

37

The dihydro derivative of benz[*j*]aceanthrylene, cholanthrane, and the fully dehydrogenated structure have each been centres of attention in the understanding of carcinogenesis. The main entries in the current literature report the synthesis of dihydrodiols, tetrahydropolyols, and epoxides derived from these systems; in most cases the same general methods of synthesis are applied, and the Reader is referred to the original literature for reports of the physical properties of these materials. A new synthesis of cholanthrene uses indanone as a source of two of the ring systems and takes advantage of the ready metallation of aryl amides (R.G. Harvey and C. Cortez, *J.Org.Chem.*, 1987, 52, 283).

Cholanthrene is formed by $HI-H_3PO_2-HOAc$ treatment which removes the acetoxyl group in the pentultimate step and, when 2,2-dideuterioindanone is used, also removes a deuterium substituent (Scheme 1).

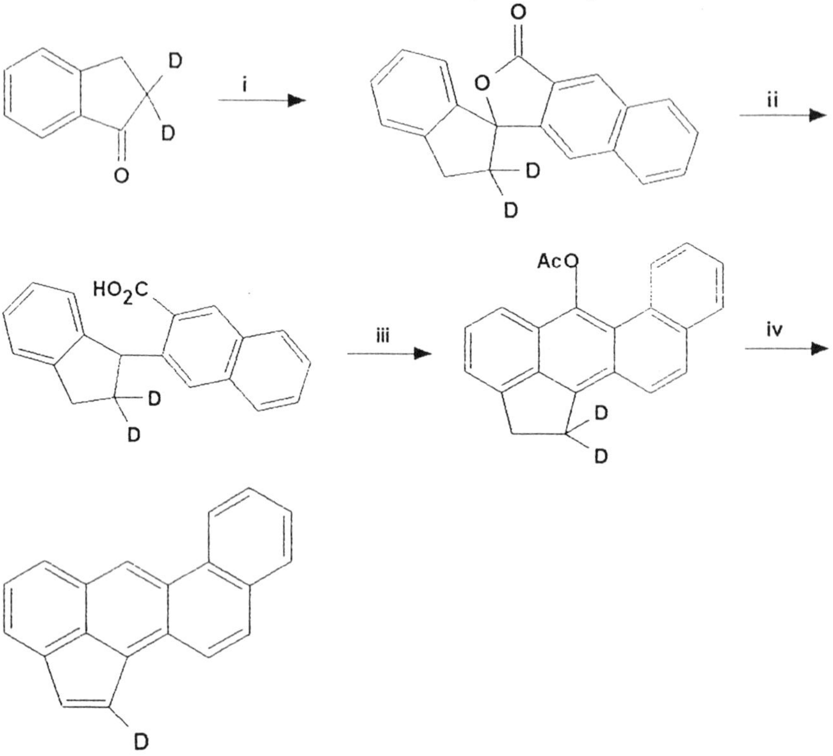

i) 2-Li-1-CONEt$_2$-naphthalene; *p*-Toluenesulphonic acid - PhH; ii) Zn - HOAc; iii) ZnCl$_2$ - HOAc; iv) HI-H$_3$PO$_2$-HOAc

Scheme 1: Cholanthrene synthesis

The synthesis of some mono-nitro derivatives of **28**, **29**, **30**, **31**, and of benz[*k*]acephenanthrylene (**38**) and their subsequent biological testing is reported (J.M. Goldring *et al.*, *Mutat.Res.* 1987, <u>187</u>, 67*)*.

38

3. The Fluorene Group

(a) Fluorene

39   **40**

Alkyl derivatives of 1*H*-fluorene (**39**) are formed in the hydrogenation of coal tar and oils (D.W. Later *et al.*, *Polynucl.Aromat.Hydrocarbons: Chem.,Charact. Carcinogen., Int.Symp., 9th*, 1984, 473 (eds. M. Cooke, A.J. Dennis. Battelle Press: Columbus, Ohio); K. Hasegawa, T. Muragishi and S. Usami, *Nippon Kagaku Kaishi*, 1988, 311; S.L.S. Sarowha, V. Ramaswami and I.D. Singh, *Erdoel Kohle, Erdgas, Petrochem.*, 1989, <u>215,</u> 97), and other analogues have been synthesised in the search for various lubrication fluids [(*a*) T. Tsubochi, N. Shimizu and K. Hata, JP 63,152,335; *Chem. Abstr.*, 1989, <u>110</u>, 118123 (*b*) H. Hata and H. Machida, EP 281,060; *Chem.Abstr.*, 1989, <u>111</u>, 60829] and for new drugs (*e.g.* Y. Oshiro *et al.*, EP 267,024; *Chem. Abstr.*, 1988, <u>109</u>, 92529). Perfluorinated polyhydro-fluorenes have also been included in blood substitutes (S.S. Davies and D.E.M. Wotton, GB 2,171,330, *Chem.Abstr.*, 1987, <u>106</u>, 38465; C.R. Sargent and D.E.M. Wotton, EP 253,529, *Chem.Abstr.*, 1988, <u>109</u>, 114754; S.K. Sharma *et al.*, *Adv.Exp.Med.Biol.*, 1987, <u>215</u>, 97). Hexa- and hepta-chloro-2,3,4,5,8,9-hexahydro-**39**, and analogous derivatives from **40**, have

Table 1:

Selected Acidity Constants for some Substituted Fluorenes and Allied
Systems

Carbon Acid	pK_{HA}	Carbon Acid	pK_{HA}
Cyclopentadiene	18.0	Indene	20.1
Cyclopenta[d,e,f]phenanthrene	22.2*	1,2,3,4,5-Me$_5$-cyclopentadiene	26.1
1,2,3,4,5-Ph$_5$-cyclopentadiene	12.5	2-Ph-indene	19.35
1,2,3-Ph$_3$-indene	15.2	Indeno[1,2,3-j,k]fluorene	10.5
Fluorene	22.60	2-Me-fluorene	23.1
3-Me-fluorene	23.4	9-Me-fluorene	22.3
9-Et-fluorene	22.7	9-^{i}Pr-fluorene	23.2
9-^{t}Bu-fluorene	24.35	9-^{t}BuCH$_2$-fluorene	20.3
9-Ph-fluorene	18.0	9-Ph$_2$CH-fluorene	20.95
9-Ph$_3$C-fluorene	20.3	9-(4-Ph.C$_6$H$_4$)-fluorene	17.7
2-MeO-fluorene	22.7	3-MeO-fluorene	23.95
2,7-(MeO)$_2$-fluorene	22.95	3,6-(MeO)$_2$-fluorene	25.3
2,3,6,7-(MeO)$_4$-fluorene	25.4	9-MeO-fluorene	22.1
9-^{i}PrO-fluorene	21.4	9-^{t}BuO-fluorene	21.3
2-F-fluorene	21.0	3-F-fluorene	22.7
2-Br-fluorene	20.0	2-PhS-fluorene	20.5
2,7-(PhS)$_2$-fluorene	18.5	2-Me$_2$N-fluorene	24.2
1,2,3,4,5,6,7,8-F$_8$-fluorene	10.8	9-CO$_2$Me	10.35
9-C$_6$F$_5$-fluorene	14.7	1,2,3,4,5,6,7,8-F$_8$-9-CO$_2$Et-fluorene	6.1
11H-Benzo[a]fluorene	20.1	11H-Benzo[b]fluorene	23.6
7H-Benzo[c]fluorene	19.5	11-Ph-11H-benzo[a]fluorene	17.9
11-Ph-11H-benzo[b]fluorene	18.9	7-Ph-7H-benzo[c]fluorene	15.6

*Streitwieser *et al.* report the value as 22.81. In other cases where pK_{HA}
have been measured by both methods the difference between reported values
is less than 0.15 pK units.

been identified as products from the hypochlorite reaction with fluorene in water but under the normal conditions of water treatment monochlorofluorenes are the observed products (S. Onodera *et al., J.Chromatogr.*, 1989, <u>466</u>, 233). 9*H*-Fluorene (**40**) shows the characteristic reactions of a polybenzenoid hydrocarbon, as 2,2'-methylenebiphenyl, together with those of the nucleophilic cyclopentadienyl anion (as 2,3:4,5-dibenzocyclopentadiene).

The Table shows selected pK_{HA} values for some fluorene analogues (F. G. Bordwell *et al., J.Org.Chem.*, 1988, <u>53</u>, 780; *J.Amer.Chem.Soc.*, 1988, <u>110</u>, 2867; *J.Org.Chem.*, 1989, <u>54</u>, 3101 and 4893; A. Streitwieser *et al., J.Org.Chem.*, 1991, <u>56</u>, 1074), Various measurements of such parameters have been reported, together with argument about the appropriateness of each parameter (*e.g.* I.S. Antipin *et al., Zhur.Org.Khim.*, 1989, <u>25</u>, 1153 and 2257; A.I. Konovalev and I.S. Antipin, *Metalloorg.Khim.*, 1989, <u>2</u>, 177; G.A. Artemkina *et al., Dokl.Akad. Nauk SSSR*, 1989, <u>304</u>, 616).

(i) Synthesis and substitution

Electrophilic substitution usually occurs at the 2- and 7-position, further or minor attack occurring at the 4- and/or 5-positions, while the ready formation of the 9-fluorenyl anion allows condensation with aldehydes and ketones to give fulvene derivatives. This tendency to form anions is encouraged by the presence of electron-withdrawing groups at C-9 (CHO, COR, CO_2R, CN); Bavin, for example, (P.M.G. Bavin, *Anal.Chem.*, 1960, <u>32</u>, 554) has proposed the use of methyl fluorene-9-carboxylate (**41**; R = H) as a reagent for the characterisation of alkyl halides through the formation of the corresponding 9-alkyl analogues (**41**; R = alkyl)

41

The most recent work on the fluorene system is usually based upon these observations, and applies the reported chemistry (see the 2nd. Edition, IIIH, pp 158-171) towards the synthesis of new compounds, often for pharma-cological use.

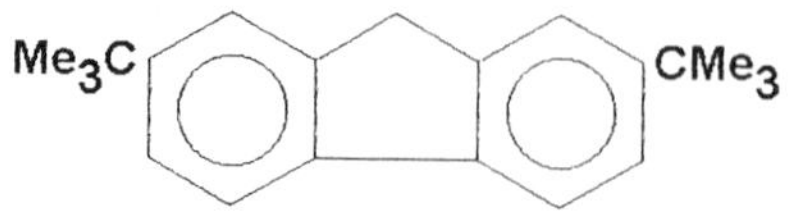

42

t-Butylation of fluorene to give the 2,7-dialkyl derivative (**42**) may be achieved using FeCl$_3$/CH$_2$Cl$_2$ (72%; L.A. Carpino, H.G. Chao and J.H. Tien, *J.Org. Chem.*,1989, *54*, 4302) or by using 2,6-di-*t*-butyl-*p*-cresol, AlCl$_3$, and CH$_3$NO$_2$ (M. Tashio, *Synthesis*, 1979, 921; S. Kajigaeshi *et al.*, *Bull.Chem.Soc.Jpn.*, 1986, *59*, 97). When Me$_3$CCl/AlCl$_3$ was used, the product is contaminated with 2-*t*-butylfluorene. The use of *t*-butyl groups as blocking agents allows the synthesis of 4-X-fluorenes (X = NO$_2$, halogen, CH$_2$Cl) *via* 2,7-di-*t*-butylfluorene; the butyl groups may be removed by trans-alkylation with benzene. In this way 4-Br, 4-CH$_3$, and 4-NHCOCH$_3$ fluorene may be prepared, but since di-substitution of **42** cannot be achieved, 4,5-substituted fluorenes are not directly accessible by this route. A similar attempt to direct further substitution to the 1- and 8-positions by electrophilic attack upon 3,6-di-*t*-butylfluorene, itself made from 2,2'-diiodo-4,4'-di-*t*-butyl-diphenylmethane by Ullmann ring-closure, was unsuccessful; attack took place entirely at C-2 and C-7 (Kajigaeshi *et al.*, *loc.cit.*).

The preparation and formylation (NaH; HCOEt) of **42** and other 2,7-disubstituted fluorenes, and of 1-chlorofluorene (**43**; itself made by the sequence shown below in 78% yield), is also described by Carpino, Chao and Tien (*loc.cit.*).

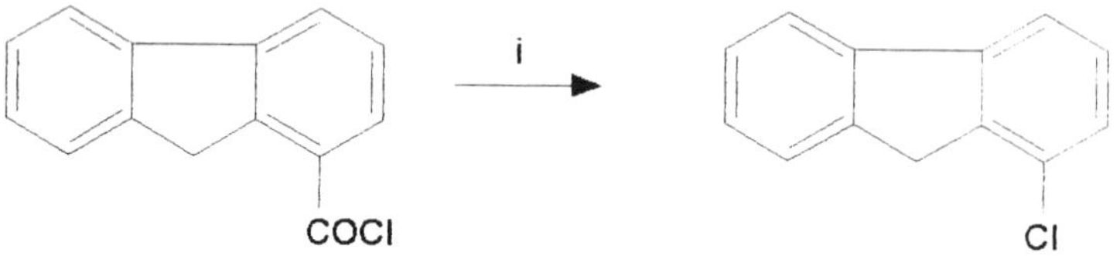

i) (Ph$_3$P)$_3$RuCl

Fluorene is attacked by PhCH$_2$NMe$_3{}^+$Br$_3{}^-$ to give 9-bromofluorene (S. Kajigaeshi *et al.*, *Chem.Express*, 1988, *3*, 347) which is best prepared by the use of N-bromosuccinimide (G. Wittig and G. Felletschin, *Liebigs Ann.*, 1944, <u>555</u>, 133). 9-Bromofluorene is also obtained by treating the complex formed from LiAlH$_4$-reduction of fluorenone with HBr, without isolating the 9-fluorenol (C. Bilger, R. Royer and P. Demerseman, *Synthesis*, 1988, 902); the method is general for the conversion of ketones to the corresponding secondary halides.

The chemistry of many 9-substituted fluorenes devolves upon the stabilities of their carbanions. Thus, the rates and products of reaction of **41** (R = H) and of 9-cyanofluorene with pentafluoropyridine, chloropentafluorobenzene and other activated aryl halides have been reported (G.A. Artamkina *et al.*, *Dokl.Akad.Nauk SSSR* 1989, <u>304</u>, 616; *ibid.*, *Izv.Akad.Nauk SSSR*, 1988, 2826). *p*-Chloronitrobenzene reacts analogously (DMSO-KOH) providing the derivative (**41**; R = *p*-nitrophenyl) in 81% yield (K. Wijciechowski, *Bull. Pol. Acad.Sci.*, *Chem.*, 1988, <u>36</u>, 235). Bordwell has used the reaction of MeI with **41** (R = H), among other fluorene derivatives, as an example of the outer-sphere SET mechanism (F.G. Bordwell and J.A. Harrelson, *J.Org. Chem.*, 1989, <u>54</u>, 4893).

Studies of the acidity of such substituted fluorenes are exemplified by the increased rate of proton loss from carbon acids with increasing DMSO content of DMSO-H$_2$O mixtures (C.F. Bernasconi *et al.*, *Stud.Org.Chem. (Amsterdam)*, 1987, <u>31</u>, 583).

The preparation and photolysis (mode-locked laser) of t-butyl 9-methyl-fluorene-9-peroxycarboxylate **43** shows the formation of the 9-methyl-9-fluorenyl radical **44** by the decarboxylation of the aroyloxy radical **45**, a slow process showing first-order decomposition of **45** (rise time, 55 picoseconds; D.E. Falvey and G.B. Schuster, *J.Amer.Chem.Soc.*, 1986, <u>108</u>, 7419)

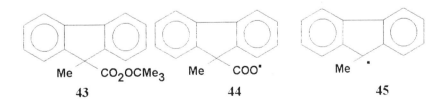

Me	CO$_2$OCMe$_3$	Me	COO·	Me	·
	43		**44**		**45**

264

The intermediacy of **45** is shown by the formation of 9,9-dimethylfluorene (6%) and 9,9'-dimethyl-9,9'-bifluorene (67%) through radical-radical reactions, and 9-(cyanomethyl)-9-methylfluorene (12%) and 9-methylfluorene by radical abstraction from MeCN.

Ring-closure is still a popular route to fluorene derivatives. The benzilic acid derivative **46** undergoes (H_2SO_4-HOAc) ring closure to give derivatives of benzo[*a*]fluorene such as **47**; the isomeric 2-naphthyl analogue may similarly provide the benzo[*c*]fluorene skeleton (A.C. Hopkinson, E. Lee-Ruff and M. Maleki, *Synthesis*, 1986, 366)

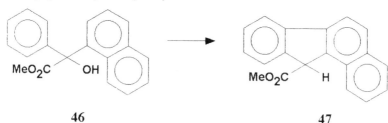

46   **47**

This process parallels the preparation of 9-phenylfluorene from Ph_3COH, and of fluorene-9-carboxylic acid from benzilic acid (H.J. Richter, *Org. Synth., Coll.Vol. IV*, 1963, 482). Fluorenylidene methanone (**48**) condenses with annulenones **49** to form derivatives by addition and the loss of CO_2 from the adduct (T. Asao *et al., Bull.Chem.Soc.Jpn.*, 1986, 59, 1713).

Snieckus (V. Snieckus *et al., J.Org.Chem.*, 1991, 56, 1682) has used the metallation (t-BuLi/LDA) of amides of *m*-teraryl and biphenyl systems to lead to fluorenone derivatives *via* a carbanionic analogue of the classical electrophilic cyclisation process. Thus, 1-phenylfluorenone (**50**) may be obtained from **51**.

48   **49**

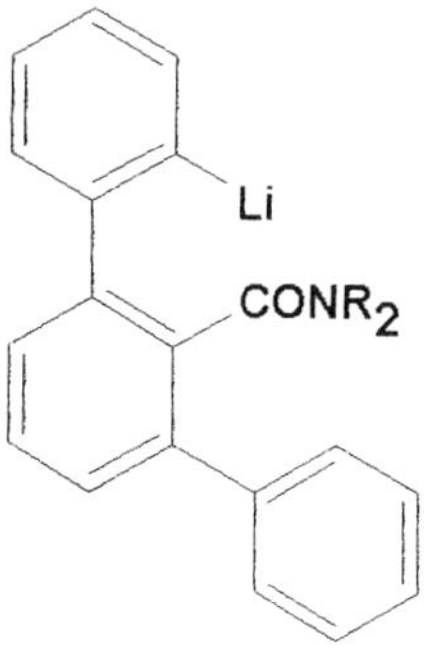

Li

CONR$_2$

51

Li

CONR$_2$

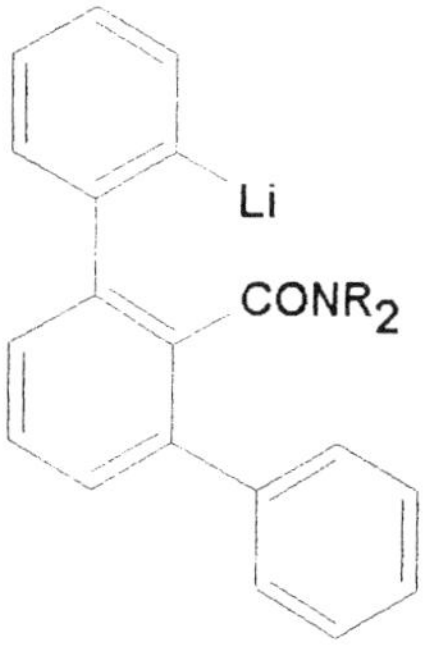

50

Restricted rotation about bonds attached to C-9 has been regularly reported for some time (M. Oki, *Angew.chem.Intern.Edn.*, 1976, 15, 87 and earlier). The more recent NMR studies have analyzed systems such as 52 (X = Me, MeO; A. Nishida *et al.*, *Kogakubu Kenkyu Kaishi (Yamaguchi Daigaku)* 1988, 38, 269).

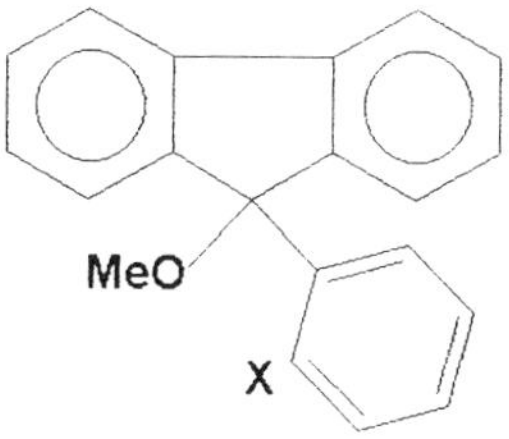

MeO

X

52

(ii) Derivatives

Fluorene may be converted in a five-step synthesis into fluorene-4-acetic acid, the acid chloride of which cyclises (AlCl$_3$) to provide 4*H*-cyclopenta[*d,e,f*]phenanthr-8-ol and the corresponding 8,9-dione. Oxidation (H$_2$O$_2$-HOAc) gives fluorene-4,5-dicarboxylic acid (S. Kajigaeshi *et al.*, *Nippon Kagaku Kaishi* 1989, 2046), which was earlier made by the ozonolysis of pyrene (H. Medenwald, *Chem.Ber.*, 1953, 86, 287). Similarly, fluorene-1,8-dicarboxylic acid is prepared in 4 steps from fluorene (S. Kajigaeshi *et al.*, *Nippon Kagaku Kaishi* 1989, 2052; cf. *J.Org.Chem.*, 1980, 45, 1847.) This, and the derived (NaN$_3$-H$_2$SO$_4$) 1,8-diamine, provide a range of 1,8-disubstituted fluorenes (*i*) by changing the functionality of the -CO_2H group and (*ii*) by diazonium processes from the -NH_2 systems.

The isomeric nitrofluorenes and -fluorenones have been synthesised by regiospecific processes (V. Snieckus *et al., Synthesis* 1989, 184) through 2-methyl-*x*-nitrobiphenyl derivatives, themselves obtained by Pd(PPh$_3$)$_4$-catalysed coupling of arylboronic acids with *x*-nitro-2-bromotoluenes. Oxidation of these methylbiaryls provided fluorenones through the PPA cyclisation of the appropriate carboxylic acids. Difficulty was experienced in the synthesis of 4-nitrofluorenone, but the low yields are justified by the selectivity of the process. Presumably the alternative route *via* the nitration of phenanthraquinone, separation of the 4-nitro isomer and the benzilic acid reaction under oxidising conditions (Scheme 2) was ignored because the isomers could not be separated completely.

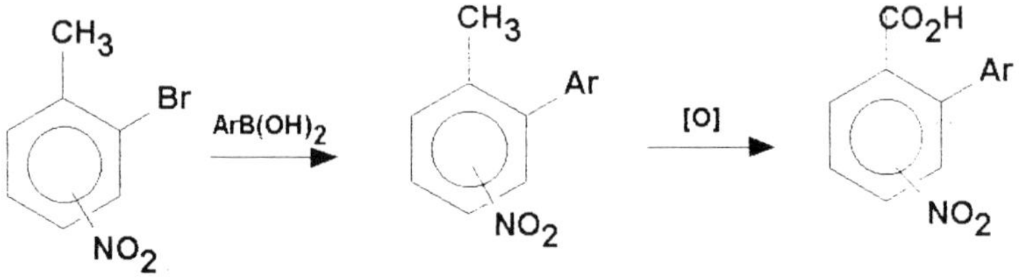

Scheme 2 - Synthesis of nitrofluorenone precursors

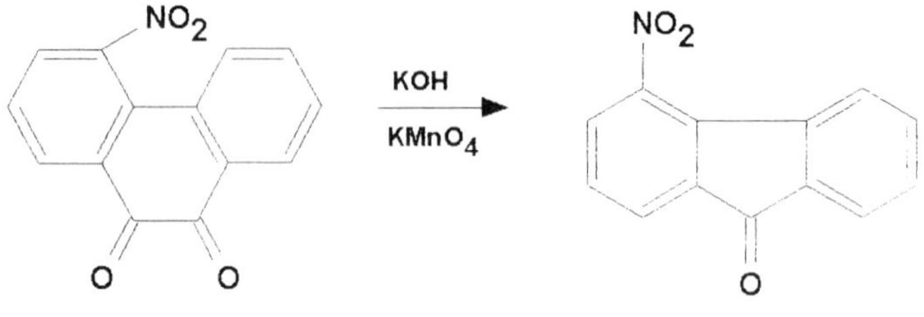

Scheme 3 - Synthesis of 4-nitrofluorenone

The nitration of fluorene yields a mixture of 2- and 4-nitro derivatives from which the 2-isomer may be isolated by crystallisation (W.E. Kuhn, *Org. Synth. Coll.Vol. 2.* 1943, 447). The corresponding nitration of 2- and 3-

methylfluorene involves competition between attack at C-7 and, where encouraged, at the site adjacent to the methyl substituent (M.J. Collins *et al.*, *Aust.J.Chem.*, 1990, 43, 1547; W.A. Vance, Y.Y. Wang, and H.S. Okamoto, *Environ.Mutagen.*, 1987, 9, 123). Thus, 3-methylfluorene gives 2-(53) and 7-nitro (54) derivatives while 2-methylfluorene provides the 3- and 7-nitro isomers.

53   54

Reduction of 2-nitrofluorene has been reported under various conditions such as the nickelocene-LiAlH$_4$ mixture (M.C. Chan *et al.*, *J Org.Chem.*, 1988, 53, 4466) leading to the formation of 2-aminofluorene or its N-hydroxy analogue (M.D. Corbett and B.R. Corbett, *Chem.Res. Toxicol.*, 1988, 1, 222); the last compound is made in the demonstration that the - NH.O.SO$_3$H system is the major ultimate carcinogenic and electrophilic metabolite in the pharmacology of 2-aminofluorene (C.C. Lai *et al.*, *Carcinogenesis (London)* 1987, 8, 471). "Unsolvated" magnesium iso-propylamide in hydrocarbons reduces nitro-arenes such as 2-nitrofluorene substantially to the amino-derivative; nitrobenzene and the simpler nitro-arenes, however, give less fully reduced derivatives, notably azo or azoxy species (R. Sanchez *et al.*, *J.Org.Chem.*, 1989, 54, 4026).

The preparation of 3-nitrofluorene by the nitration of fluorene, elaboration of the 2-nitro-isomer to 2-acetamidofluorene, separation of the 3- (45%) and 7-nitro derivatives obtained by nitration, and hydrolysis, diazotisation and deamination (Me$_3$CONO-THF) has been reported in yields of 20-25% overall (J.R. Moran and G. Emmett, *Org.Prep.Proc.Int.*, 1990, 22, 639). This improves upon the yield of earlier, but similar, routes.

9-Bromo-*N*-(9*H*-9'-fluorenylidene)-9-aminofluorene (55) may be prepared from 9,9-dibromofluorene and 9-(*N*-trimethylsilylimino)fluorene; Lewis acids, by removing the halogen, provide an azaallenium cation (H. Frey, A. Mehlhorn and K. Ruehlmann, *Tetrahedron*, 1987, 43, 2945).

268

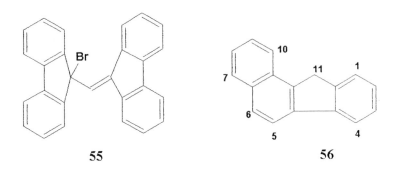

55

56

(b) Benzofluorenes

57

58

Benzo[*a*]-, [*b*]-, and [*c*]-fluorene (**56-58**) have each received attention recently, and again the main areas of interest focus upon the acidity of their derivatives or their application to the cancer problem.

11*H*-Benzo[*a*]fluoren-11-one (**61**) has been synthesised by an apparent rearrangement of 11*H*-benzo[*b*]fluoren-11-one (**59**). This is obtained by the condensation of phthalaldehyde with 1-indanone, and cleavage (KOH-PhMe) to give *o*-(2-naphthyl)-benzoic acid (**60**) which cyclises (PPA) to form the isomeric benzofluorenone **61** (A Streitwieser and S.M. Brown, *J.Org.Chem.*, 1988, **53**, 904).

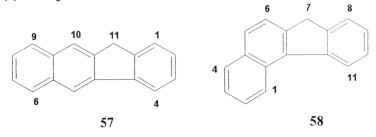

59

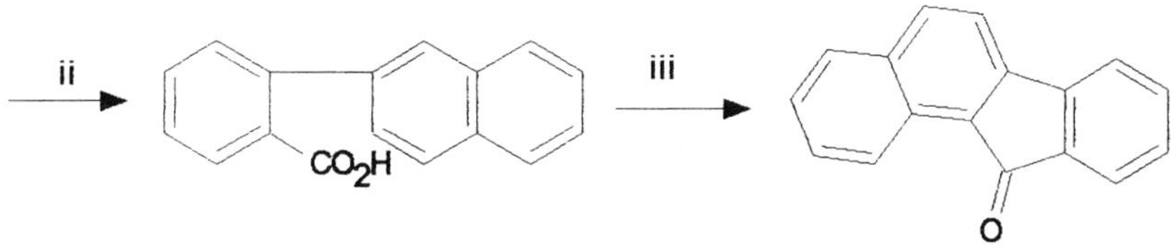

60   61

i) Base; ii) KOH - PhMe; iii) PPA

Scheme 4: Synthesis of 11*H*-benzo[*a*]fluorenone

E. Krogh and P. Wan (*Can.J.Chem.*, 1990, <u>68</u>, 1725) repeated this preparation of **61** and reported syntheses of the corresponding alcohol and hydrocarbon by reduction.

Snieckus' method of metalling arylamides (V. Snieckus *et al.*, *J.Org.Chem.*, 1991, <u>56</u>, 1682) also provides 7*H*-benzo[*c*]fluoren-7-one.

The Stobbe condensation between p-R.C$_6$H$_4$.CO.CH$_3$ (**62**; R = Me, OMe, Cl) and diethyl (cyclohex-1-enyl)succinate leads to an unsaturated diester which cyclises to provide the corresponding 2-(cyclohexenyl)naphthalene system **63**. Saponification and electrophilic acylation of the double bond provides derivatives of tetrahydrobenzo[*b*]fluorenone (**64**; Scheme 5) (H. A. A. Regaila, A.-K. M. N. Gohar and O. M. M. El-Roddi,, *Egypt J.Pharm. Sci.*, 1988, <u>29</u>, 207; Regaila and El-Roddi, *J.Chem.Soc.Pak.*, 1988, <u>10</u>, 79). The analogous process may be carried out with benzaldehydes (Regaila *et al. Egypt J.Pharm.Sci.,*, 1988, <u>29</u>, 219).

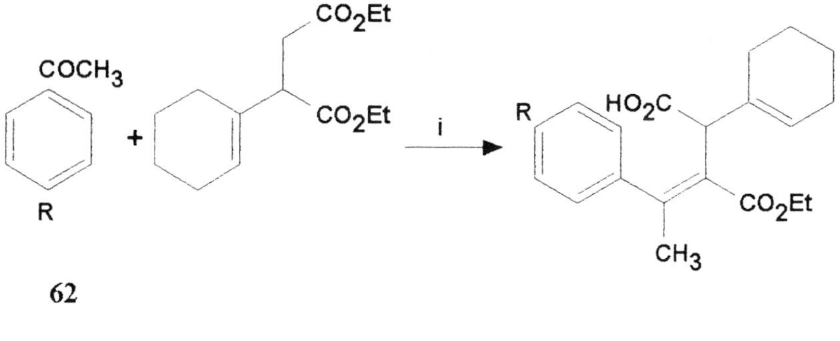

62

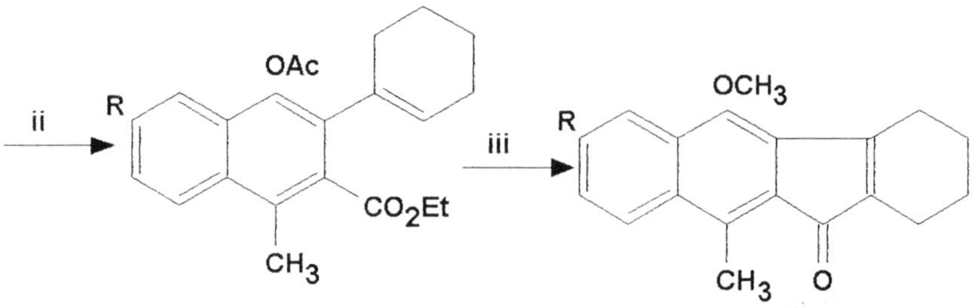

i) NaOEt; ii) NaOH; PPA; Ac$_2$O; iii) acid; methylation

Scheme 5: General synthesis of 11*H*-benzo[*b*]fluorenone

7*H*-Benzo[*c*]fluorene (3,4-benzofluorene, **58**) is formed by the photochemical rearrangement of 1-(bromomethyl)dibenzosemibullvalene (**65**, 1-bromomethyl-3,4;6,7-dibenzotricyclo-[3,3,0,0^{2,8}]octa-3,6-diene) in HOAc (S.J. Cristol and B.J. Vanden Plas, *J.Org.Chem.*, 1989, **54**, 1209); the 9,9-D$_2$ derivative gave 6-deuterio-**58**. This complements earlier studies on the corresponding reaction of the CH$_2$Cl analogue of **65** which gives 15% **58** and acetolysis products (S. J. Cristol *et al.*, *Can.J.Chem.*, 1986, **64**, 1081), and studies of the thermal reaction with AgOAc-HOAc and of the deamination of 1-NH$_2$-dibenzosemibullvalene in HOAc (L. A. Paquette *et al.*, *J.Org. Chem.*. 1972, **37**, 3852), but leaves the mechanism(s) of these processes still unsure..

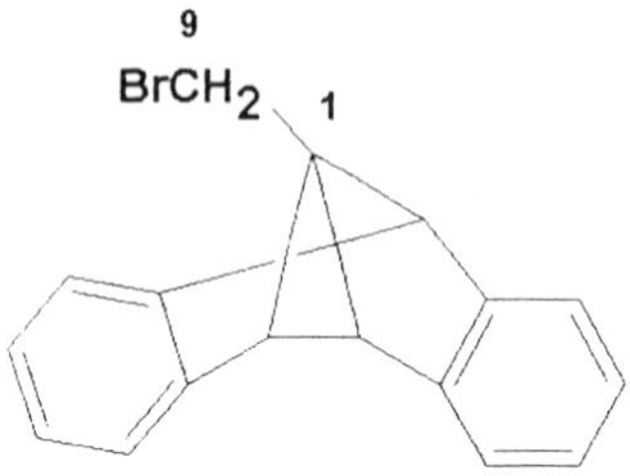

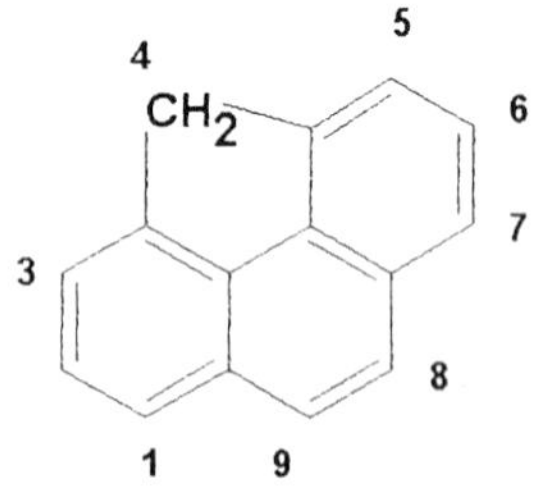

4*H*-Cyclopenta[*d,e,f*]phenanthrene (**66**) may also be regarded as benzo-[*d,e,f*]-fluorene; its chemistry falls directly between that of fluorene, with its acidic methylene protons, and phenanthrene and its ready addition reactions.

Its synthesis and reactions have been reviewed (M. Minabo and K. Suzuki, *Yuki Gosei Kagaku Kyokaishi* 1986, <u>44</u>, 421) and the Japanese workers have expanded their studies of the substitution reactions of 66 in a number of areas.

A total synthesis of **66** uses the application of *t*-butyl groups to block attack at C-2 and C-7 of fluorene, so that 2,7-di-*t*-butylfluorene (**42**) reacts with $ClCH_2OMe-TiCl_4$ to insert the CH_2Cl group. This gives the corresponding acid chloride *via* $ArCH_2CN$ and $ArCH_2CO_2H$; cyclisation ($AlCl_3$), reduction (HI/P in HOAc) and finally dealkylation ($AlCl_3$-PhMe) gives **66** in 13% yield overall (Scheme 6: M. Minabe, M. Yoshida and T. Takayanagi, *Bull.Chem.Soc.Jpn*, 1988, <u>61</u>, 995).

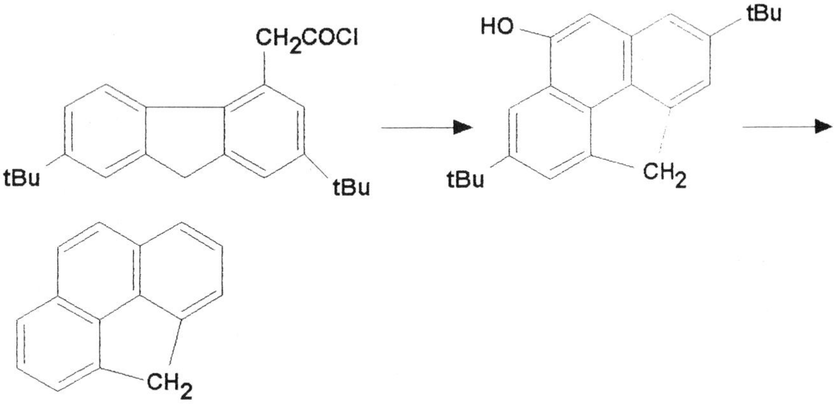

Scheme 6: Synthesis of 4*H*-Cyclopenta[*d,e,f*]phenanthrene

Electrophilic substitution of the parent hydrocarbon has been studied in detail earlier, but the directive effect of substituents is still of interest. The acetamido compounds are much more selective than the amino compounds in electrophilic bromination, and much less prone to encourage disubstitution (Table 2).

The Friedel-Crafts acylation of the four isomeric methoxy-66 systems, and of 2-methoxy-8,9-dihydro-4*H*-cyclopenta[*d,e,f*]phenanthrene shows a competition between the OMe substituent effect and the inherent influence of the parent hydrocarbon. 1-Methoxy-4*H*-cyclopenta[*d,e,f*]phenanthrene gives five mono-acetyl products in which the 2- and 8-isomers predominate

Table 2:

Orientation of bromination of amino-4H-cyclopenta[def]phenanthrenes

Substituent	Site of attack	Yield (%)	Disubstitution	Yield (%)
1-NH$_2$	2-Br	89	-	-
2-NH2	1-Br	80	1,3-Br$_2$	12
2-NHAc	1-Br	99		
3-NH$_2$	8-Br	20	2,8-Br$_2$	34
3-NHAc	8-Br	61	-	-
8-NH$_2$	9-Br	45	3,9-Br$_2$	31
8-NHAc	9-Br	84	-	-

(2.2 mol. eq. AlCl$_3$, 20°); increased amounts of AlCl$_3$ encourage the formation of the 2-isomer but at 75° the 3-ketone comprises 66% of the mixture. 2-MeO-**66** only suffers attack at C-1 (83%) and the 3-isomer correspondingly undergoes exclusive attack at C-8 (92%). 8-Methoxy-**66** gives a mixture of 1- and 3-acetyl derivatives in the ratio of 37:63 (2.2 mol. eq. AlCl$_3$, 20°); this distribution is almost exactly reversed when 3.3 mol. eq. AlCl$_3$ is used (M. Minabe *et al.*, *Bull.Chem.Soc.Jpn.*, 1988, 61, 729).
Although the orientation of acylation of **66** has been carefully studied the products of the Fries rearrangement (often regarded as mechanistically allied with the Friedel-Crafts acylation) have only recently been identified. The relative amounts of isomeric products are dependent upon the reaction conditions, as with the rearrangement of simpler aryl systems such as phenyl acetate (Table 3; M. Minabe *et al.*, *Bull.Chem.Soc. Jpn.*, 1988, 61, 879). The preparations of the isomeric isopropenyl and isopropyl analogues made through the sequence Ar.COCH$_3$ → ArC(CH$_3$)=CH$_2$ → ArCHMe$_2$ are also reported.
Photochemical carboxylation of 4H-cyclopenta[d,e,f]phenanthrene takes place when solutions in DMSO are irradiated in the presence of PhNMe$_2$. This general process, which gives fluorene-2-carboxylic acid (37%) or acenaphthene-5-carboxylic acid (22%) from the appropriate ArH's, gives 8,9-dihydro-4H-cyclopenta[d,e,f]phenanthrene-2- and -8-CO$_2$H as well as 4H-cyclopenta[d,e,f]phenanthrene-8-carboxylic acid. In the absence of CO$_2$ coupling occurs to give *meso*- and racemic 8,8'-bis(8,9-dihydro-4H-cyclo-penta[d,e,f]phenanthren-yl); the stereochemistry was defined because both

Table 3:

Orientation of Fries rearrangement of acetoxy-4*H*-cyclopenta[*d,e,f*]-
phenanthrene

Ester function **Product (%)**
1-OCOCH$_3$ 2-CH$_3$CO-1-OH (85%)
2-OCOCH$_3$ 7-CH$_3$CO-2-OCOCH$_3$ (34%) and a little phenol (C$_2$H$_4$Cl$_2$;
r.t.)
1-CH$_3$CO-2-OH (65%, PhNO$_2$; 40°; 33%, C$_2$H$_4$Cl$_2$, b.p.)
3-OCOCH$_3$ 2-CH$_3$CO-3-OH (62%)
8-OCOCH$_3$ 9-CH$_3$CO-8-OH (51%) and deacylated starting material

products gave the same product (**67**) upon dehydrogenation (M. Minabe *et al.*, *Bull.Chem.Soc.Jpn.*, 1988, <u>61</u>, 2063).

The alkene **68**, an analogue of 9,9'-bifluorenylidene, shows similar twisting about the central double bond, but surprisingly to a smaller degree (30.8°; *cf.* 40.7°) which was rationalised in terms of syn pyramidalisation (J. J. Stezowski *et al.*, *Struct.Chem.*, 1990, <u>1</u>, 123).

67   **68**

4. Cyclopentaphenanthrenes and cyclopentaphenalenes

The recent chemistry of 4*H*-cyclopenta[*def*]phenanthrene has been discussed in Section 3.

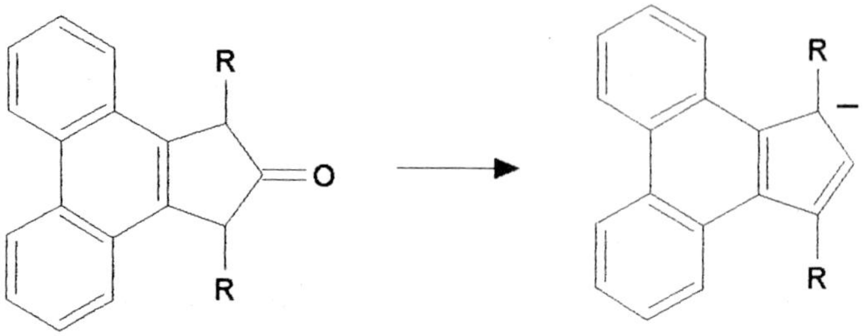

72

Scheme 7: Synthesis of 1*H*-cyclopenta[*l*]phenanthrene carbanions

Dibenzo[*a,c*]cyclononatetraenyl anion (**73**; R = H) readily ring-closes to form the anion of **69**; in contrast, and as a result of steric interactions, no such ring-closure is found with the 5,9-diphenyl analogue (**73**; R = Ph) (B. Eliasson *et al., J.Org.Chem.*, 1989, <u>54</u>, 171).

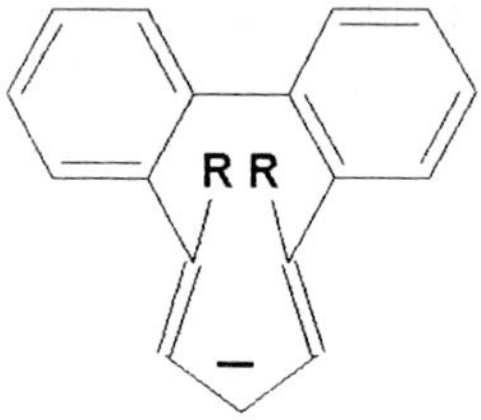

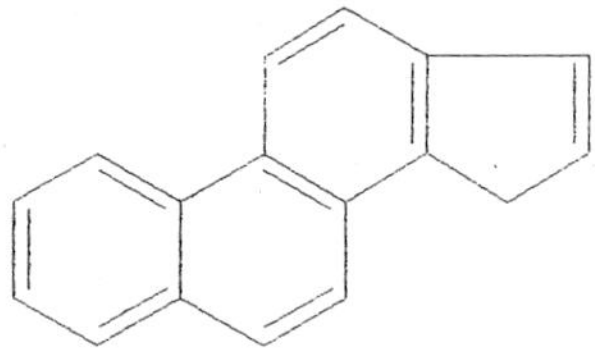

73   **74**

Substituent and structural effects upon the formation of the cyclopenta-[*l*]phenanthrene skeleton through photodehydrocyclisation of 1,2-diaryl-cyclopent-1-enes have been reported (J.B.M. Somers and W.H. Laarhoven, *J.Photochem.Photobiol. A*, 1987, <u>40</u>, 125; 1989, <u>48</u>, 353). The thermal cyclisation of 9,10-bis(1,2,2-trifluorovinyl)phenanthrene gives 1,1,3-trifluoro-2-(CF_3)-**69** through a series of rearrangements (W.R. Dolbier *et al.*, *J.Amer.Chem.Soc.*, 1991, <u>113</u>, 1059).

Cyclopenta[*a*]phenanthrene (**74**) has been competently reviewed (M.M. Coombes, *Polycyclic Aromat.Hydrocarbons Carcinog.: Structure-Act. Relat.*, 1988, <u>1</u>, 59).

5. The Fluoranthene Group

(a) Fluoranthene

(75)

The fluoranthene system (75) arises in a number of pyrolytic processes (G. Perez, A. Cristalli and E. Lilla, *Chemosphere*, 1991, 22, 279; E.J. Soltes and S.C.K. Lin, *Prepr.Pap. - Am.Chem.Soc.,Div.Fuel Chem.*, 1987, 32, 178) including laser angioplasty (J.M. Kokosa and D.J. Doyle, *Chem.Oggi*, 1987, 19; *Chem.Abstr.*, 1988, 110, 100947) and the pyrolysis of benzene-naphthalene mixtures (which also provides the two isomeric phenylnaphthalenes) (R.J. Evans and T.J. Milne, *Energy Fuels*, 1987, 1, 123) as well as the β-particle decay of tritiated naphthalene in benzene solution (G. Angelini *et al.*, *Radi-ochim.Acta*, 1990, 51, 173). Health hazards are reviewed (Anon. *Dangerous Prop.Ind.Mater.Rep.*, 1987, 7, 80). Some substitution reactions of 75 have been well described previously; they may be applied to the synthesis of even more elaborate hydrocarbon systems as in the general method described by B.P. Cho and R.G. Harvey (*J.Org.Chem.*, 1987, 52, 5668).

Nitration of fluoranthene by NO_2-N_2O_4 has been identified to occur under two separate conditions. An electrophilic process, favoured at low temperatures, gives the 2-nitro compound (76); a free-radical reaction by which the 3-nitro-isomer (77) is preferred prevails at higher temperatures. The confusion which this dichotomy presents is exemplified by the formation of 76 as the major product (53.7%) in CCl_4 although 77 is the major (69%) product in CH_2Cl_2 under similar conditions. In both cases, only low conversions of 75 were found ((a) G.L. Squadrito, D F Church and W A Pryor, *J.Amer. Chem. Soc.*, 1987, 109, 6537; (b) Squadrito *et al.*, *J.Org.Chem.*, 1990, 55,

2616). Under classical electrophilic conditions, 3-nitrofluoranthene is the major isomer; 1,2- and 1,3-dinitrofluoranthene are held to arise from further addition of NO_2 to the adduct 78 (B.S. Shane *et al., Environ.Mol. Mutagen* 1991, 17, 130). The preparation of 7- and 8-nitrofluoranthenes has also been described (Shane *et al., loc.cit.*)

The further nitration of 77 by fuming nitric acid is shown to give 3,7-, 3,9-, and 3,4-dinitrofluoranth-enes together with 3,4,7- and 3,4,8-trinitrofluoranthene (R. Nakagawa *et al., Mutat.Res.* 1987, 191, 85). In a report of the mass spectrometry of 18 dinitro- and 4 trinitro-fluoranthenes, which focuses upon electron-impact fragmentation, the further nitration of 3-7- and 8-nitrofluoranthenes is described (T Ramdahl *et al., Biomed.Environ.Mass Spectrom.* 1988, 17, 55).

Formylation of fluoranthene occurs with Cl_2CHOMe (Cho and Harvey, *loc. cit.*; K.W. Bair *et al., J.Med.Chem.*, 1991, 34, 1983) to give a mixture of the 3- (61%), 7- (2%) and 8-formylfluoranthenes (22%. K. W. Bair, US 4,720,587; *Chem.Abstr.*, 1987, 109, 170069); 3-X- and 4-X-fluoranthenes (X = Cl, Br, Et) behaves similarly (Bair, *loc.cit.*). Halomethylation and hydroxymethylation products of 75 are described in studies of the carcinogenic effects of the products. The techniques used did not involve the isolation of these compounds, and the usual method of synthesis was the free-radical attack upon the corresponding methylfluoranthenes (J. C. Ball and W. C. Young, *Chem.Biol Interact.* 1991, 77, 291).

Iodination has been reported under a number of reaction conditions (see R. Pfaender and H. Hoecker, *Makromol.Chem.Macromol.Symp*, 1986, 4, 119). Iodine, followed by HNO_3-H_2SO_4, provides 3-iodo-8-nitrofluoranthene; 3,4-diiodo-8-nitrofluoranthene is also formed in the corresponding reaction of sodium fluoranthene-3-sulphonate (V. M. Zinchenko and N. A.

Ivleva, *Vopr.Khim.Khim.Tekhnol.* 1986, 82, 16; *Chem.Abstr.*, 1988, 110, 212320). Radner (F. Radner, *Acta Chem.Scand.*, 1989, 43B, 481) reports the formation of iodofluoranthene(s) by ICN but without isolation or the separation of possible isomers.

The flash-vacuum pyrolysis of 7,10-bis(ethynyl)fluoranthene provides a route to corannulene (**79**). The starting diacetylene was obtained from dimethyl fluoranthene-7,10-dicarboxylate by reduction ($LiAlH_4$) to the diol, oxidation (PCC) to the dial and conversion (Ph_3P, CBr_4, Zn) to the bis (2,2-dibromoethenyl) derivative. Pyrolysis of this (1000°) provides **79** in 40% yield; 4 equivalents of LDA give 7,10-diethynylfluoranthene which on flash pyrolysis gives 10% yields of **79**, but considerable amounts of the diacetylene are lost by thermal decomposition before sublimation into the reaction chamber.

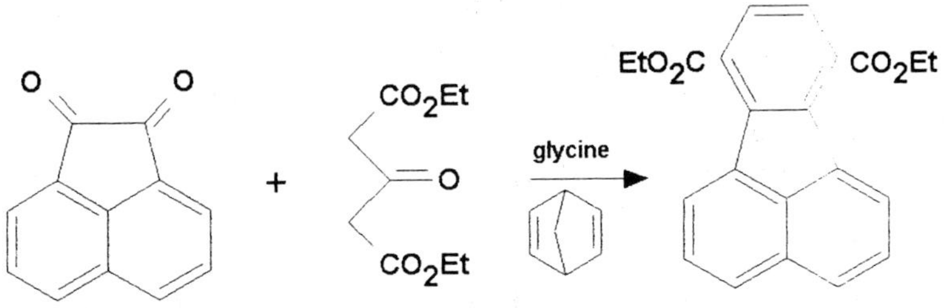

Scheme 8: Synthesis of dimethyl fluoranthene-7,10-dicarboxylate

The parent diester is obtained by a one-pot synthesis in which norborna-diene acted both as solvent and reagent in a Diels-Alder addition with an intermediate cyclopentadienone (Scheme 8) (L. T. Scott *et al.*, *J.Amer. Chem.Soc.*, 1991, 113, *7082*).

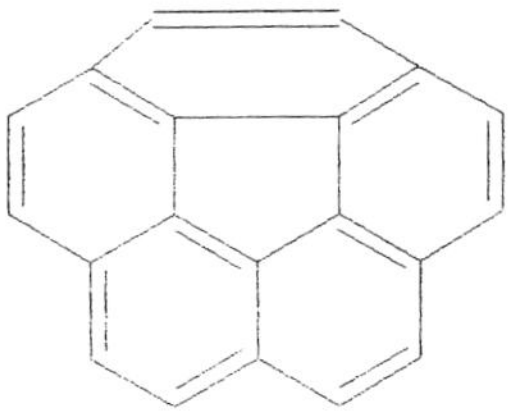

79

(b) Benzofluoranthenes

80   81   82

Much of the recent chemistry of the benzofluoranthenes has been directed towards their carcinogenicity and therefore their metabolites. The carcinogenicity has been evaluated in terms of their ^{13}C-NMR chemical shifts (Y. Sakamoto and Y. Sakamote, *Bull.chem.soc.Jpn.*, 1989, 62., 330) and by CASE-SAR analysis (A. M. Richards and Y. T. Woo, *Mutat.Res.*, 1990, 242, 285). Mathematical models for the predicity of carcinogenicity have also been described (L. Wang *et al.*, *Huanjing Kexue Xuebao*, 1987, 7, 340; *Chem. Abstr.*, 1987, 108. 126345).

Benzo[j]fluoranthene (**81**) is among the polycyclic aromatic systems whose orientation and rate of electrophilic substitution has been linked with DEWAR-PI parameters (M.J.S. Dewar and R.D. Dennington, II, *J.Amer.Chem.Soc.*, 1989, 111, 3804). *trans*-4,5-Dihydroxy-4,5-dihydro-benzo-[j]-fluoranthrene (J.E. Rice *et al.*, *J.Org.Chem.*, 1987, 52, 849) and the 9,10-analogue (J.E. Rice *et al.*, *loc.cit.*; J.E. Rice, N.G. Geddie and E.J. Lavoie, *Chem.Biol.Inter-act.* 1987, 63, 227) were identified as metabolites, along with the 3-, 4-, 6-, and 10-hydroxy derivatives of benzo[j]fluoranthene. .

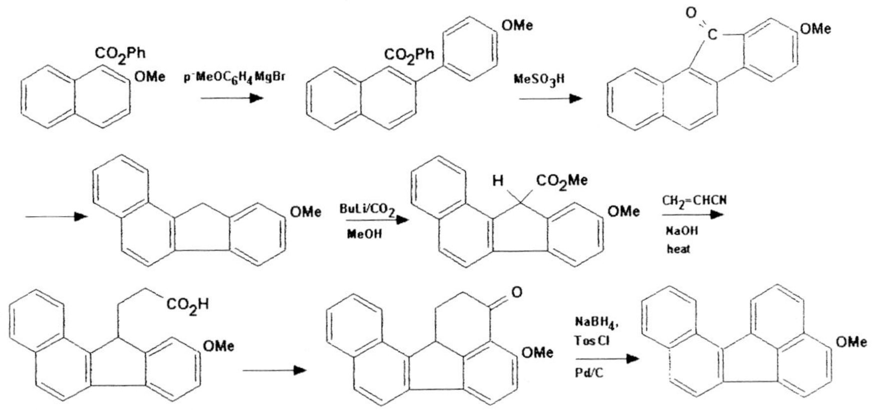

Scheme 8: Synthesis of 4-methoxybenzo[j]fluoranthene

The synthesis of 4-methoxybenzo[*j*]fluoranthene proceeds by conventional chemistry involving the elaboration of the fluoranthene skeleton from a similarly substituted fluorene system. (Scheme 8) Cleavage of the methyl ether (BBr_3) and oxidation (Fremy's salt) provides the 4,5-quinone which, with KBH_4/O_2 yields the *trans*-4,5-dihydro-4,5-diol. This sequence of reactions is common in the preparation of metabolites of carcinogenic hydrocarbons. Thus, 1,2,3,12c-tetrahydro-3-keto-benzo[j]fluoranthene is converted into 2,3-dihydro-2,3-dihydroxybenzo[j]fluoranthene (Scheme 9)

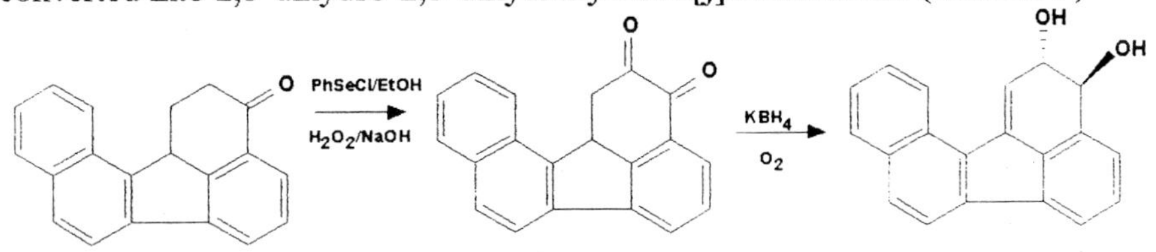

Scheme 9: Synthesis of *trans*-2,3-dihydro-2,3-dihydroxybenzo[*j*]-fluoranthene

and 10-methoxybenzo[j]fluoranthene (Scheme 10) provides the corresponding 9,10-diol.

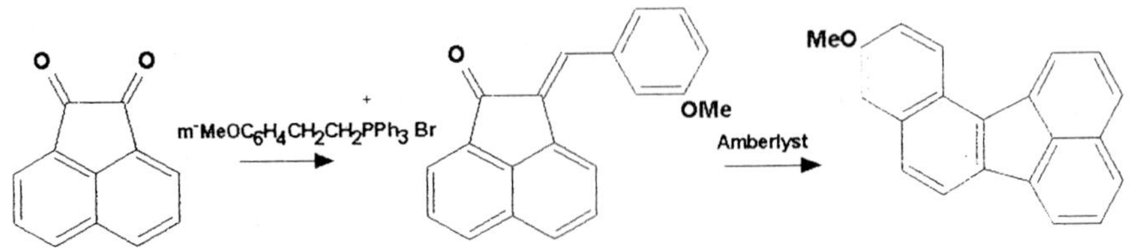

Scheme 10: Synthesis of 10-methoxybenzo[j]fluoranthene

4-Fluorobenzo[j]fluoranthene (83) has been obtained from 9-fluoro-11H-benz[a]fluorenone, itself prepared by conventional chemistry. Reaction with the protected Grignard reagent from BrCH$_2$CH$_2$CHO (as the acetal) quantitatively provides the hydroxyacetal which, with PPA, undergoes cleavage and cyclisation to give 83. (J. E. Rice and Z. M. He, *J.Org.Chem.*, 1990, 55, 5490)

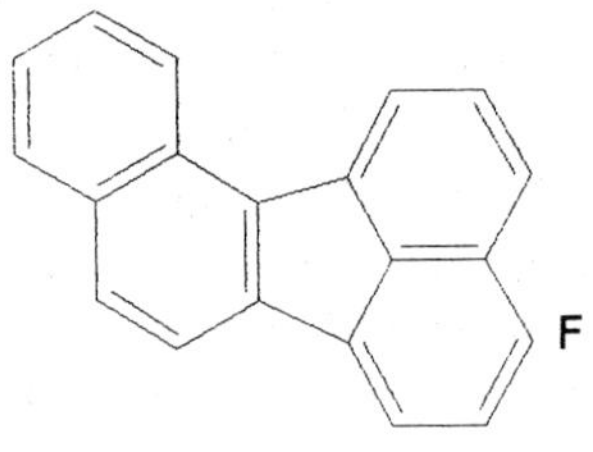

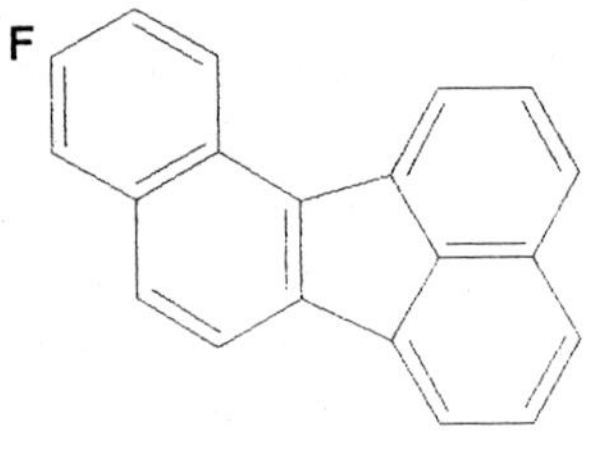

83

84

10-Fluorobenzo[j]fluoranthene (84) is obtained from 1-(p-fluorophenyl)acenaphthylene (85) *via* a carbene addition and fragmentation of the three-membered ring to provide the ester (86). Reduction to ArCHO, and cyclisation (PPA) gives (84) which may also be obtained in the same yield (58%) by the direct reduction of the carbene addition product to the aldehyde, when PPA brings about ring opening and cyclisation in one step (Rice and He, *loc.cit.*)

anti:syn, 2:1

85   86

Benzo[*k*]fluoranthene (**82**) may be obtained by the double Wittig reaction between acenaphthene quinone (**20**) and the triphenylphosphonium salt from 1,2-bis(bromo-methyl)benzene (Papageorgiou, Nicolaides and Stephanidou-Stephanatou, *v.s.*); under phase-transfer conditions, the corresponding 3- and 4-fluoro-*o*-xylene deriva-tives gave 8-fluoro- (15%) and 9-fluoro-benzo[*k*]-fluoranthene (2%; J.E. Rice *et al*., *J.Org.Chem.*, 1988, 53, 1775). Compared with the metabolism of benzo[*k*]fluoranthene itself, the 3-, 8-, and 9-fluoro derivatives are rather more resistant.   Fluorine is not lost during metabolism; 3-fluorobenzo[*k*]fluoranthene gives the 10,11-dihydro-10,11-diol and either the 2,3- or the 4,5-analogues.   The 8-fluoro isomer did not give either 8- or the 11-hydroxy-derivatives, and 9-fluoro-**82** did not give 9-hydroxy-**82**, nor yet the 8,9-dihydro-8,9-diol nor the 10,11-analogues.   The dihydrodiols formed from this compound appeared to be derived from either the 2,3-dihydro-2,3-diol system or the 4,5-analogue.   In contrast, **82** itself provided 8,9-dihydro-8,9-diol, benzo[*k*]fluoranthene-2,3-quinone, and 3-, 8- and 9-hydroxy-**82** (E. H. Weyand *et al.*, *Carcinogenesis (London)*, 1988, 9, 1277).

Similarly, 11*H*-benzo[*b*]fluorenone affords a route to the metabolism products of **82**, making use of the readily formed nucleophilic carbanion derived from **86** and its Michael addition to methyl acrylate. Decarboxylative hydrolysis, and cyclisation of the acid chloride, provides the benzo[*b*]fluor-anthene skeleton (**87**) which by Prevost oxidation gives the dihydrodiol **88** (S. Amin *et al.*, *Polynucl.Aromat.Hydrocarbons [Pap.Int.Symp.] 8th*, 1983,

99. Eds. M. Cooke, A. J. Dennis; Battalle Press, Columbus, Ohio, 1985). The isomeric diol **89** was also obtained.

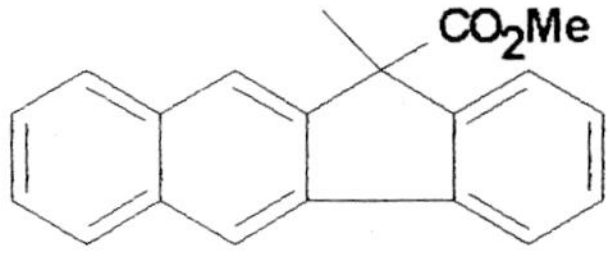

86 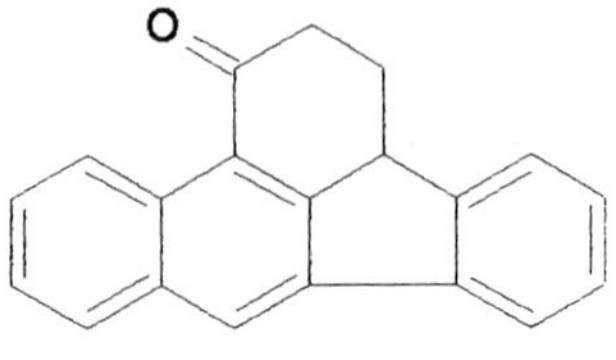   **87**

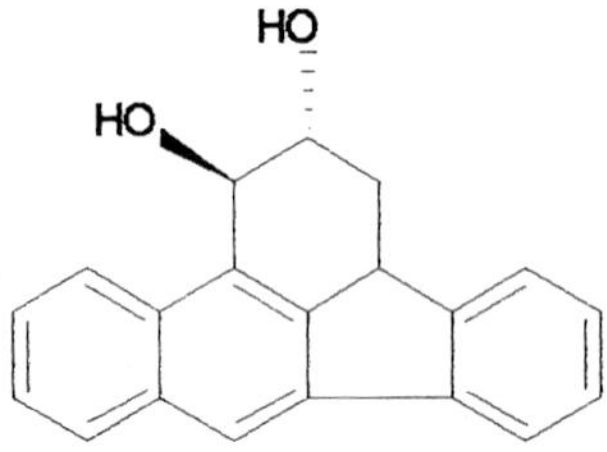

88 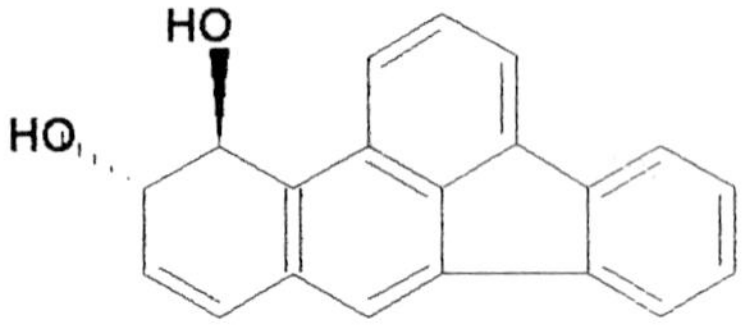   **89**

6. Larger systems containing five-membered rings

(a) Synthesis

The construction of multi-ring systems is, in general, an extension of the methods already described. In this Section, therefore, new general synthetic methods are reported.

R.G. Harvey *et al.* (*J.Org.Chem.*, 1991, 56, 1210) show the wide use of enamines such as **90** and **91** as reactants towards benzylic or 2-arylethyl halides. In the following Scheme the synthesis of benzo[c]fluorene (**56**) is shown; by using 1-naphthyl-methyl chloride, benzo[a]fluorene (**58**) results. Analogously, **91** with 1- or 2-naphthylmethyl chloride leads to the formation of 7H-dibenzo[c,g]fluorene (**92**), 13H-dibenzo[a,i]fluorene (**93**) or 13H-dibenzo[a,g]fluorene (**94**).

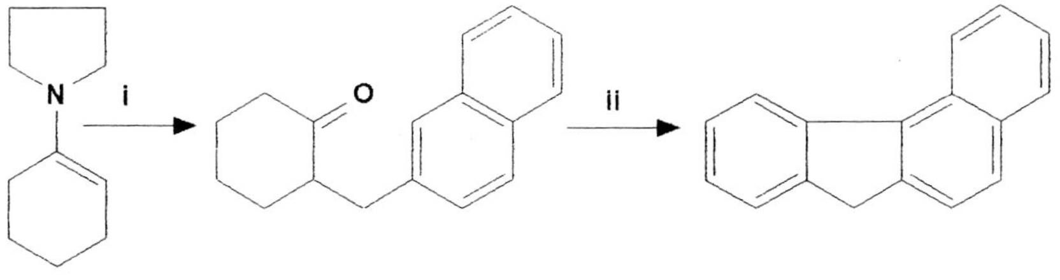

(i) 2-$C_{10}H_7CH_2Cl$ (ii) PPA; DDQ

Scheme 11: Synthesis of fluorenone analogues

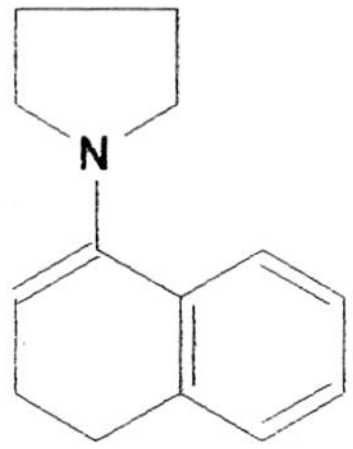

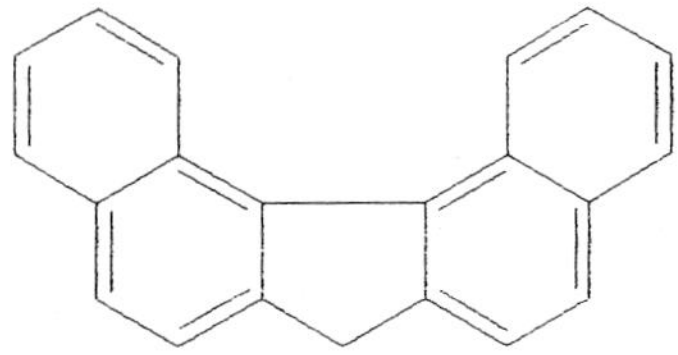

91   92

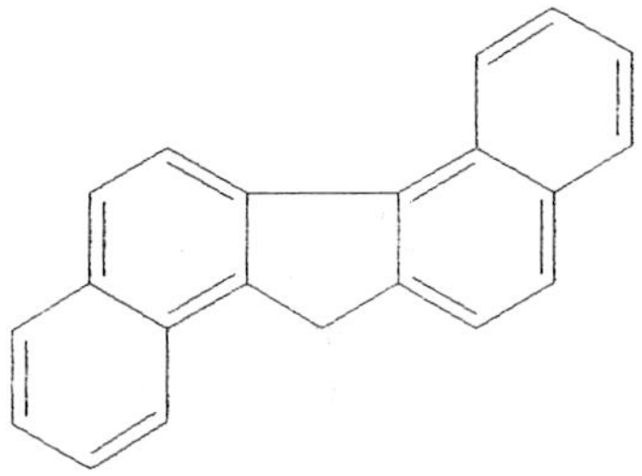

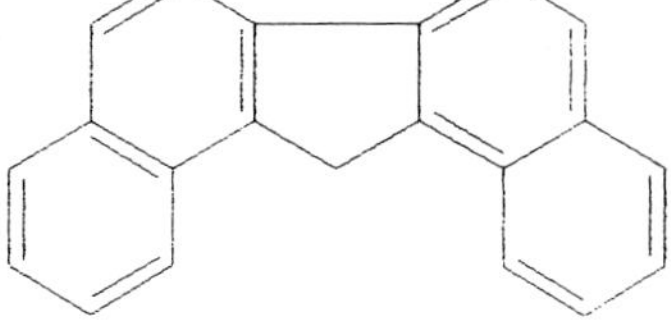

93   94

1-Bromoacenaphthene correspondingly gives 2-(1'-acenaphthenyl)cyclohex-anone which, *via* the epoxide ($Me_2S=CH_2$) provides the 2-arylcyclohexane-carboxaldehyde, and this then gives 95. The reaction between 96 and 90 provides the 2-arylcyclohexanone which (PPA) leads to 4*H*-cyclopenta-[*d,e,f*]chrysene (**97**).

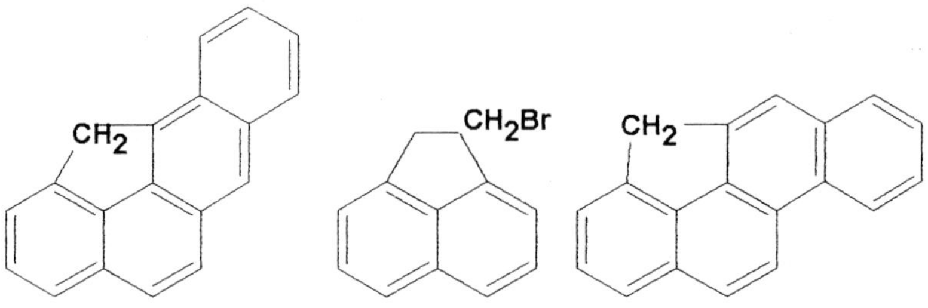

95   96   97

B.P. Cho and R.G. Harvey (*J.Org.Chem.*, 1987, <u>52</u>, 5668) have used cyclo-hexene epoxide as a source of the new six-membered ring; aryl-lithiums then provide the 2-arylcyclohexanol oxidation (PDC) of which provides the 2-arylcyclohexanone whose cyclisation (PPA) and aromatisation (DDQ) then provides the range of structures shown below (Table 7).

Table 7:

Aryl halides as sources of polyannular hydrocarbons

Aryl halide	ArH product
9-Bromophenanthrene	Benz[*e*]acephenanthrylene(**98**)
1-Bromopyrene	Indeno[1,2,3-*c,d*]pyrene (**100**)
4-Bromopyrene	Indeno[1,2,3-*c,d*]pyrene (**100**)
6-Bromochrysene	Indeno[1,2,3-*h,i*]chrysene (**99**)

Table 7 (contd.):

Aryl halides as sources of polyannular hydrocarbons (*contd.)*

6-Bromobenz[*a*]pyrene	Benz[*d,e,f*]indeno[1,2,3-*h,i*]-chrysene (**101**)
6-Bromo-1,2,3,4-H$_4$-benz[*a*]pyrene	Fluoreno[3,2,1-*d,e,f,g*]chrysene (**102**)
7-Bromobenz[*a*]anthracenyl	Dibenz[*a,e*]aceanthrylene (**103**; 46%)
	Dibenz[*a,j*]aceanthrylene (**104**; 40%)
9-Bromoanthracene	Benz[*a*]aceanthrylene (**28**)
6-Cl-1-Br-benz[*a*]pyrene	Benz[*d,e,f*]indeno[1,2,3-*q,r*]-chrysene
4-Bromobenz[*a*]anthracene	Dibenz[*e,k*]acephenanthrylene (**105**)

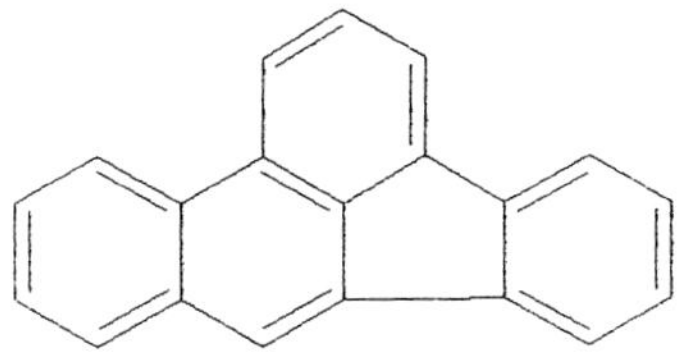

98 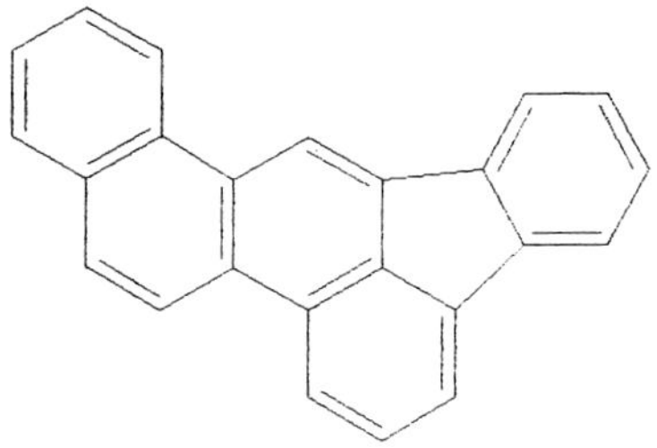   **99**

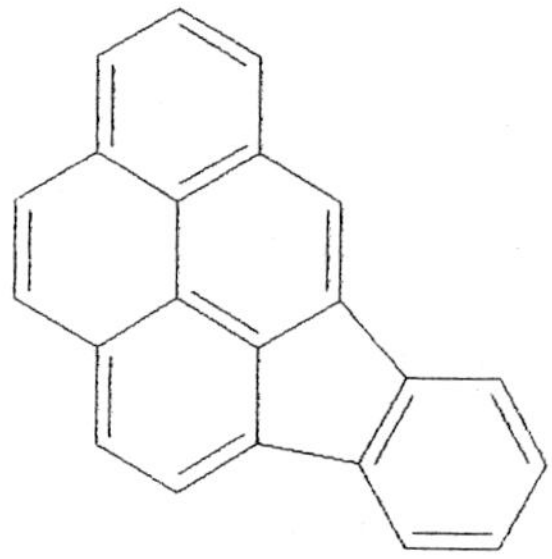

100

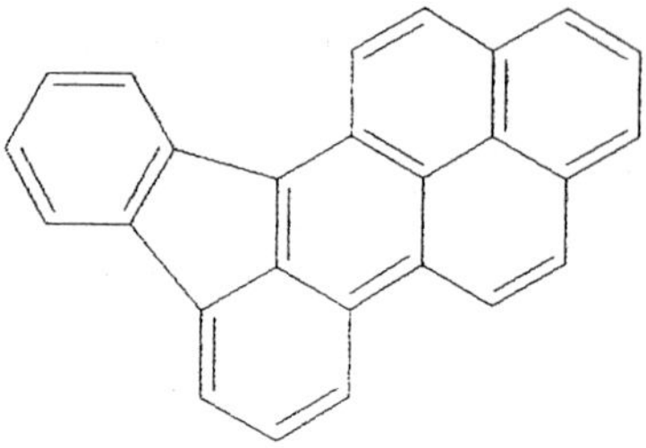

101

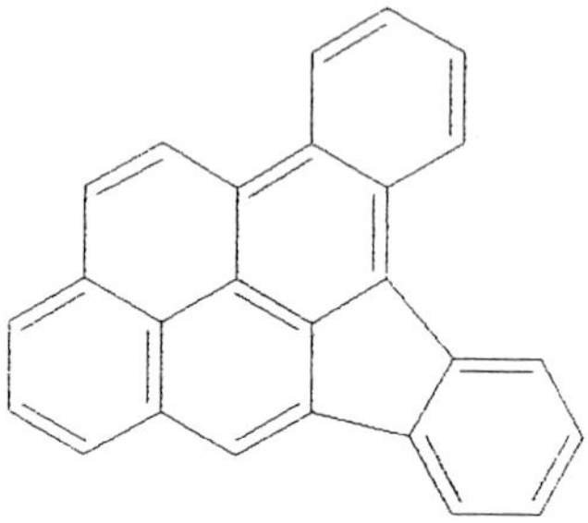

102

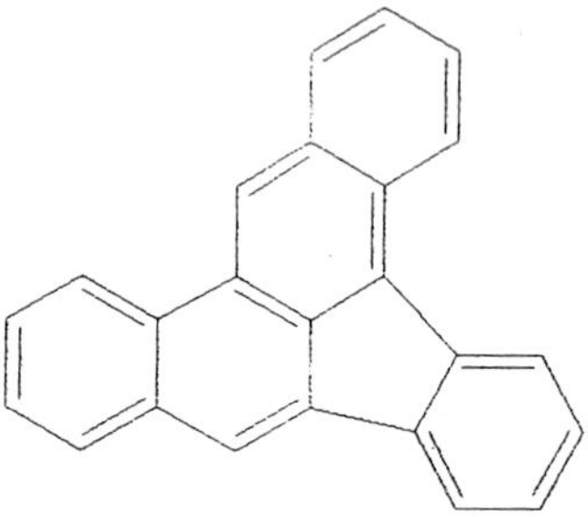

103

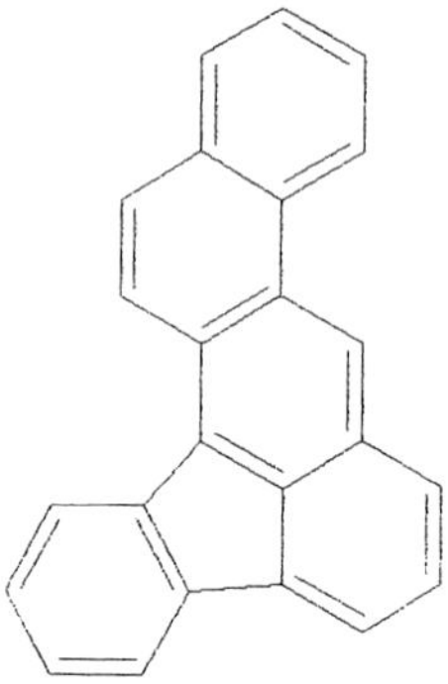

104

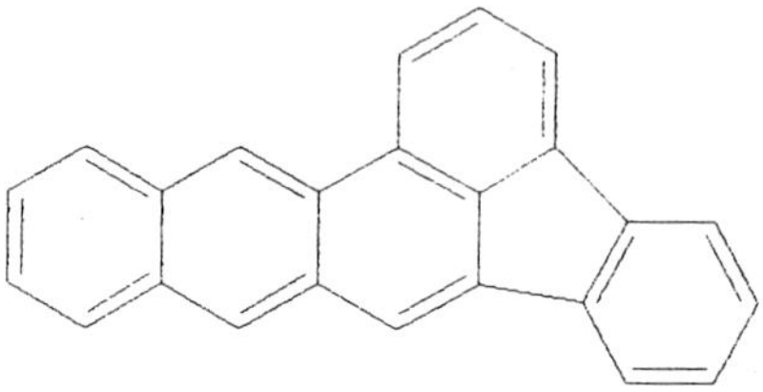

105

C.S. Rao and his co-workers (*Tetrahedron*, 1991, 47, 3499) report the preparation of many poly-annulated structures from the common use of structures such as (106) which also behaved as masked ketones. Thus, 106 reacts with benzyl magnesium chloride to provide a route to 11*H*-benzo[*b*]-fluorene (57). The method is general, and may be extended by the use of various $ArCH_2MgCl$ and other oxoketene dithioacetals (Scheme 12).

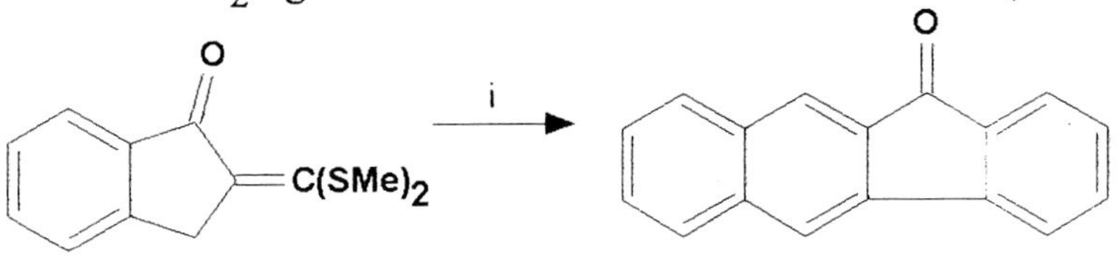

106   57

(i) $NaBH_4$/AcOH, then $PhCH_2MgCl$; then $BF_3.Et_2O$/PhH

Scheme 12: Synthesis of 11*H*-benzo[*b*]fluorenone

(b) Structures with two or more five-membered rings

Indeno[1,2-*b*]fluorene (107; R.G. Harvey *et al.*, *J.Org.Chem.*, 1991, 56, 1210) and indeno[1,2-a]fluorene (108) have been the subjects of physical organic chemical measurements (D. Bethell, P. Gallagher and D.C. Bott, *J. Chem.Soc.,Perkin Trans.2*, 1989, 1097); the synthesis of the latter ring system was based upon reported classical chemistry such as the cyclisation of 3,6-diphenylphthalic acid.

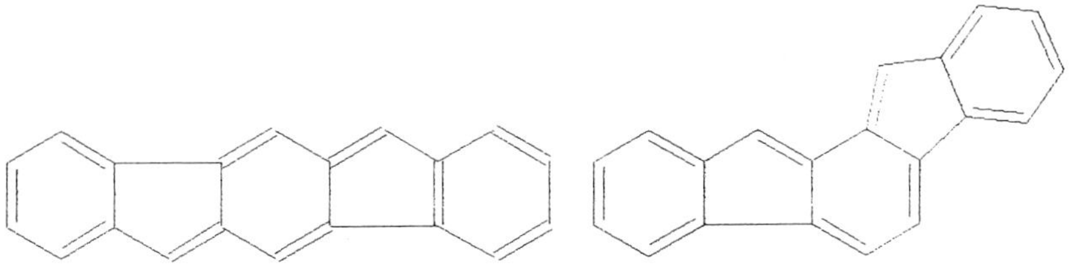

107   108

8*H*-Indeno[2,1-*b*]phenanthrene (**109**) was obtained by C.S. Rao *et al.* (*loc. cit.*).

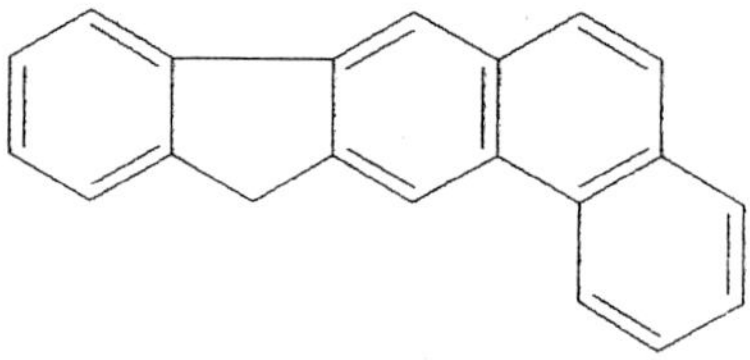

109

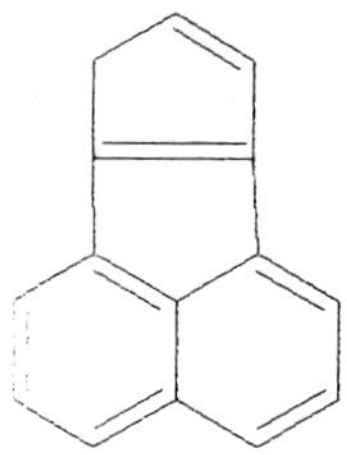

110

7*H*-Cyclopenta[*a*]acenaphthylene (**110**) was obtained as the 8,9-dihydro derivative through condensation reactions of acenaphthylene quinone which reflect the formation of the cyclopenta[*l*]phenanthrene system (**69**) from phenanthraquinone (Scheme 6).

The cyclopent[*h,i*]aceanthrylene **111** is obtained by conventional chemistry through the dialkylation of anthracene using lithium bromoacetate in liquid NH$_3$ forms *cis*-9,10-dihydroanthracene-diacetic acid (**112**). Ring-closure gives, after reduction, 2,7-dihydro-**111** (B.B. Boere *et al.*, *Recl.Trav.Chim. Pays-Bas*, 1990, 109, 463).

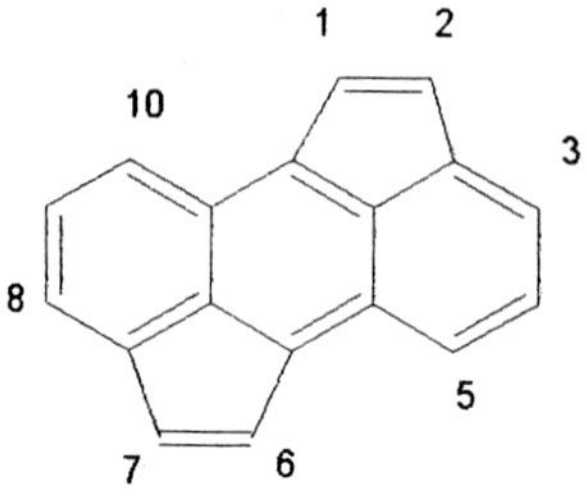

111

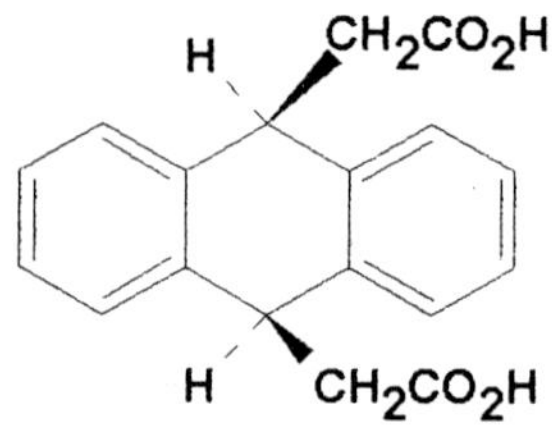

112

(c) Fullerenes

The concept of a hollow, stable, all-carbon molecule derived from 60 carbon atoms arranged in five- and six-membered rings - an icosahedron with 12 pentagonal and 20 hexagonal faces - had been suggested by Japanese (E. Osawa, *Kagaku (Kyoto)* 1970, 25, 854; Z. Yoshida and E. Osawa, *Aromaticity* (Kagakudojin, Kyoto) 1971, 174) and by Russian scientists (D.A. Bochvar and E.G. Gal'pern, *Proc.Acad. Sci.USSR*, 1973, 209, 239) over twenty years ago. Although polymeric carbon systems became experimentally available in 1984 (E.A. Rohlfing, D.M. Cox and A. Kaldor, *J.Chem.Phys.*, 1984, 81, 3322) the breakthrough occurred with the preparation of C_{60} (12%) and C_{70} (2%) in sub-gram quantities by the evaporation of graphite in an argon or helium atmosphere and separation by chromatography or Soxhlet extraction ((a) W. Krätschmer, K. Fostiropoulos and D. Huffman, *Chem.Phys.Lett*, 1990, 170, 167; (b)W. Krätschmer *et al.*, *Nature*, 1990, 347, 354; A. Koch, K.C. Khemani and F. Wudl, *J.Org. Chem.*, 1991, 56, 4543; (c) K.C. Khemani, M. Prato and F. Wudl, *J.Org. Chem.,*, 1992, 57, 3254). This allowed an explosive interest in such systems, and especially of C_{60}; over 1500 papers have been written in 1990-92.

Substitution reactions of C_{60} cannot occur, but electrophilic, nucleophilic and free-radical addition processes have been well documented. Cyclic voltammetry shows the incorporation of up to 6 electrons reversibly (Q. Xie, E. Perez-Cordero and L. Echegoyen, *J.Amer. Chem.Soc.*, 1992, 114, 3978; in contrast, electron-loss to form radical cations (C_{60}^{n+}; n = 1-3) can only occur in non-nucleophilic media (*e.g.* (a) FSO_3H-SbF_5; G.P. Miller *et al.*, *Mat.Res.Soc.Symp. Proc.* 1992, 347, 293; (b) H_2SO_4-SO_3; S.G. Kukolich and D.R. Huffman, *Chem.Phys.Lett.*, 1991, 182, 263; (b) H. Thoman, M. Bernardo and G.P. Miller, *J.Amer.Chem.Soc.*, 1992, 114, 6593) to avoid trapping.

Fluorination at *ca.* 70°C occurs (J.H. Holloway *et al.*, *J.Chem.Soc.*, *Chem. Commun.*, 1991, 966). While up to 76 fluorine atoms may be incorporated in the C_{60} molecule, such extensive addition occurs only with u.v. irradiation.

Although $C_{60}F_{60}$ might be expected to be the stable product, less fully fluorinated systems such as $C_{60}F_{48}$ have been reported to lose fluorine upon prolonged storage (A.A. Tuinman *et al.*, *J.Phys.Chem.*, 1992, <u>96</u>, 7584). Similarly, free-radicals such as allyl and cyclopentadienyl add readily; repeated addition is also observed to the extent to which the fullerene may be justly described as a "free-radical sponge" (P. Kusic *et al.*, *J.Amer.Chem. Soc.*, 1991, <u>113</u>, 6274).

The chemistry of this and other fullerenes expands with such speed that the Reader is referred to the plethora of reviews on these fascinating systems.

Chapter 30

POLYANNULAR AROMATIC COMPOUNDS CONTAINING FOUR OR MORE SIX-MEMBERED RINGS

R. BOLTON

The carcinogenicity of polybenzenoid arenes has been the main driving force in the recent interest in their chemistry. In many instances, a wide range of arenes, their alkyl-, amino-, and nitro- analogues, and their polyhydrogenated analogues have been the subject of extensive studies. In other cases, the careful and systematic identification of the metabolites is reported, since the physiological activity of the hydrocarbons is associated with epoxides derived by formal addition processes. This Review deals with the chemistry of such polybenzenoid systems. To avoid repetition, the following reports deal with aspects of carcinogenicity of a large number of compounds which include derivatives of each of the groups dealt with in this Chapter.

Computer Automated Structure Evaluation of mutagenicity [(*a*) G. Klopman, M.R. Frierson and H.S. Rosenkranz, *Mutat.Res.*, 1990, 228, 1; (*b*) G. Klopman and C. Raychaudhury *J.Chem.Inf.Comput.Sci.*, 1990, 30, 12; (*c*) Klopman and Rosenkranz, *Mutagenesis*, 1990, 5, 425; (*d*) Rosenkranz and Klopman, *Teratog.,Carcinog.,Mutagen.*, 1990, 10, 73] and an extensive CASE-SAR assessment (A.M. Richards and Y.T. Woo, *Mutat. Res.*, 1990, 242, 114) make important contributions. In these papers a wide range of substances is considered, including most of the polybenzenoid hydrocarbons and some of their hydrogenated derivatives; for example, Richards and Woo report studies of benz[*a*]anthracene and its 7,12-dihydro analogue, the 12 isomeric monomethyl derivatives, 1,12- and 7,12-dimethylbenz[*a*]anthracene, 5,6- and 7,12-dihydro-7,12-dimethylbenz[*a*]anthracene and the

8,9,10,11-tetrahydro analogue. Other treatments emphasise polarisability (B.K. Park, M.C. Suh and U.H. Paek, *Bull.Korean Chem.Soc.*, 1986, 7, 183) and a barrage of calculations (*e.g.* (*a*) L. Wang *et al.*, *Huanjing Kexue Xuebao*, 1987, 7, 340; *Chem.Abstr.*, 1988, 108, 126345 (*b*) J.W. Flesher and S.R. Myers, *Teratog.,Carcinog.,Mutagen.* 1991, 11, 41) is directed towards the identification of parameters which reflect the tendency towards attack upon the arene. This, it is thought, provides the ultimate carcinogens during metabolism. Other treatments of the carcinogenicity problem emphasise the significance of metabolic products (*e.g.* benz[*a*]anthracene and methyl derivatives, S.K. Yang, *Polycycl. Aromat. Hydrocarbon Carcinog.:Struct.-Act.Relat.* 1988, 1, 129. Ed. S.K. Yang, B.D. Silverman CRC: Boca Raton, Fla.). The observation that the crystallisation of cholesterol is highly and specifically influenced by the presence of carcinogens (B. Contag, *Z.Naturforsch., C: Biosci.* 1991, 46, 663) may suggest another mechanism of physiological action of cholesterol.

1. Naphthacene (Tetracene)

1

Naphthacene, like anthracene, is widely studied in applications for electrically conducting materials and photochromic dyes. Structure-enthalpy relationships between polybenzenoid systems include naphthacene among many polyannular systems (W.C. Herndon. D.A. Connors and P. Lim, *Pure Appl.Chem.*, 1990, 62, 435). The early synthesis of rubrene by the self-condensation of arylpropynols (*e.g.* ArPhC(OH)C≡CPh) has been improved by a two-step process. The alcohol is treated with RSO_2Cl in the presence of a base such as NEt_3 at -30° to +30° and the product is then heated with a hindered (non-nucleophilic) base to encourage elimination. The resulting 5,11-diphenyl-6,12-diarylnaphthacenes (**2**) have wide uses as electrochromic compounds and in charge-development in conducting systems (A.P. Essenfeld, EP 302,195; *Chem.Abstr.*, 1989, 111, 114894).

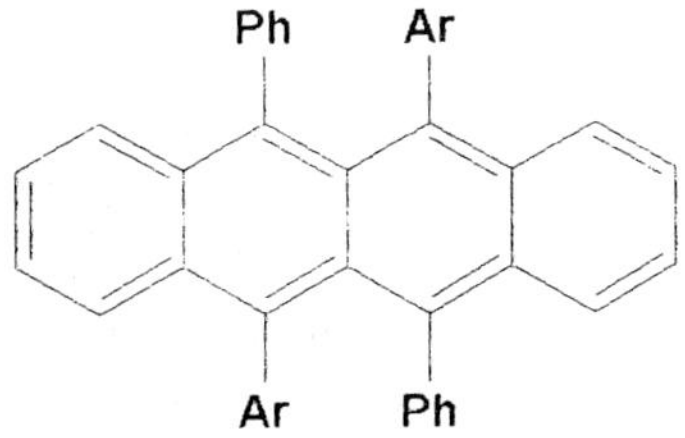

2

A comparable reaction of 9-(arylethynyl)-9-fluorenols gives the four-membered ring system **3** as the major product, with small amounts of the fluoranthene derivative **4** (M. Minabe *et al., Bull.Chem.Soc.Jpn.,* 1988, <u>61</u>, 2067).

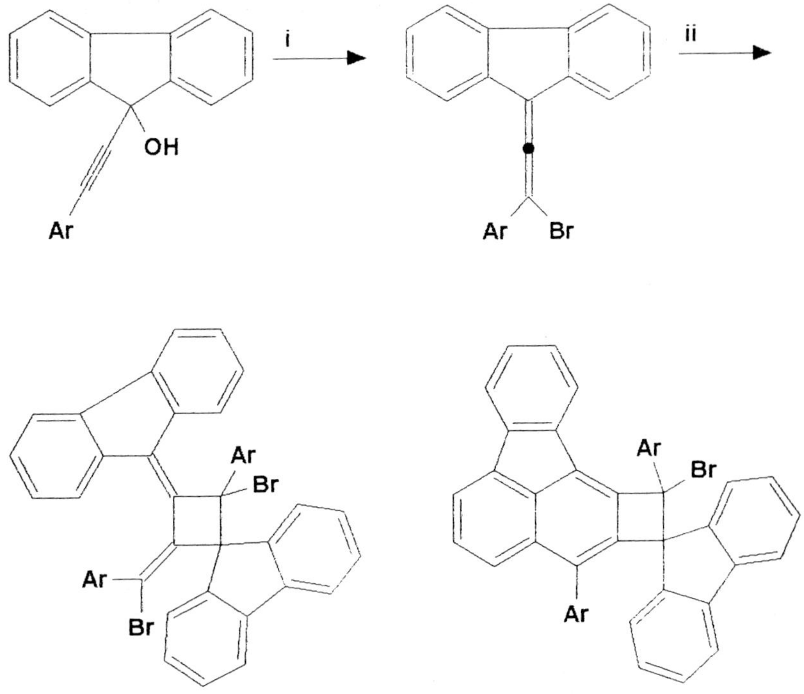

3   **4**

i) HBr-HOAc ii) 2,4,6-Me$_3$-pyridine

<u>Scheme 1: Cumulene dimerisation processes</u>

Arene-1,4-endoxides may be often easily obtained through the addition of arynes to furan. They undergo cycloaddition to benzocyclobutene (which reacts as *o*-xylylene) to give adducts dehydration and dehydrogenation of which provides catenated polybenzenoid systems; naphthalene-1,4-endoxide provides tetracene (J. Luo and H. Hart, *J.Org.Chem.*, 1987, <u>52</u>, 4833) and the method is amenable to the synthesis of substituted tetracenes also. Friedel-Crafts acylation of naphthalene-1,4-diol by 4,5-difluorophthalic anhydride (B_2O_3, 200°) gives 2,3-difluoro-6,11-dihydroxy-5,12-naphtha-cenedione (5). The derived hexachloride (6) on reductive aromatisation gives (7) leading to the dibridged system (8; X = F; Y = Se) [(*a*) B. Hilti and C.W. Mayer, DE 3,635,124; *Chem.Abstr.*, 1987, <u>107</u>, 154320; (*b*) Hilti *et al.*, EP 285,564; *Chem.Abstr.*, 1989, <u>110</u>, 155481). By similar processes T Maruo *et al.* (*Chem.Mater.*, 1991, <u>3</u>, 630) report the synthesis of 2,3-dimethyl-5,6:11,12-tetrathiatetracene (8; X = Me; Y = S).

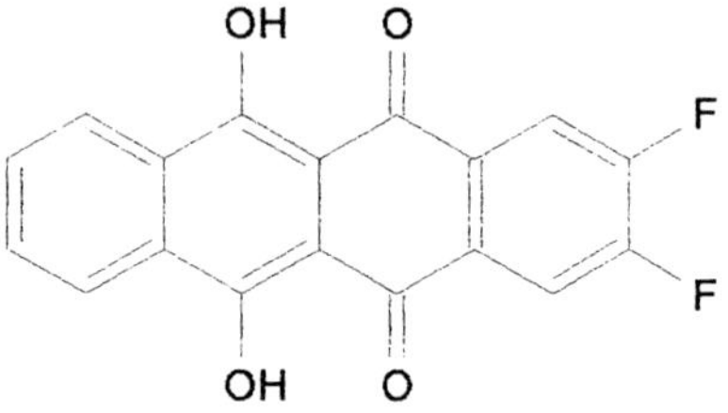

5

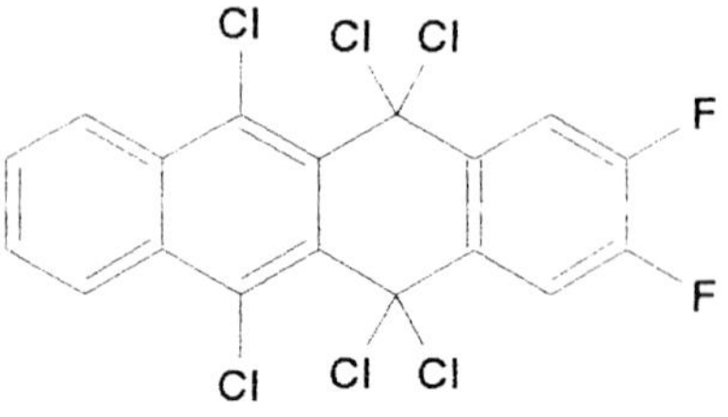

6

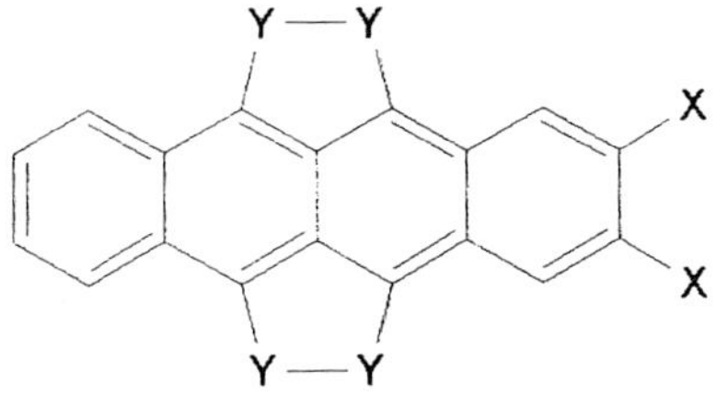

8

In making the acene systems phthaloyl chloride is advocated in the acylation (AlCl$_3$-PhNO$_2$) of p-C$_6$H$_4$X$_2$ (X = Me, OMe, OH) since the intermediate ketoacid cyclises concurrently (G. Sartori *edt al., Gazz.Chim.Ital.,* 1990, 120, 13). 5,12-Diacetoxy-2-CF$_3$-naphthacene has been prepared (M. Baumann *et al.,* EP 344,111; *Chem.Abstr.* 1990, 113, 115288) by standard chemistry involving the 5,12-dione. K.W Bair *et al.,* [(*a*) *J.Med.Chem.,* 1991, 34, 1983; (*b*) US 4,882,358; *Chem.Abstr.,* 1990, 113, 40196] report the synthesis of the 3-aldehyde. Hydrogenation of arenes, including naphthacene, may be achieved using tetraalkyldiboranes which add to give borylated hydroarenes; hydrolysis then removes the borane function (R. Koester, W. Schuessler and M. Yalpani, *Chem.Ber.,* 1989, 122, 677; *cf. ibid., idem* 1990, 123, 719). The aromatisation of 1,2,3-tris-(*t*-butyl)-6,11-dihydronaphthacene to give 9 is achieved by O$_2$ in the presence of KO^{t}Bu. Upon photoirradiation of 9 the remarkable isomer 10 is formed (Z. Yoshida *et al.,* JP 01,238,546; *Chem.Abstr.,* 1990, 112, 197890).

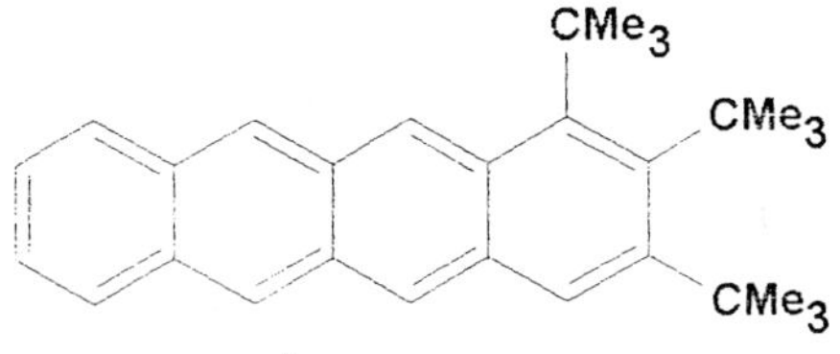

9

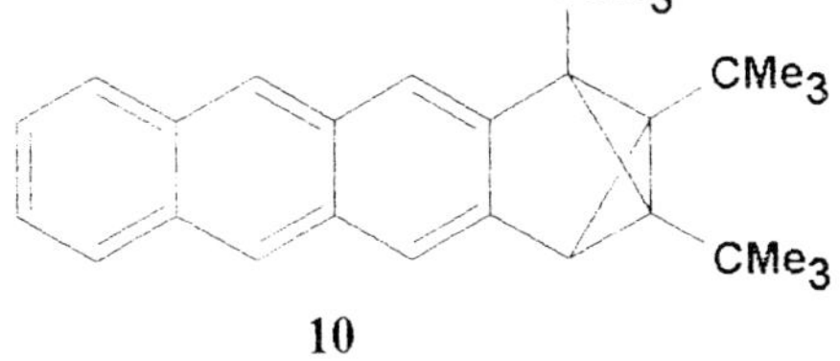

10

2. Chrysene

Chrysene, derived from 4b,10b-dihydrochrysene, may be obtained by the flow pyrolysis $(500°)$ of the adduct between benzyne and benzobarrelene (M.Banciu et al., Rev.Roum.Chim.,1986, 31, 503).

Like most aromatic hydrocarbons, chrysene (12) is hydrogenated in a number of well-defined steps (e.g. B. Fixari and P. Le Perchec., Fuel., 1989, 68, 218). 5,6-Dihydrochrysene is found as the product of the addition of rare-earth metals (Ce, Nd, Sm, Yb) to chrysene followed by protonation (Y. Chauvin, H. Oliviera and L. Saussine, Inorg.Chim.Acta, 1989, 161, 45). 1,2,3,4-tetrahydro-12, 1,2,3,4,5,6-hexahydro-12, and 1,2,3,4,4a,7,8,9,10,-11,12,12a-dodecahydro-12 are formed by the catalysed $(^iPr_2BB^iPr_2)$ hydrogenation of chrysene (M. Yalpani and R. Koester, Chem.Ber., 1990, 123, 719).

Various isomers of octadecahydro-12 have been described and their NMR spectra reported (N.S. Vorob'eva et al., Neftekhimiya, 1987, 27, 750; Chem. Abstr., 1987, 109, 169637). A range of hydrogenated chrysenes appear in the literature; dihydroalkyl derivatives and other partially hydrogenated analogues arise in the aromatic fraction of solvent-refined coals (J. Chen and S. Qian, Fuel Sci.Technol.Int., 1988, 6, 687). Methyl- (M. Barfield et al., J.Amer.Chem.Soc., 1989, 111, 4285) and dimethyl-chrysenes (S. Amin et al., Chem.Res.Toxicol., 1988, 1, 349) have been prepared, sometimes by routes which involve the previous synthesis of their polyhydro-analogues. For example, T.L. Gilchrist and R.J. Summersell (J.Chem.Soc., Perkin Trans. 1, 1988, 2595) describe the coupling $(Pd(PPh_3)_4)$ of 13 with some arylzinc halides to provide species which behave as conjugated trienes and cyclise thermally to give derivatives of 9,10-dihydrophenanthrene or of chrysene (Scheme 2).

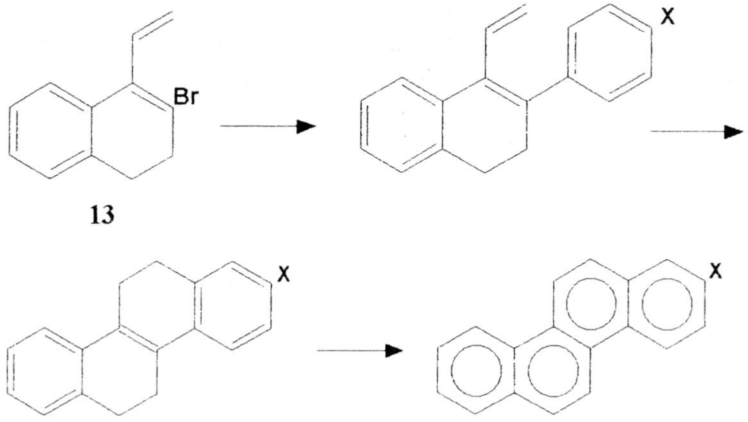

13

14

Scheme 2: Synthesis of chrysene systems

In this way, 5,6,11,12-tetrahydro-2-methoxychrysene (**14**; X = MeO) was obtained. The 2,8-(MeO)$_2$ analogue, and 4b,5,6,12-tetrahydro-2,8-dimethoxy-4b-methylchrysene ((*a*) D.J. Collins and J.D. Cullin, *Aust.J. Chem.*, 1988, 41, 735 (*b*) D.J. Collins, J.D. Cullin and G.M. Stone, *idem.*, 1988, 41, 746) have been prepared by more lengthy methods. The preparations of 1- and 2-MeO-6-methyl-**12** as routes to the isomeric *syn*- and *anti*-1,2-dihydroxy-3,4-epoxy-1,2,3,4-tetrahydro-6-methylchrysenes use 1,4-dimethylnaphthalene. Mono-bromination (NBS), reaction with Ph$_3$P, and treatment of the resulting phosphonium salt with *m*-methoxy-benzaldehyde gives 1-(4-methylnaphthyl)-2-(*m*-methoxyphenyl)ethene; photochemical ring closure in the presence of a mild oxidising agent (I$_2$-PhH) gives 1- or 2-methoxy-6-methyl-**12** (S Amir *et al.*, *Carcinogenesis (London).*, 1986, 7, 2067). Similarly, the synthesis of 2-MeO-5-propyl-**12** (S. Amir *et al.*, *Carcinogenesis (London)*, 1988, 9, 2305) and of 8-MeO-1,11-Me$_2$-**12** (S. Amir *et al.*, *Chem.Res.Toxicol.*, 1988, 1, 349) follow standard procedures and lead to the synthesis of potentially carcinogenic metabolites. Thus, the isopropyl and ethyl substituents were obtained by the

elaborations ArCHO → ArCHOHCH$_3$ → ArCOCH$_3$ → ArEt, and
ArCOCH$_3$ → ArC(Me)=CH$_2$ → ArCHMe$_2$.

1,2,3,4-Tetrahydro-11-methylchrysene (16; R = Me) and the demethylated
analogue have been reported from a one-pot synthesis between silyl enol
ethers such as 15 and arylacetaldehydes. Scheme 3 shows the synthesis of
tetrahydrochrysene in a process which is intrisincally generally applicable (P.
Di Raddo and R.G. Harvey, *Tetrahedron Lett.*, 1988, <u>29</u>, 3885). This
compound, and the 11-methyl system, are also reported (*loc.cit*) to undergo
dehydrogenation (aromatisation) using DDQ.

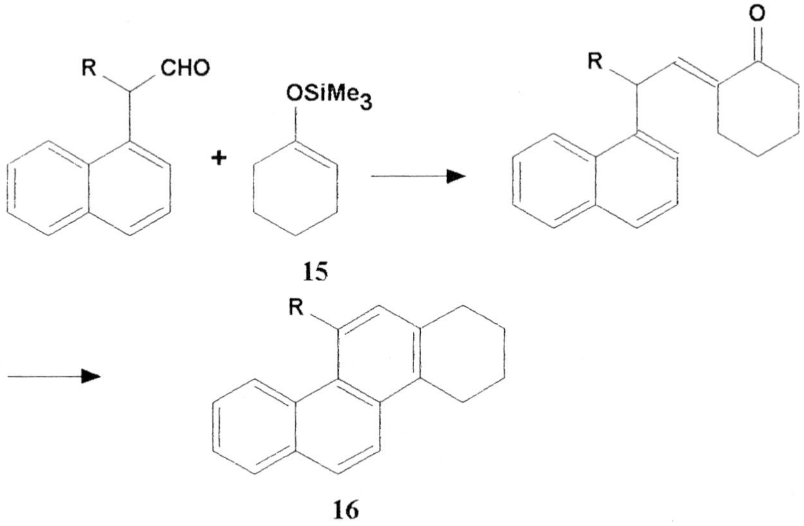

Scheme 3: Aldol synthesis of 11-R-1,2,3,4-tetrahydrochrysenes

Side-chain bromination (NBS) of 4-Me-chrysene is reported in the synthesis
of ArCH$_2$NHR derivatives (K.W. Bair *et al.*, *J.Med.Chem.*, 1991, <u>34</u>,
1983). In the same report the synthesis of a number of aryl derivatives is
described; 2-, 3-, and 6-chrysenyl systems were obtained by the acylation of
12 and separation of the three isomeric acetyl derivatives (S.H. Tucker and
M. Whalley, *J.Chem.Soc. (C)*, 1949, 3213).

Substituted chrysenes (*e.g.* 6-Ph-12) have been obtained (15-47%) by the
coupling of acetylenes R.C≡CH (R = Ph, SiMe$_3$, CO$_2$Et or CMe$_2$OH) with

2-Ph-1-naphthalene diazonium ion in pyridine (R. Leandini *et al.*, *Synthesis*, 1988, 333); phenanthrenes may similarly be made from 2-biphenylamines. Substitution of alkylchrysenes has an evident effect upon their carcinogenicity. 5-Methyl-6-nitrochrysene, obtained by the nitration (HNO_3-H_2SO_4-HOAc) of 5-methylchrysene along with 5-methyl-12-nitrochrysene (K. El-Bayoumy *et al.*, *Carcinogenesis (London)*, 1989, 10, 1685) is reported ((*a*) G.H. Shive, K. El-Bayoumy and S.S. Hecht, *idem*, 1987, 8, 1327 (*b*) K. El-Bayoumy, S. Amin and S.S. Hecht, *idem*, 1986, 7, 673) to show much weaker activity than those of either methyl- or nitro-chrysenes. A similar effect is reported as a result of the *ortho*-nitration of other alkylarenes. 6-Nitrosochrysene, a metabolite of the nitro-derivative which is significantly less tumerogenic than its precursor (K. El-Bayoumy, G.H. Shive and S.S. Hecht, *Carcinogenesis (London)*, 1989, 10, 369), has been prepared by the oxidation (MCBA) of 6-NH_2-12; its reduction provides the very active 6-NHOH-12 (K.B. Delclos *et al.*, *idem*, 1987, 7, 1703).

3. Benz[a]anthracene

17

Benz[*a*]anthracene (**17**) is a common pollutant and a well-known carcinogen. Much of the recent reported chemistry of this and its substitution products rests upon these two observations. For example, aspects of the oxidation of 7,12-dimethyl-**17** by H_2O_2 with horse-radish peroxidase (P.J. O'Brien, *Microsomes Drug Oxid.,Proc.Int.Symp.6th 1984* 284. Ed. A.R. Boobis, Taylor and Francis, London UK) and the formation of epoxides (S.K. Balani *et al.*, *J.Org.Chem.*, 1987, 52, 137) have their links with carcinogenicity, as has the photochemical autoxidation of **17** (L. Shevchuk and M. Gubergrits, *Polynucl.Aromat.Hydrocarbons: Meas.Means,Metab. Int.Symp, 11th* 1987 831. Eds. M. Cooke, K. Loening and J. Merritt,

Battelle Press, Columbus, Ohio) and the cocarcinogencity shown by K$_2$Cr$_2$O$_7$ (P.A. Zhuravlev, *Eksp.Onkol.*, 1991, 13, 19; *Chem.Abstr.*, 1991, 115, 129778); the kinetics of the radical-promoted autoxidation of 17 (T Caceres, S. Guaiquil and E.A. Lissi, *Bol.Soc.Chil.Quim.*, 1988, 33, 177; *Chem.Abstr.*, 111, 38701) may also have bearing on the observed mechanism of its metabolism.

It arises from a range of pyrolytic processes, so that occupational exposures to bitumen fumes in road-paving operations [(*a*) S. Monarca *et al.*, *Int.Arch. Occup.Environ.Health* 1987, 59, 393; (*b*) Z. Braszczynska *et al.*, *Med.Pr.*, 1987, 38, 259; *Chem.Abstr.*, 1988 108, 155783] and industrial hygiene in relation to laser angioplasy (J.M. Kokosa and D.J. Doyle, *Chim.Oggi*, 1987, 19; *Chem.Abstr.*, 1989, 110, 100947) are matters of concern at its occurrence.

The identification of the metabolic products has been driven by the chemical reactivity of epoxides towards aminoacid structures, and much of the reported chemistry of carcinogenic systems is directed towards the synthesis, identification, and chemistry of such metabolites, and their incorporation into DNA structures. As with anthracene, the aromaticity of 17 has been discussed (Z. Chen, *Shanxi Daxue Xuebao, Ziran Kexueban*, 1985, 30, 53; *Chem.Abstr.*, 1987, 107, 22723) in relation to its benzenoid character (M. Randic, S. Nikolic and N. Trinajstic, *Gazz.Chim.Ital.*, 1987, 117, 69; *Chem.Abstr.*, 1987, 108 5330r). Calculations have been made of the heat of formation (J. Kao, *J.Amer. Chem.Soc.*, 1987, 109, 3817) and, by AM1 MO methods, the geometry and resonance energy of the system (W. C. Herndon, D. A. Connor and P. Liu, *Pure Appl.Chem.*, 1990, 62, 435). The oxidation of benz[*a*]anthracene by Tl(III) trifluoroacetate provides the radical cation of 7-CF$_3$COO-benz[*a*]anthracene. The identification of this by ESR spectroscopy, and the analogous reactions of the 12 monomethyl and the 7,12-dimethyl derivatives, are also reported (X. Chen and P.D. Sullivan, *J.Magn.Reson.*, 1989, 83, 484). Cerium (IV) ammonium sulfate in acid provides 7-oxo-12-hydroxy-7,12-dihydrobenz[*a*]anthracene and the derived 7,12-quinone, presumably by similar electron-transfer processes (G. Balanikas *et al.*, *J.Org.Chem.*, 1988, 53, 1007). Epoxidation takes place when NaOCl reacts with 17 in a two-phase (CH$_2$Cl$_2$-H$_2$O) phase-transfer catalysed (Bu$_4$N$^+$HSO$_4^-$) process in which HCl is initially added, followed by NaOH to maintain pH 8.5. Addition to form the chlorohydrin and subsequent cyclisation seems probable (H. Mitsui, T. Hagashi and I. Maeda,

JP 61,109,784; *Chem.Abstr.*, 1987, <u>106</u> 20291). Biological oxidation by *Beijerinckia* strain 1 gives **18**, **19**, and **20** reflecting a deep-seated and extensive attack in which labelled C-12 is lost as CO_2 (W. R. Mahaffey *et al.*, *Appl.Environ.Microbiol.* 1988, <u>54</u>, 2415).

18

19

20

al., *Chemosphere* 1990, <u>20</u>, 525) or in the presence of dimethylbutene (R.W. Murray, S.N. Rajadhyaksha and R. Jeyaraman, *Polycyclic Aromat. Compd.*, 1990, <u>1</u>, 213), brings about similar oxidation of **17**.

The rate of phenylation ($PhN_2^+BF_4^--I^-$; P. R. Singh and A. Kumar, *Aust.J. Chem.* 1972, <u>25</u>, 2133; M Tilset and V.D. Parker, *Acta Chem.Scand.*, 1982, <u>B36</u>, 281) of benz[*a*]anthracene and a number of other polybenzenoid systems has been measured (CH_2Cl_2, 25°) by competition methods (V.D. Parker, K.L. Handoo and B. Reitstoen, *J.Amer. Chem.Soc.*, 1991, <u>113</u>, 6218) taking the known k_2 for phenylation of benzene ($4.5 \times 10^5 \, M^{-1}s^{-1}$) as standard. Benz[*a*]anthracene reacts (k_2, $1.8 \times 10^8 \, M^{-1}s^{-1}$) at a similar rate to that of anthracene (k_2, $2.05 \times 10^8 \, M^{-1}s^{-1}$) but, surprisingly, slower than phenanthrene (k_2, $5.5 \times 10^8 \, M^{-1}s^{-1}$).

(a) Substituted benz[a]anthracenes

The hydrogenation of benz[a]anthracene in the presence of tetrapropyl-diborane provides 8,9,10,11-tetrahydro-, 5,6,8,9,10,11-hexahydro- and 1,2,3,4,8,9,10,11-octahydrobenz[a]anthracenes (M. Yalpani and R. Koester, *Chem.Ber.*, 1990, <u>123</u>, 719).

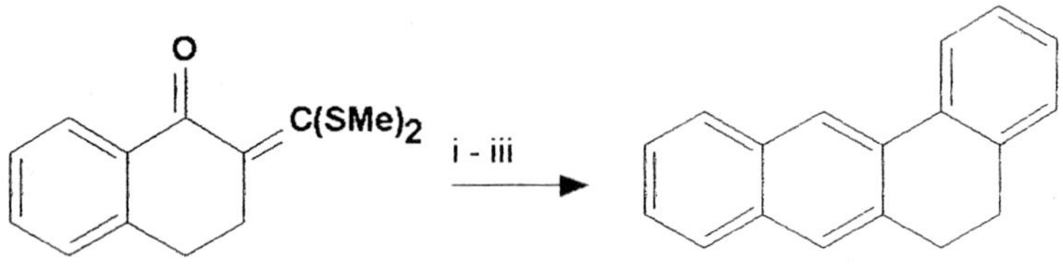

21

(i) NaBH$_4$; (ii) PhCH$_2$MgCl; (iii) BF$_3$.Et$_2$O

Scheme 4: Synthesis of 5,6-dihydrobenz[a]anthracene

5,6-Dihydrobenz[a]anthracene may be prepared by the addition of PhCH$_2$MgCl to the β-oxo-dithioacetal **21** and subsequent cycloaromatisation by BF$_3$.Et$_2$O. This process may be applied generally to the synthesis of a number of polybenzenoid systems; α-oxo-ketene dithioacetals may be similarly used (C.S. Rao *et al.*, *Tetrahedron*, 1991, <u>47</u>, 3499). Monofluoro-benz[a]anthracenes have been made by conventional synthetic methods. Thus, the reaction between fluoroaryl Grignard reagents and phthalic anhydride (D.T. Winiak *et al.*, *J.Org.Chem.*, 1986, <u>51</u>, 4499) or with oxazolines (D.T. Winiak, S. Goswami and G. E. Milo, *J.Org.Chem.*, 1988, <u>53</u>, 345) provides substituted phthalides which then lead to cyclised structures with acids such as PPA. The Friedel-Crafts acylation of substrates such as 6-fluorotetralin similarly provides 2-aroylbenzoic acids the cyclis-ation of which gives fluorobenz[a]anthraquinones. Both 4- and 7-bromo-benz[a]anthracenes have been prepared (B.P. Cho and R.G. Harvey, *J.Org.Chem.*, 1987, <u>52</u>, 5668). Direct electrophilic halogenation leads to the 7-isomer; the 4-bromo isomer is prepared by the bromination of benz[a]anthraquinone and reduction of the bromoquinone obtained (G.M. Badger and A.R.M. Gibb, *J.Chem.Soc.*, 1949, 799); however, HI is reported as a better reducing agent since both the purity and the yield of the product is improved (Cho and Harvey, *loc.cit*).

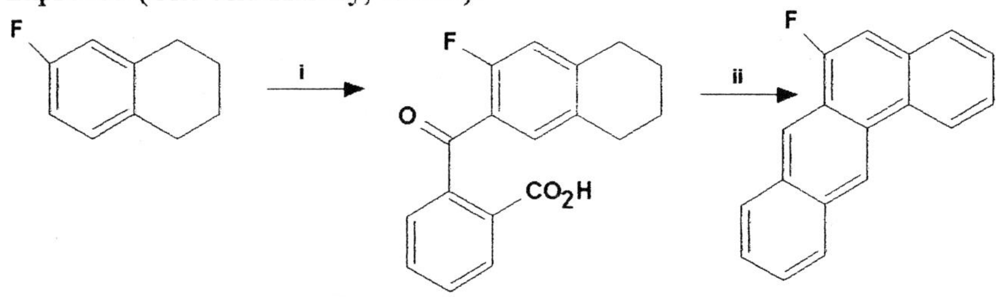

i) phthalic anhydride/$AlCl_3$; ii) PPA, aromatisation

Scheme 5: Synthesis of 6-fluorobenz[*a*]anthracene

5-Methylbenz[*a*]anthracene undergoes side-chain bromination with NBS in CCl_4, but gives the 7-bromo- derivative in DMF when conventional electrophilic substitution is preferred. (J. Che and T. C. Yang, *Proc.Arkansas Acad.Sci.*, 1987, 41, 24; *Chem.Abstr.*, 1987, 110, 94659). Methylbenz[*a*]-anthracenes may be prepared from *p*-toluoyl chloride and 1- or 2-naphth-aldehydes (C.C.Lai *et al.*, *J.Chin.Chem.Soc.(Taipei)*, 1991, 38, 207) by using the oxazoline substituent as a masked carboxamide function; metall-ation takes place at the site *ortho* to the group by displacement either of halogen to form a Grignard reagent, or of hydrogen by BuLi. The secondary alcohol which arises from subsequent reaction with ArCHO appears as a lactone which cyclises under acidic (Friedel-Crafts) conditions (Scheme 6; compare the similar use of oxazolines in Grignard syntheses of phthalides). Friedel-Crafts reactions are also used in the synthesis of methylbenz[*a*]anth-racenes; thus, the spiro-anhydride (22) reacts with benzene, toluene, or *p*-xy-lene to give the keto-acid (23; Ar = C_6H_5, *p*-Me.C_6H_4, or 2,5-$Me_2C_6H_3$) and this leads to the spiro-hydrocarbon (24) which rearranges during dehydrogenation to provide 17, its 2-methyl derivative or the 1,4-dimethyl analogue (Scheme 7). Side-chain brominated derivatives of 7- and 12-methyl-17, and of 7,12-dimethyl-17, have been prepared in a study of their mutagenicity (E.G. Rogan *et al.*, *Chem.-Biol.Interact.* 1986, 58, 253). The introduction of fluorine into such systems profoundly affects the direction of metabolism, and may therefore inhibit carcinogenic or mutagenic behaviour; for example, 9- or 10-CF_3 derivatives of 7,12-dimethylbenz[*a*]anthracene

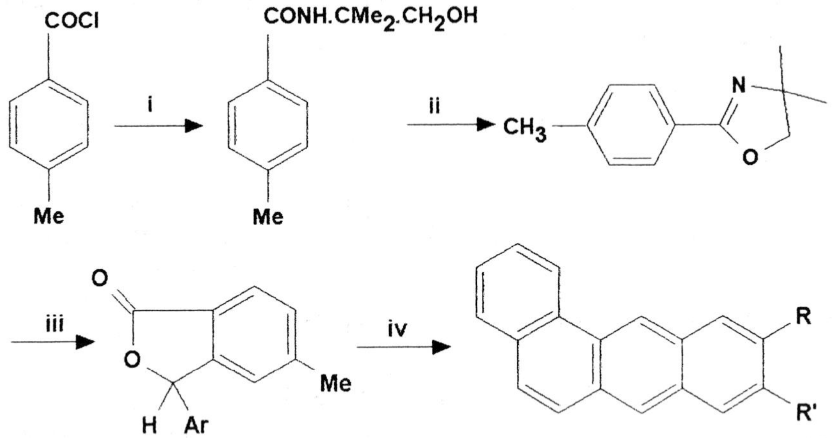

(Ar = 1- or 2-naphthyl, giving 9- or 10-methylbenz[a]anthracene; R = Me, R' = H or R = H, R' = Me)

i) Me$_2$C(NH$_2$)CH$_2$OH; ii) SOCl$_2$; iii) BuLi/ArCHO; iv) HI/HOAc

Scheme 6: Metallation of aryloxazolines in benz[a]anthracene synthesis

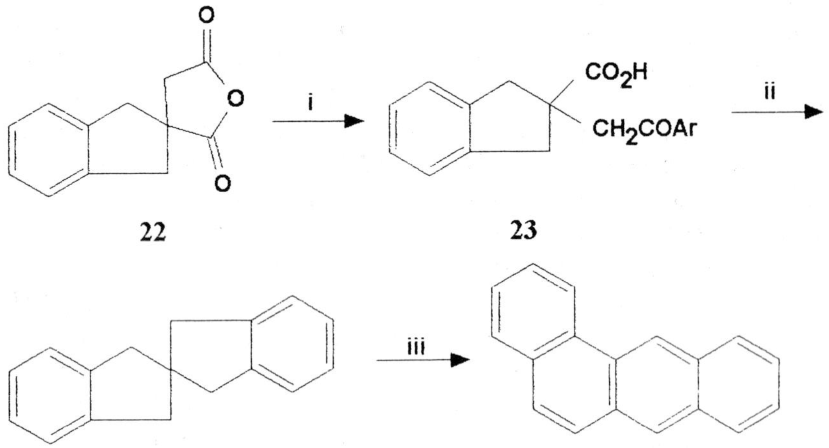

i) ArH/AlCl$_3$; ii) hydrogenolysis, PPA, Clemmensen reduction iii) Pd/C

Scheme 7: Spiroanhydride acylation in benz[a]anthracene synthesis

show no carcinogenicity at the levels tested (T.W. Sawyer, E.P. Fisher and J. DiGiovanni, *Carcinogenesis (London)*, 1987, <u>8</u>, 1465).

(b) Metabolites of benz[a]anthracene derivatives

Metabolism of polycyclic hydrocarbons often provides polyols and epoxides both formally obtained through electrophilic attack upon >C=C< systems, such products are responsible for the alkylation of DNA which brings about the carcinogenic or mutagenic effect. In consequence the stereochemistry, site of addition, and syntheses of these compounds have been widely investigated. The synthesis of such metabolites from arenes often relies upon the presence of a methoxy or hydroxy function in the arene. For example, the demethylation of -OMe (BBr$_3$) gives -OH. Oxidation of the phenol by Fremy's salt [(KSO$_3$)$_2$NO] provides an *ortho*-quinone (R.G. Harvey, *Synthesis* 1986, 605), appropriate reduction of which may give *cis* or *trans* dihydrodiols. A remaining alkene function may be further functionalised to provide an epoxide (*via* MCBA epoxidation or a bromohydrin through addition by Br$_2$-H$_2$O) or by further hydroxylation can provide tetrahydrotetraols (Scheme 8).

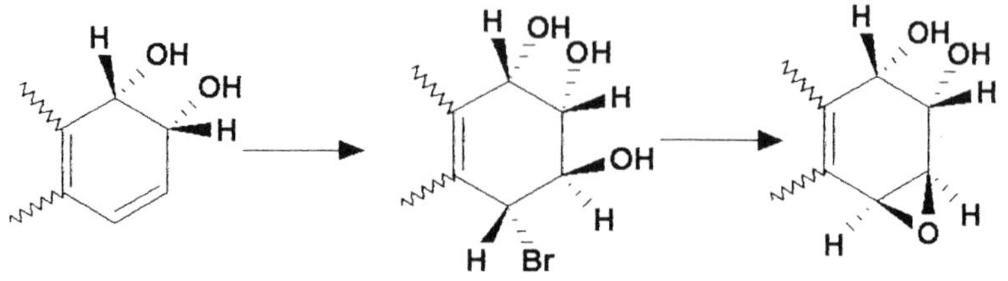

Scheme 8: General synthesis of some arene metabolites

Thus, 7-acetoxy-1,2,3,4-tetrahydrobenz[*a*]anthracene is further acetoxylated (DDQ-AcOH, r.t.) to give a separable mixture of the 1- (24%) and 4-acetoxy (38%) derivatives, each of which (HCl-HOAc) eliminate HOAc from the alicyclic ring and hydrolyse to give isomeric dihydrobenzanthrones which may be reduced and dehydrated to give the dihydrobenzanthracenes (MeLi followed by dehydration similarly provides the 7-Me analogues). These may then yield dihydrodiols and tetrahydroepoxy derivatives by appropriate chemistry (R.E. Lehr *et al..*, *J.Org.Chem.*, 1989, 54, 850). The synthesis of the phenols necessary for following sequences such as Scheme 7 therefore becomes important.

Cyclodehydration of the product of the reaction between 6-lithio-1,4-dimethoxycyclohexa-2,5-diene and 2-(2-naphthyl)ethyl iodide gives 3-methoxybenz[*a*]anthracene contaminated with large amounts of 3-methoxy-benzo[*c*]phenanthrene (ratio, 1:3, based upon the acetoxyarenes formed by reduction, acetylation and dehydrogenation of the ketones shown in Scheme 9) (R.G. Harvey, *J.Med.Chem.*, 1988, 31, 154); however, 6-methoxy-2-naphthylacetaldehyde (25) reacts with the silyl enol ether (26) in the presence of $TiCl_4$ (1 mol) to give a mixture of the alkyl- and alkylidene-cyclohexanones 27 and 28 (Scheme 10). Hydrogenation of 28 gives the saturated ketone which, with $MeSO_3H$, yields an isomeric mixture containing 80% of 10-methoxy-1,2,3,4-tetrahydro-benz[*a*]anthracene (J. Pataki, P. di Roddo and R.G. Harvey, *J.Org. Chem.*, 1989, 54, 840). The synthesis of 1,4-dimethoxybenz[*a*]anthracene and its 7- and 12-methyl analogues is reported (M.S. Newman, A.R. Chaudhury and A. Kumar, *Org.Prep. Proced.Int.*, 1990, 21, 37).

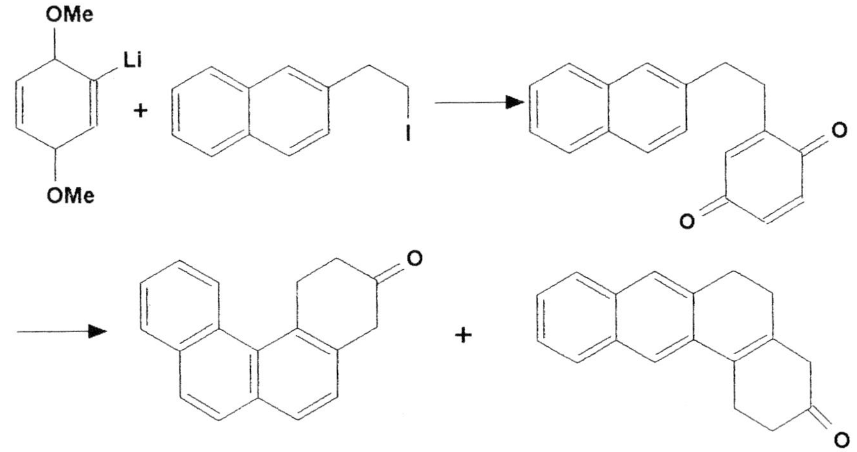

Scheme 9: Synthesis of 3-oxo-1,2,3,4-tetrahydro-benz[*c*]phenanthrene and -benz[*a*]anthracene.

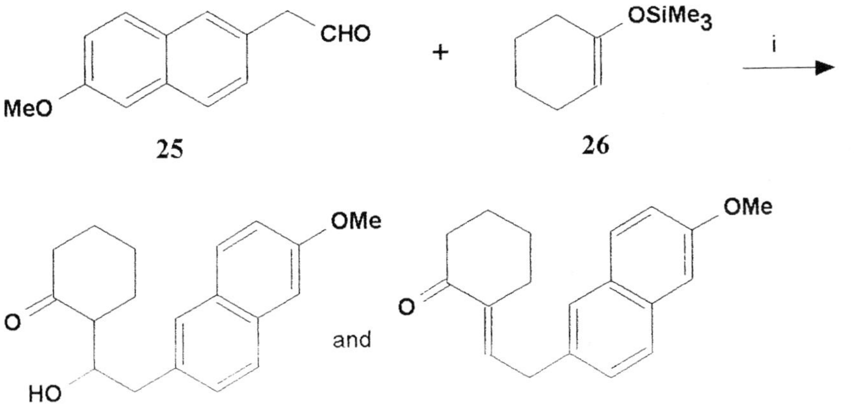

25 26 27 28

i) $TiCl_4$

Scheme 10: Synthesis of 10-methoxybenz[*a*]anthracene precursors

Hydroperoxide-dependent epoxidation of 3,4-dihydroxy-3,4-dihydrobenz-[*a*]anthracene (**29**) with arachidonic acid in methanol gives four isomeric tetrahydrotetraols and a methyl ether arising from methanolysis of the *anti*-diol epoxide (T.A. Dix, J.R. Buck and L.J. Marnett, *Biochem.Biophys.Res. Commun.*, 1986, 140, 181).

29

4. Benz[d,e]anthracene

The 1*H*-benz[*d,e*]anthracene (**30**) skeleton is reported in the product of fluorine addition and substitution of the aromatic hydrocarbon, in which 28 fluorine atoms have been attached. Such materials, and the corresponding products of reaction of benz[*a*]anthracene, benzofluorenes, cyclopentaphenanthrenes and cyclopentanthracenes have been proposed as fluxes in the working of solders and other low-melting metal mixtures (C.R. Sargent and D.E.M. Wotton, EP 253,529; *Chem.Abstr.*, 1988, 109, 114754)

30

31

The isomeric 7*H*-benz[*d,e*]anthracene (**31**) system is mainly seen as the readily available ketone benzanthrone (**32**) which results, not surprisingly, from the oxidation (MnO_4^-) of **31** (S.M. Gannon and J.G. Krause, *Synth.*, 1987, 915). Purification of **32** by recrystallisation from solvent mixtures such as $PhCH_2OMe-H_2O$ is advocated (H. Hoch, G. Kilpper and P. Miederer, DE

3,811,635; *Chem.Abstr.*, 1990, 112, 157894). Grignard processes allow the formation of 7-Et-**31**, for example (P.W. Rabideau, J.L. Mooney and K.B. Lipkowitz, *J.Amer.Chem.Soc.*, 1986, 108, 8130). In the same way that benzanthrone arises from the reaction between anthraquinone and glycerol, appropriately substituted anthraquinones lead to 6,8-, 6,11- and 8,11-dichloro-7*H*-benz[*d,e*]anthrones. Simple nucleophilic substitution by PhO⁻, aided by the electron-withdrawing carbonyl functions, provides the diphenoxy analogues (Yu.E. Gerasimenko, N.T. Sokolyuk and L.P. Pisulina, *Zh.Org.Khim.*, 1986, 22, 632; *Chem.Abstr.*, 1987, 106, 66871). Similarly, Cu-catalysed aminodebromination of 3-Br and 3,9-Br$_2$-**31** is reported (K. Kato and M. Yoshida, JP 02,243,662; *Chem.Abstr.*, 1991, 114, 121786).

32

Self-condensation of **32** and its derivatives provides some large molecular systems (S. Iwashima and H. Honda, *Res.Bull.Meisei Univ.,Phys.Sci.Eng.*, 1985, 21, 31; *Chem.,Abstr.*, 1989, 111, 153751) under Scholl-type conditions. Thus in the presence of Zn, tetrabenzoperylene (**33**), benzo-phenanthropentaphene and dibenzonaphthopentaphene (**34**) systems are formed; the major product from the alkaline fusion of **32** is violanthrene, along with dinaphthoperylene. Mixtures of 1-phenalenone systems such as **32** lead to highly condensed systems, identified predominantly by u.v. and mass spectrometry [(*a*) J.C. Fetzer, *Adv.Chem.Ser.*, 1988, 217, 308; (*b*) J.C. Fetzer and W.R. Biggs, *Org.Prep.Proced.Int.*, 1988, 20, 223].

33   34

5. Benzo[c]phenanthrene

Hydrogenation of benzo[c]phenanthrene gives 1,2,3,4,7,8- and *cis*-5,6,6a,7,
8,12b-hexahydro-35 together with 1,2,3,4,4a,5,6,7,8,12a,12b-H_{12}-35, which
arise from initial attack at the 5,6-positions. (I. Amer *et al.*, *J.Mol.Catal.*,
1987, 39, 185; see also R. Koester and M. Yalpani, *Chem. Ber.*, 1990, 123,
719). The synthesis of 1,2,3,4,4a,5,7,8-octahydro-35 is reported (T.L.
Gilchrist and R.J. Summersell (*J.Chem.Soc., Perkin Trans. 1*, 1988, 2595).

35

Bromination of 35 provides the 5-bromo derivative, which undergoes amination as a route (Balz-Schiemann) to 5-F-35.   6-Fluorobenzo[c]phenan-
threne arises from the addition of BrF to 1-(2-naphthyl)-2-phenylethene,
followed by dehydrobromination and photochemical cyclisation of the
resulting alkene. The 5,7-, 5,8-, and 6,7-difluoro analogues were made
correspondingly (S. Mirsadeghi *et al.*, *J.Org.Chem.*, 1989. 54, 3091).
Addition to 35 provides *trans*-5-azido-6-hydroxy-5,6-H_2-35. Reduction
(LiAlH$_4$) of the derived (SOCl$_2$) chloride provides the benzophenanthro-

aziridine from formal nucleophilic cyclisation of the aminochloro intermediate (E. Abu-Shqara and J. Blum, *J.Heterocycl.Chem.*, 1989, 26, 377).

The synthesis of 6-methylbenzo[c]phenanthrene is reported (M.Barfield *et al*, *J.Amer.Chem.Soc.*, 1989, 111, 4285). The reaction of 6-lithio-1,4-dimethoxycyclohexadiene with 2-(2-naphthyl)ethyl iodide, after cyclodehydration and dehydrogenation, gives 3-methoxybenzo[c]phenanthrene contaminated with 3-methoxybenz[a]anthracene (R.G. Harvey, *J.Med.Chem.*, 1988, 31, 154)

4-Methoxybenzo[c]phenanthrene is obtained (B. Misra and S. Amin, *J.Org. Chem.*, 1990, 55, 4478) from 1-(2-naphthyl)-2-(2-methoxyphenyl)ethene by photochemical ring-closure; demethylation (BBr_3) leads to the phenol, and conventional chemistry then provides *anti*-1,2-epoxy-3,4-dihydroxy-1,2,3,4-tetrahydrobenz[c]phenanthrene.

6,7-Difluorobenzo[c]phenanthrene undergoes oxidative metabolism to give 7-fluorobenzo[c]phenanthrene-5,6-dione. An unusual, but by no means unique, reaction occurs when the oxidation is carried out in the presence of acetone, when either one of the >C=O systems is reduced to >CH-OCH_2COCH_3 (M.A. Patrick *et al.*, *J.Org.Chem.* 1991, 56, 888)

6. Pyrene

36   37   38

The hydrogenation of pyrene (36) has received regular study. 4,5-Dihydro-, 4,5,9,10-tetrahydro- (37), 1,2,3,6,7,8-hexahydro- (38), and 1,2,3,3a,4,5,5a,-6,7,8- and 1,2,3,3a,4,5,9,10,10a,10b-decahydro-pyrene have each been identified [(a) A. Berg, J. Lam, and P.E. Hansen, *Acta Chem.,Scand.*,

Ser.B, 1986, <u>40</u>, 665; (*b*) G. Mann *et al.*, *J.Prakt.Chem.*, 1989, <u>331</u>, 267; (*c*) with ^{i}Pr$_2$BB^{i}Pr$_2$, R. Koester and M. Yalpani, *Chem.Ber.*, 1990, <u>123</u>, 719).

Hydrogenation of pyrene provides at least two synthetically useful derivatives. 4,5,9,10-Tetrahydropyrene (**37**) and 1,2,3,6,7,8-hexahydropyrene (**38**) have been used to obtain 2- and 4-substituted pyrenes, in contrast with the strong directive effect found in the electrophilic attack of pyrene itself, when substitution occurs almost exclusively at C-1. Thus, **37** was first used (R. Bolton, *J.Chem.Soc.*,, 1964, 4637) to obtain 2-nitro-, 2-benzoyl- and 2-acetyl-**37** which were successfully dehydrogenated to the correspondingly substituted pyrenes. Analogously, **38** gave 4-nitro-**38** by direct substitution; dehydrogenation provides 4-NO$_2$-**36** (P.M.G. Bavin, *Can.J.Chem.*, 1959, <u>37</u>, 1614). In both cases, diazonium chemistry converted the derived amines into halogeno-analogues.

This chemistry has been extended to the formation of bromopyrenes from **37** [(*a*) H. Lee and R.G. Harvey, *J.Org.Chem.*., 1986, <u>51</u>, 2847; (*b*) R.G. Harvey *et al.*, *idem.*, 1988, <u>53</u>, 3936] and **38** respectively to give 2- and 4-methylpyrenes (R. Lapouyade, J. Pereyre and P. Garrigues, *C.R.Acad. Sci.*, *Ser. 2.*, 1986, <u>303</u>, 903) and of 4,9-dibromopyrene [(*a*) S. Yamaguchi, K. Nagareda and T. Hanafusa, *Synth.Met.*, 1989, <u>30</u>, 401; (*b*) K. Blatter and A.D. Schlueter, *Synthesis* 1989, 356] and 4,9-diiodopyrene from **38**. The 1,2,3,6,7,8-hexahydro-4,9-diiodopyrene also leads to the hexahydropyreno-quinodimethane **39** (Yamaguchi *et al.*, *loc.cit.*). 4-Methylpyrene has also

39

been made (M. Barfield *et al.*, *J.Amer.Chem.Soc.*, 1989, <u>111</u>, 4285) by the chloromethylation of **38**. The syntheses of 2- and 4-nitropyrenes have been repeated (A.M. Vanden Braken - Van Leersum *et al.*, *Recl.Trav.Chem.*

Pays-Bas 1987, 106, 120), and the preparation of 2-fluoropyrene is reported by standard aromatic chemistry (L. Rodenburg *et al.*, *Recl.Trav.Chem. Pays-Bas* 1988, 107, 1). A similar synthesis of cyclopenta[*c,d*]pyrene 3,4-oxide uses 1,2,3,6,7,8-hexahydropyrene as a route to 2-(4-pyrenyl)ethanol which is converted first to the aldehyde (N-chlorosuccinimide-Me$_2$S) and then to the carboxylic acid (Ag$_2$O) which yields to *trans*-3,4-dihydro-3,4-dihydroxycyclopenta[*c,d*]pyrene monotosylate and then (powdered KOH) to the epoxide (Y. Sahali, P.L. Skipper and S.R. Tannenbaum, *J.Org. Chem.*, 1990, 55, 2918). The formation of 2-*t*-butylpyrene and 2,7-di-*t*-butylpyrene from the 4,5,9,10-tetrahydro analogues reflects the observation that metacyclophanes undergo cyclisation to 4,5,9,10-tetrahydropyrenes when electrophilic attack occurs upon one of the aryl rings. The resulting carbocation attacks the second ring, forming the pyrene skeleton. This process, first reported in 1961 (N.L. Allinger, M.A. da Rooge and R.B. Herman, *J.Amer.Chem.Soc.*, 1961, 83, 1974), was observed again in the attempted trans-*t*-butylation of some metacyclophanes and was advocated as a route to the pyrene system [(*a*) T. Yamato, T. Arimura and M. Tashiro, *J.Chem.Soc.*, *Perkin Trans.1* 1987, 1; (*b*) T. Yamato *et al.*, *J.Org.Chem.*, 1987, 53, 3196] in cases where the initial adduct contained substituents at C-10a and C-10b (-F, -OMe) which were labile under the reaction conditions and led to the 4,5,9,10-tetrahydropyrene structure. In forcing conditions, using the halide as solvent, up to five alkyl groups (CHMe$_2$, cyclopentyl., cyclohexyl) can be substituted into pyrene (A. Berg, J. Lam, and P.E. Hansen, *Acta Chem.Scand.*, *Ser.B*, 1986, 40, 665); cyclodehydrogenation processes could yield some new polycyclic arenes. 2-Nitropyrene is attacked by Pb(OAc)$_4$ in HOAc to give, after hydrolysis, a mixture of 1-hydroxy-2-nitropyrene and 1-hydroxy-7-nitropyrene (1:1.7). Nitration of 2-nitropyrene similarly gives a mixture of 1,2- (38%) and 1,7-dinitropyrene (57%); expectedly, the former isomer undergoes ready nucleophilic displacement of NO$_2$⁻ (A.M. van den Braken-van Leersum, J. Cornelisse and J. Lugtenburg, *J.Chem.Soc.,Chem.Commun.*, 1987, 1156).

7. Triphenylene

The crowding between C-4 and C-5 in triphenylene is more evident in 1,2,3,4-tetraphenyltriphenylene where the structure is twisted by *ca* 30°; this angle is scarcely changed when the 5,6,7,8,9,10,11,12-octadeuterio analogue is studied (R.A. Pascal *et al.*, *J.Org.Chem.*, 1988, 53, 1687).

40   **41**

Hydrogenation of **40** yields 1,2,3,4-tetrahydro-**40** (M. Yalpani and R. Koester, *Chem.Ber.*, 1990, 123, 719); using Li-NH$_3$ provides also the 1,4,4a,12-isomer, *cis*- and *trans*-1,2,3,4,4a,12b-hexahydro-**40** derivatives. These appear to arise from the initial formation of 1,4-dihydro derivatives, which is at variance with the prediction of calculations (Z. Marcinow, A. Sygula and P.W. Rabideau, *J.Org.Chem.*, 1988, 53, 3603). Acetylation of triphenylene (AcCl-AlCl$_3$-PhNO$_2$) provides good yields of the 2-isomer (L.H. Klemm *et al.*, *J.Heterocycl.Chem.*, 1989, 26, 1241); this leads to the construction of a number of 2-triphenylenyl compounds such as ArCO$_2$H and ArCHO in the synthesis of an azadibenz[*a,c*]anthracene (M.J. Tanga, R.F. Davis and E.J. Reist, *J.Heterocycl.Chem.*, 1987, 24, 39). The formation of 1- and 2-nitrotriphenylene in similar amounts by treatment of **39** with H$_2$SO$_4$-MeNO$_2$-(CF$_3$CO)$_2$O (L.H. Klemm, E. Hall and S.K. Sur, *J.Heterocycl.Chem.*, 1988, 25, 1427) may reflect the age of the nitro-methane used; unfortunately the details of this surprising process have not been studied.

Many of the reports about triphenylene systems deal with hexaalkoxy systems such as 2,3,6,7,10,11-hexamethoxytriphenylene **41** which arises from the anodic (J.M. Chapuzet and J. Simonet, *Tetrahedron*, 1991, 47, 791) or chemical (chloranil-H$_2$SO$_4$: H. Levanon and Y. Mignon, IL 70,572; *Chem.Abstr.*, 1987, 107, 188249) trimerisation of veratrole; Chapuzet and

Simonet report electrolysis conditions in which trimerisation and dimerisation are separately favoured.

The synthesis of metabolites from benzo[g]chrysene (**42**) and formally derived from 11,12,13,14-tetrahydro-**42** involve some conventional chemistry based upon **40** (D.R. Bushman *et al.*, *J.Org.Chem.*, 1989, <u>54</u>, 3533). Thermal decomposition of 2-lithiononafluorobiphenyl provides 1-C_6F_5-undecafluoro-**40**. 1,2,3,4-Tetra- and 1,2,3,4,5,6,7,8-octa-fluorotriphenylene may be made by the interaction of 2-BrC_6H_4I with BrC_6F_5 or 2-$BrC_6F_4C_6F_5$ respectively in the presence of BuLi (C.A. Beaumont *et al.*, *J.Organomet.Chem.*, 1988, <u>344</u>, 1).

8. Benzo[*g*]chrysene

42

Benzo[*g*]chrysene (**42**) is made by the photochemical cyclisation of the ester of acid **43** followed by acid-catalysed cyclisation of the resulting arylbutyric acid function (Scheme 11) (C.M. Utermoehlen, M. Singh and R.E. Lehr, *J.Org.Chem.*, 1987, <u>52</u>, 5574). 1-Naphthyl and 2-naphthyl systems lead eventually to benzo[*g*]chrysene, but the former gives 4-(5-chrysenyl)-butyric acid and hence, on acid-cyclisation, derivatives of 1,2,3,4-tetrahydro-**42** whereas the latter gives the 5,6,7,8-tetrahydro-**42** system through the analogous benzo[*c*]phenanthrene compounds.

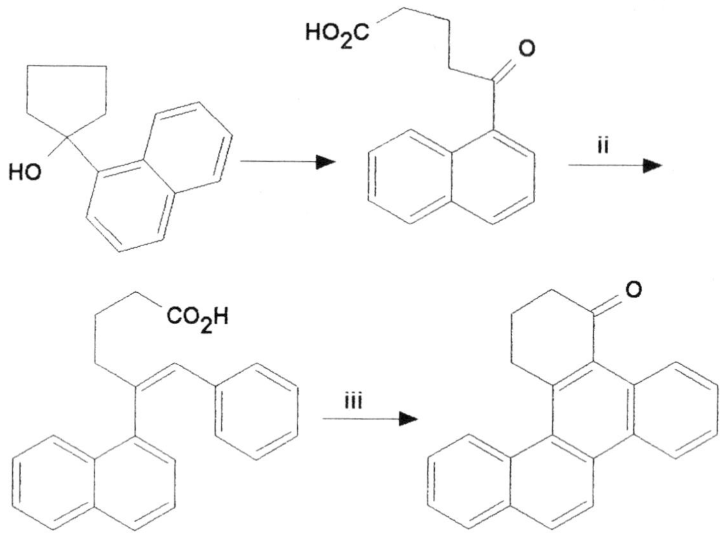

i) TsOH; $KMnO_4$; ii) $PhCH_2PPh_3{}^+Cl^-/NaOEt$; iii) $h\nu-I_2-PhH$; acid

Scheme 11: Synthesis of benzo[*g*]chrysene skeleton

9. Benzopyrenes

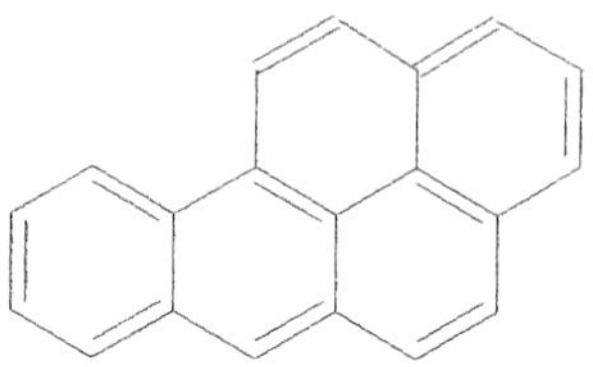

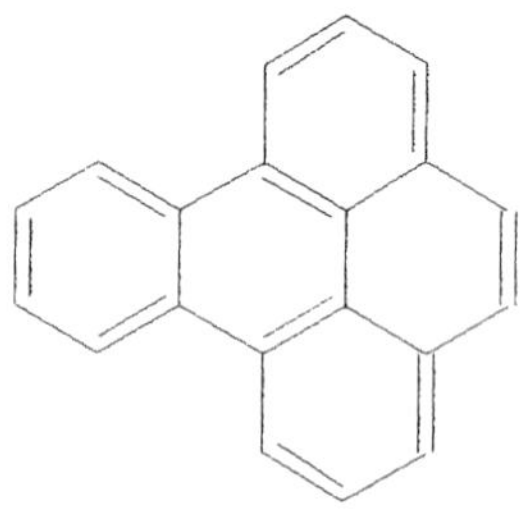

Benzo[*a*]pyrene (**44**) and benzo[*e*]pyrene (**45**) are the subject of a text (M.R. Osborne and N.T. Crosby, *Benzopyrenes*, Cambridge University Press. Cambridge, U.K., 1987). Both are extensively reported in the nitration of **42** (K. Fukuhara *et al.*, *Chem.Pharm.Bull.*, 1990, <u>38</u>, 3158) provides a mixture of 1,6- and 3,6-dinitro-**44** which may be separated by reduction (NaSH) to the corresponding 1- and 3-NH$_2$ derivatives. These may be obtained pure by chromatography; diazotisation and standard chemistry provides single, pure, dinitro compounds. In contrast, Z. Zhang *et al.* (*Huaxue Xuebao* 1990, <u>48</u>, 73; *Chem.Abstr.* 1990, <u>113</u>, 77885) report the synthesis of these compounds by the nitration of benzo[*a*]pyrene The preparation of 10-deuterio-6-substituted benzo[*a*]pyrenes (**46**; X = Cl. Br. NO$_2$, CHO) by direct substitution, and some consequent elaboration or replacement of these groups, is also reported (Y. Kim *et al.*, *J.Chin.Chem.Soc. (Taipei)* 1988, <u>35</u>, 387).

46

Similarly, B.P. Cho and R.G. Harvey (*J.Org.Chem.*, 1987, <u>52</u>, 5668) have reported the chlorination of **44** by SO$_2$Cl$_2$ to give 6-Cl-**44**, and the analogous synthesis of 1- and 3-bromo-6-chloro-**44**. The cyclisation of 4-(1-pyren-yl)butyric acid (HF; H. Schulde, *Chem.Ber.*, 1971, <u>104</u>, 3995) provides 7,8,9,10-tetrahydrobenz[*a*]pyren-7-one (**47**). J. Pataki and R.G. Harvey (*J.Org.Chem.*, 1987, <u>52</u>, 2226) used this as a starting material for the preparation of 1-substituted-**44** *via* 1-Br-9,10-H$_2$-**44**. Halogenation (Br$_2$ in CCl$_4$-CH$_2$Cl$_2$) gave 1-bromo-**47**, although the subsequent addition of AgF before work-up suggests that substitution and addition occurred concurrently. The chemistry is shown in Scheme 12.

4,5-Difluorobenz[*a*]pyrene (**48**) has been obtained from the *cis*-4,5-dihydro-4,5-diol (**49**) through oxidation (DDQ in dioxan) to the quinone and the formation of 4,5-dihydro-4,4,5,5-tetrafluorobenz[*a*]pyrene through the action of DASF (Me$_2$NSF$_3$) in PhH. Aromatisation was then brought about with LiAlH$_4$ (S.C. Agarwal *et al.*, *Carcinogenesis (London)* 1991, <u>12</u>, 1647).

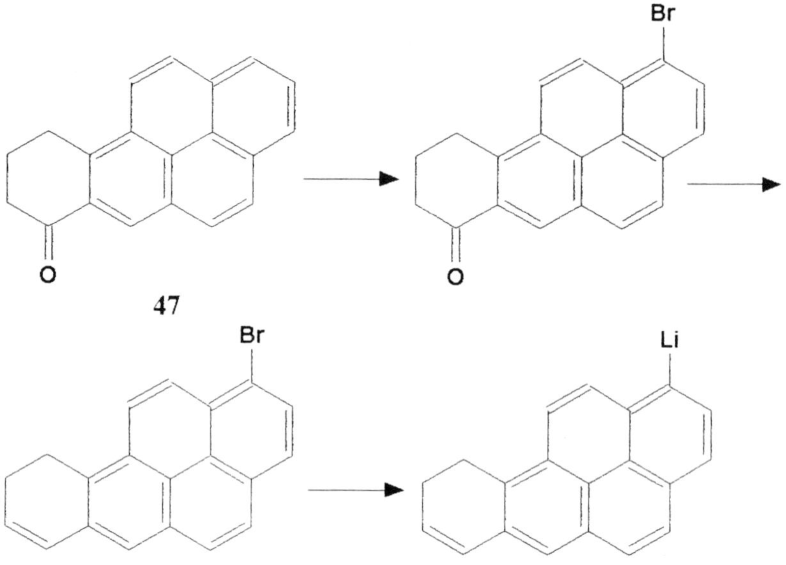

318

<u>Scheme 12: Synthesis of 1-lithio-9,10-dihydrobenz[a]pyrene</u>

Benzo[e]pyrene chemistry is reflected in the application of polyhydro-derivatives towards directing electrophilic attack (cf pyrene). The synthesis

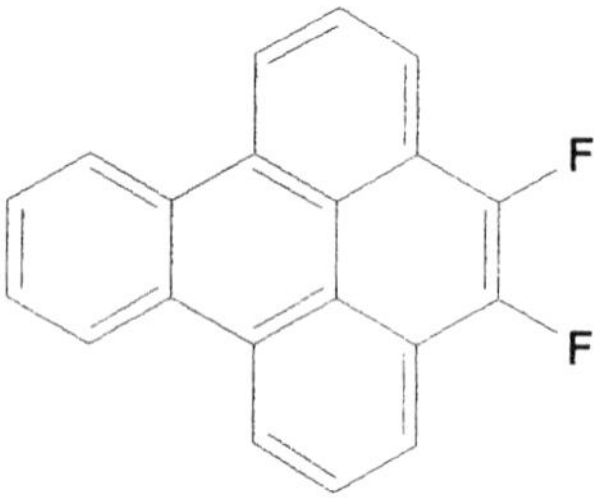

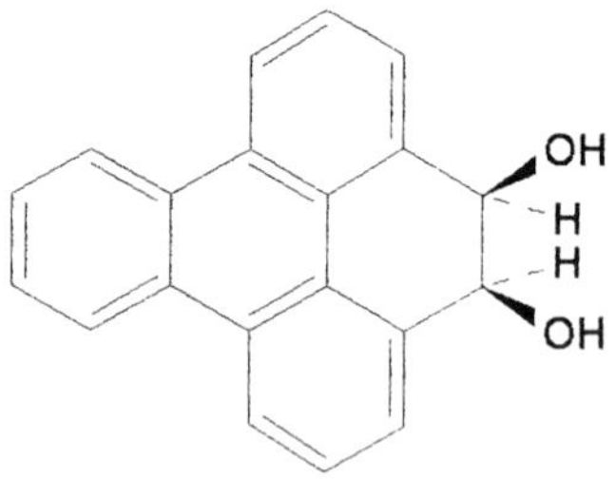

48   49

and mutagenicity of 1,6-, 1,3- and 1,8-dinitro-9,10,11,12-tetrahydrobenz[e]-pyrenes, and 1,6- and 1,8-dinitro-**45** are reported; the biological activity falls in the order given. Of the mononitrobenzo[a]pyrenes, the 3-isomer shows the highest activity (P.P. Fu *et al.*, *Mutat.Res.*, 1991, <u>17</u>, 169).

The Friedel-Crafts acetylation (AcCl-AlCl$_3$-PhH) of 1,2,3,6,7,8,9,10,11,12-decahydro-**45** provides **50** after dehydrogenation (DDQ). R. Sangaiah and

A. Gold (*J.Org.Chem.*, 1988, <u>53</u>, 2620) report the synthesis of naphtho-[1,2,3-*m,n,o*]acephenanthrylene (**52**) from 4-acetylbenz[*e*]pyrene (**50**) (Scheme 13), and the corresponding formation of cyclopenta[*c,d*]pyrene (**51**) from 4-acetylpyrene. The unusual aspect of the chemistry is the conversion ArCOCH$_3$ → ArCH$_2$CO$_2$H. This variant of the Willgerodt reaction is achieved under much milder oxidising conditions (equimolar Tl(NO$_3$)$_3$ in MeOH/CH$_2$Cl$_2$ with 70% HClO$_4$ 4:2:1); provided that it is general both in the length of the chain along which the migration occurs and the nature of the attached aryl group this rearrangement will have considerable applications in arene chemistry.

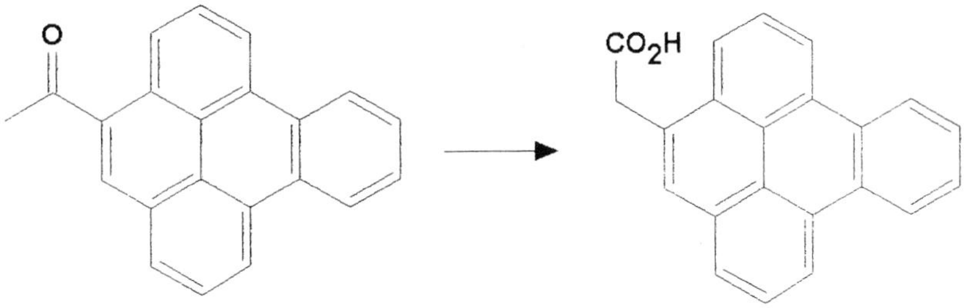

50

51   **52**

Scheme 13: Synthesis of cyclopenta[*c,d*]pyrene and naphtho(1,2,3-*m,n,o*)-acephenanthrylene

10. Dibenzanthracenes

Dibenz[*a,c*]anthracene (**53**) metabolites, such as the dihydrodiols and phenolic precursors, have been synthesised (P.L. Kole, S.K. Dubey and S. Kumar, *J.Org.Chem.*, 1989, <u>54</u>, 845).

53

Dibenz[*a,h*]anthracene, **54**

Dibenz[*a,j*]anthracene, **55**

A range of methoxydibenz[*a,h*]- and -[*a,j*]-anthracenes have been made by a regiospecific preparation in which 2,5- or 2,6-dichlorobenzoquinone adds to *o*-, *m*- or *p*-methoxystyrene. The resulting chloromethoxyphenanthrene diones are then further treated with 1-vinylcyclohexene and the resulting adducts are deoxygenated and demethylated by treatment with HI-HOAc. In such a way 1,2,3,4-tetrahydro-9-, -10-, and -11-methoxydibenz[*a,h*]anthracene and 1,2,3,4-tetrahydro-10-, -11-, -12-, and -13-methoxydibenz[*a,j*]-anthracene are made (G.M. Muschik *et al.*, *Polynucl.Aromat.Hydrocarbons Chem.*, *Charact.Carcinog.,Int.Symp. 9th 1984* 657. Eds. M. Cooke, A.J. Dennis. Battelle Press. Columbus, Ohio). Dibenz[*a,j*]anthracene is produced by Rao's α-oxoketene dithioacetal method (C.S. Rao *et al.*, *Tetrahedron*, 1991, <u>47</u>, 3499).

11. Dibenzophenanthrenes

The photochemical cyclisation of a number of fluoro-derivatives of *cis*-di(2-naphthyl)ethene (**56**) gives fluoro derivatives of dihydrodibenzo-[*c,g*]phenanthrene (**57**) provided that neither site of attack bears a fluorine substituent (Scheme 14; X and Y = F or H). Coloured intermediates which were interconvertible were observed in the course of the study (Y. Ittah *et al.,. J.Photochem.Photobiol. A*, 1991, 56, 239). Optically active crown ether systems have been made which rely upon the distortion of the dibenzophenanthrene (or [5]helicene) system for these properties. The starting material, 2,13-X_2-1,4,11,14-tetramethyldibenzophenanthrene (**58**; X = Br; Scheme 15), underwent the reactions ArBr →ArCHO → ArCH$_2$OH by successive treatment with BuLi/DMF and then LiAlH$_4$. The resulting diol was alkylated by tetraethyleneglycol ditosylate to provide a crown ether ring system (K. Yamamoto *et al., J.Chem.Soc., Perkin Trans.1*, 1990, 271).

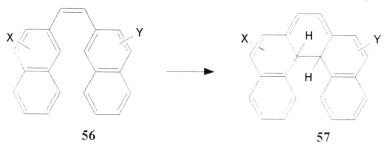

56   **57**

Scheme 14: Dibenzo[*c,g*]phenanthrenes from 1,2-diarylethenes

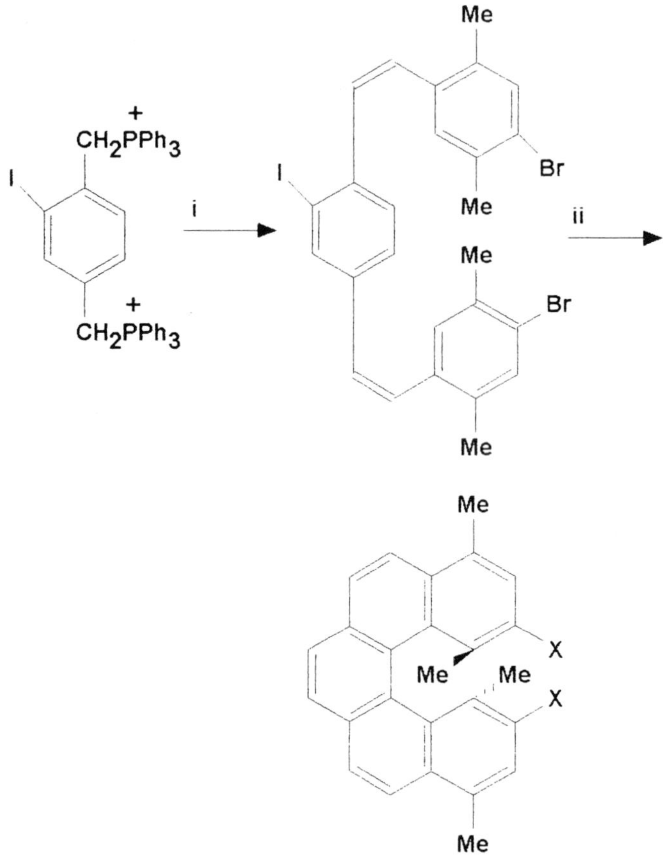

i) 2,5-Me$_2$-4-BrC$_6$H$_2$CHO (2 moles); ii) hν

Scheme 15: Dibenzo[c,g]phenanthrene synthesis

L Liu and T.J. Katz (*Tetrahedron Lett.*, 1990, **31**, 3983) have made the dibenzophenanthrene system by condensing 1,4-divinylbenzene and benzo-quinone. The 1:2-adduct (**59**), a bis(quinone), may be separated into its enantiomers by acetylating the hemiketal, selectively hydrolysing the product with bovine pancreas acetone powder to give a mixture of optically active hemiketal and acetate, separating the acetate and oxidising it to provide both

(+)- and (-)-**59**. A variant upon this method uses 1,4-dimethoxybenzene and
1-vinylnaphthalene *via* mercuration (Hg(OAc)$_2$-AcOH; NaCl-H$_2$O) and
Heck coupling (LiPdCl$_3$-MeCN) with the vinylnaphthalene. The first
addition provides 1,4-dimethoxybenzo[*c*]phenanthrene (**60**). Mercuration
again takes place at C-2 and C-3. The isomeric organomercurials may be
separated, when a second addition of 1-vinylnaphthalene gives 9,18-dimeth-
oxydinaphtho[*a,h*]- and -[*a,j*]-anthracenes **61** and **62**. The methoxyl group

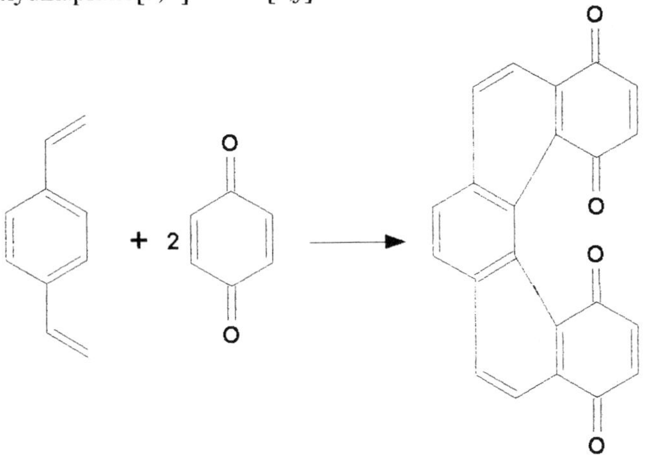

59

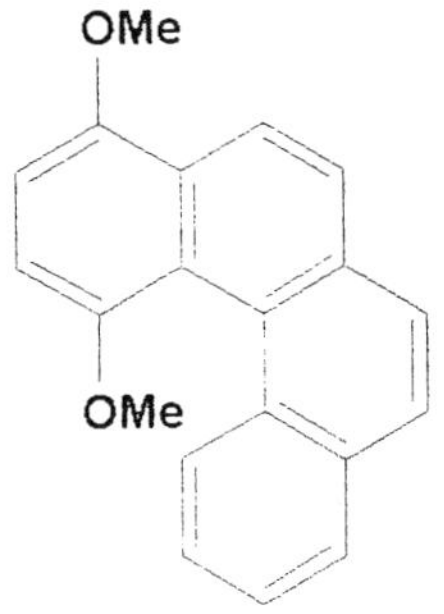

60

attached to C-18 of **61** is greatly distorted away from the expected planarity
(R.B. Gupta *et al.*, *J.Amer.Chem.Soc.*, 1991, <u>113</u>, 359).

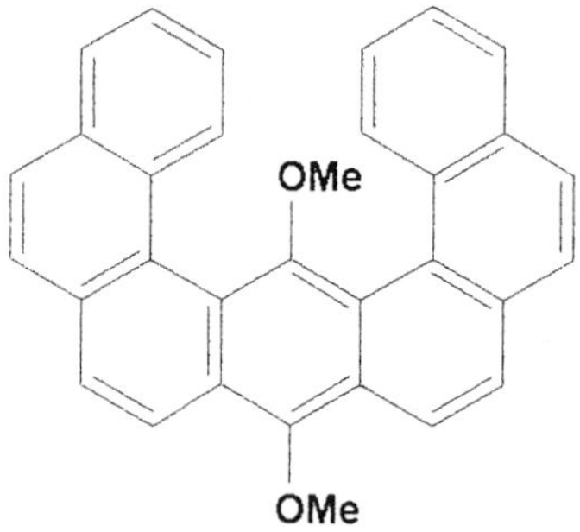

OMe

61

OMe

OMe

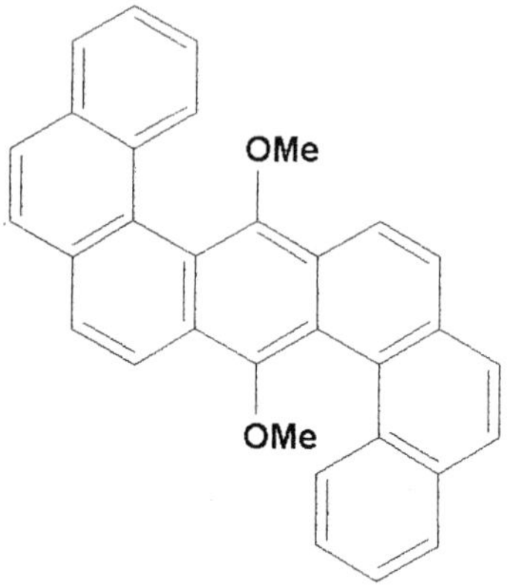

62

12. Perylene

With NBS in DMF perylene (63) gives 3-bromoperylene (R. Lapouyade, J. Pereyre and P. Garrigues, *C.R.Acad.Sci., Ser.2*, 1986, 303, 903). Electrophilic attack of perylene derivativates by NO_2^+ and RCOCl is reported (V.I. Rogovik and L.F.Gutnik, *Zh.Org.Khim.*, 1988, 24, 635). 3,4,9,10-Perylene-tetracarboxylic acid dianhydride (64) provides the 1,6,7,12-tetrachloro derivative 65 (Cl_2-3% oleum). The 1,7-dinitro derivative of 64 results from conventional substitution, but chlorination in DMF at 80° causes chloro-denitration to give 1,7-Cl_2-64. This may be further attacked to give 65.

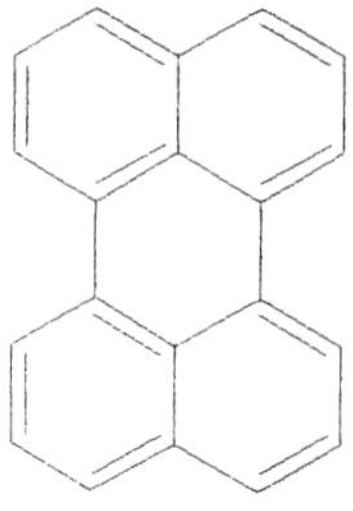

63

Gas-phase chlorodenitration has been regularly reported; it occurs under milder conditions than chlorodeprotiation and often appears to proceed by a free-radical mechanism, but selective and specific displacement of -NO_2 in solution is unusual. The Russian authors also report reductive dehalogenation (Cl or Br) by KOH-ethylene glycol, or by aqueous KOH in the presence of NH_2OH and light; the latter process may involve radicals, but the former process is not so easily understood.

13. Dibenzonaphthacenes

The janusene structure (5,5a,6,11,11a,12-hexahydro-5,12[1',2']:6,11[1',2']-dibenzonaphthacene; 66) formally arises from the 1,4-addition of benzene structures across naphthacene, and similarly for larger structures. Filler first prepared 1,2,3,4-tetrafluoro- and 1,2,3,4,5,6,7,8-octafluorojanusenes (R.

66   67

Filler and G.L. Cantrell, *J.Fluorine Chem.*, 1987, 36, 407) by the addition of dibenzobarrelene (67) to tetra- or octa-fluoroanthracene respectively. Strong support was found for donor-acceptor interaction between the two facing six-membered rings, and this was supported by MMX calculations and by the formation of a persistent radical-cation which arose from single-electron transfer *in FSO₃H-SbF₅* when no proton-uptake occurred. Two different dimers were identified from this species (K.K. Laali, E. Gelerinter and R.Filler, *J.Fluorine Chem.*, 1991, 53, 107).

14. Iptycenes

A range of iptycenes have been obtained by Hart using difficult but fundamentally similar chemistry. The simplest member of the group, 68, is obtained by the addition of benzyne across the 9,10-position of anthracene. As similar addition may take place across *meso*-positions of other acenes such as naphthacene (as in the formation of the janusenes) extended systems may be elaborated. The following serve as examples:

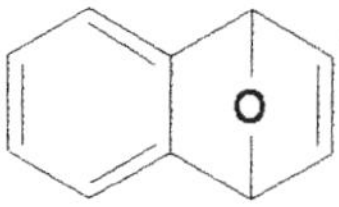

73

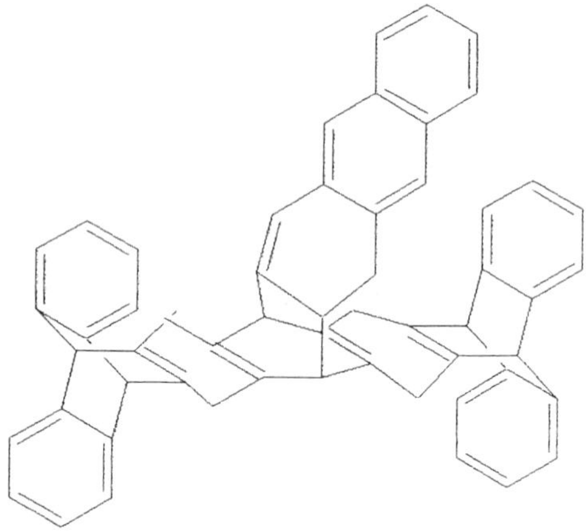

74

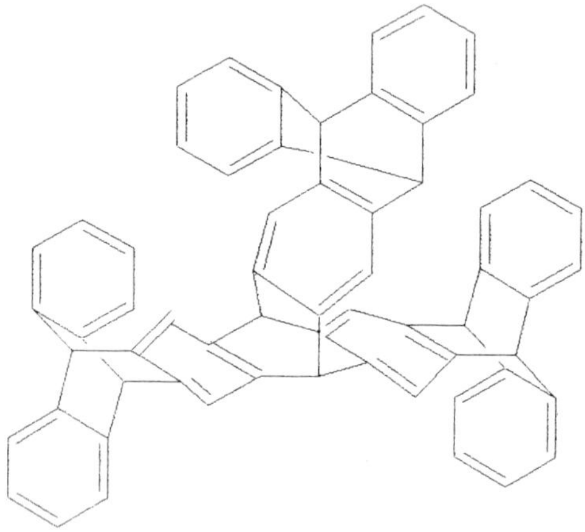

75

Aromatisation (Pd/C) to the anthracene system then allows a further addition using benzyne (o-$N_2^+C_6H_4CO_2^-$ with propene oxide to remove acid) and the final synthesis of tritriptycene **75** (5,7,9,14,16,18,28,33-octahydro-28, 33-[1',2']benzeno-7,16[2'.3']anthraceno-5,18[1',2',]:9,14[1",2"]dibenzeno-heptacene) (A. Bashir-Hashemi, H. Hart and D.L. Ward, *J.Amer.Chem. Soc.*, 1988, <u>110</u>, 5237; *ibid., idem*, 1986, <u>108</u>, 6675).

15. Larger systems

The literature contains occasional references to many extremely large aromatic systems. Many of these, however, reflect either reviews of health hazards (*e.g.* Dibenzo[*a,i*]pyrene: Anon, *Danger.Props.Ind.Mater.Rep.*, 1987, 7, 66) or the extensions of a calculation with little chemical support for the structures invented. Examples include the use of graph theory to support the proposal of various large polybenzenoid systems [J.R. Dias, *THEOCHEM* 1987, 34, 213; (*b*) *ibid., idem*, 1989, 54, 57], the invention of algorithms and functions to determine the number of Kekulé structures possible among the more extensive structures (S.J. Cyvin, B.N. Cyvin and J. Brunvoll, *J.Mol.Struct.*, 1989, 198, 31; see also Yu.A. Kruglyak and M.E. Dokhtmanov, *Zhur.Org.Khim.*, 1989, 25, 1817), the application of simple MO parameters to identify potential carcinogens [(*a*) U.R. Kim *et al.*, *Taehan Hwahakhoe Chi.*, 1987. 31, 153; *Chem.Abstr.*, 1988, 108, 36872; (*b*) E. Rachin, *Izv.Khim.*, 1988, 21. 63; *Chem.Abstr.*, 1988, 109, 87870], and the calculation of chemical thermodynamic properties [(*a*) R.A. Alberty, M.B. Chung and *Chem.Abstr.*, 1989, 110, 22925; (*b*) R.A. Alberty, M.B. Chung and A.K. Reif, *J.Phys.Chem.Ref.Data* 1989, 18, 77] and structure-enthalpy relations in polycyclic catenated aromatic hydrocarbons (W.C. Herndon, D.A. Connor and P. Liu, *Pure Appl.Chem.*, 1990, 62, 435).

As a generalisation, condensation or polymerisation processes are popular routes to large aromatic systems. The application of the Scholl reaction of benzanthrone to forming aromatics such as benzo[*r,s,t*]phenanthro[1,10,9-*c,d,e*]pentaphene (76) containing up to 11 ring systems has already been cited (J.C. Fetzer and W.R. Biggs, *loc.cit.*) S. Obenland and W. Schmidt [*NATO ASI Ser., Ser. C* 1987, 191 (*Polycyclic Aromat.Hydrocarbons Astrophys.*) 165] have reported the synthesis of tribenzo[*a,g,m*]coronene (77) in a five-step process from 1,5-bis(chloromethyl)tetralin (78); trimerisation is a logical beginning.

76

77

78

Guide to the Index

This index is constructed in a similar manner to the volume indexes of the first edition of the Chemistry of Carbon Compounds. However, to make the index easier to use, more descriptive entries have been made for the commonly occurring individual, and groups of chemicals.

The indexes cover primarily the chemical compounds mentioned in the text, and also include reactions and techniques, where named, and some sources of chemical compounds such as plant and animal species, oils, etc.

Chemical compounds have been indexed alphabetically under the names used by authors, editing being restricted to ensuring uniformity of entries under the same heading. In view of the alternative nomenclature that can often be used, a limited amount of cross-referencing has been done where it is considered to be helpful, but attention is particularly drawn to Convention 2 below.

For this and the succeeding volumes, the indexing conventions listed below have been adopted.

1. Alphabetisation

(a) A letter by letter alphabetical sequence is followed for entries, firstly for the main entry, followed by the descriptive entry.

(b) The following prefixes have not been counted for alphabetising:

n-	*o-*	*as-*	*meso-*	*C-*	*E-*
	m-	*sym-*	*cis-*	*O-*	*Z-*
	p-	*gem-*	*trans-*	*N-*	
	vic-			*S-*	
		lin-		*Bz-*	
				Py-	

Some prefixes and numbering have been omitted in the index, where they do not usefully contribute to the reference.

(c) The following prefixes have been alphabetised:

Allo	Epi	Neo
Anti	Hetero	Nor
Bis	Homo	Pseudo
Cyclo	Iso	

2. Cross references

In view of the many alternative trivial and systematic names for chemical compounds, the indexes should be searched under any alternative names which may be indicated in the main body of the text. Only a limited amount of cross-referencing has been carried out, where it is considered that it would be helpful to the user.

3. Derivatives

Simple derivatives are not normally indexed if they follow in the same short section of the text.

4. Collective and plural entries

In place of "– derivatives" the plural entry has normally been used. Plural entries have occasionally been used where compounds of the same name but differing numbering appear in the same section of the text.

5. Main entries

The main entry of the more common individual compounds is indicated by heavy type. Multiple entries, such as headings and sub-headings over several pages are shown by "–", e.g., 67–74, 137–139, etc.

Index